Proceedings in Life Sciences

A Specialized Symposium Organized for the
European Society for
Comparative Physiology and Biochemistry
in Crans-sur-Sierre, Switzerland, September 16–18, 1986

Conference Organization

Scientific Organizer: Ch. Gerday, Liège (Belgium)
General Organizers: L. Bolis, Milano (Italy), R. Gilles, Liège (Belgium)

Under the Patronage of

Fidia Research Laboratories
European Society for Comparative Physiology and Biochemistry

Ch. Gerday L. Bolis R. Gilles (Eds.)

Calcium and Calcium Binding Proteins

Molecular and Functional Aspects

With 157 Figures

Springer-Verlag
Berlin Heidelberg New York
London Paris Tokyo

Dr. Charles Gerday
Laboratory of Muscle Biochemistry, Institute of Chemistry B6,
University of Liège, Sart Tilman, 4000 Liège, Belgium

Professor Dr. L. Bolis
Institute of General Physiology, University of Messina, Via dei Verdi 85,
98100 Messina, Italy

Professor Dr. R. Gilles
Laboratory of Animal Physiology, University of Liège, 22,
quai Van Beneden, 4020 Liège, Belgium

ISBN 3-540-18434-1 Springer-Verlag Berlin Heidelberg New York
ISBN 0-387-18434-1 Springer-Verlag New York Berlin Heidelberg

Library of Congress Cataloging in Publication Data. Calcium and calcium binding proteins: molecular and functional aspects/Ch. Gerday, L. Bolis, and R. Gilles (eds.). p. cm. – (Proceedings of life sciences) "A specialized symposium organized for the European Society for Comparative Physiology and Biochemistry in Crans-sur-Sierre, Switzerland, September 16–18, 1986" – P. facing t. p. Includes index. 1. Calcium-binding proteins – Congresses. 2. Calcium-Physiological effect – Congresses. I. Gerday, Charles. II. Bolis, Liana. III. Gilles, R. IV. European Society for Comparative Physiology and Biochemistry. V. Series. [DNLM: 1. Calcium-Binding Proteins – physiology – congresses. 2. Calcium – metabolism – congresses. 3. Calmodulin – Binding Proteins – physiology – congresses. 4. Cells – physiology – congresses. QU 55 C1438 1986] QP552.C24C34 1988 599'.019214 – dc 19 DNLM/DLC for Library of Congress 87-32207 CIP

Printing and Binding: Brühlsche Universitätsdruckerei, Giessen
2131/3130-54321

Preface

The idea of Professors Bolis and Gilles to gather together for a 3 days' meeting in the splendid environment of Crans-Montana in Switzerland a limited number of people around the subject of calcium and calcium binding proteins seemed at first particularly attractive, and when they asked me to take charge of the scientific organization of the symposium, I accepted with enthusiasm.

It rapidly became clear that the major problem would be the selection of the topics, since it was impossible to cover completely and in depth such a broad and dynamic area of research. In our view, one imperative was to associate as intimately as possible the structural and the functional aspects of the areas covered. Apart from one whole day focused on the fascinating roles played by calmodulin in cellular activities, the other sessions were devoted to calmodulin-related calcium binding proteins in muscle and non-muscle tissues and to some selected biological systems such as mitochondria, secretory cells or sarcoplasmic reticulum in which calcium also plays a crucial role.

The presentations were made by leading investigators in their field. Some of them do not, however, appear in the present volume, for which there are two reasons: first, some of the contributions were somewhat outside the scope of the book; second, three speakers, for valid reasons, simply found no opportunity to write a manuscript in the allotted time.

When analyzing the topics developed herein, one can obviously marvel at the progress which has been made recently at the level of the molecular characterization of calcium binding proteins, but also deplore that their role in cell functioning often remains incompletely understood, evasive, or simply unknown, as in the case of several EF hand calcium binding proteins.

Although the meeting has only covered a limited number of topics and has probably raised many more questions than it gives answers, it has offered us the opportunity of evaluating the path we have still to follow to reach an understanding of the role of calcium and its targets in cellular activities.

I should like to thank all the participants, invited speakers, and those who presented posters, for the high level of their contribution.

I also express my gratitude to Liana Bolis, Professor at the University of Milano, for providing the essential support and raising the necessary funds that rendered the organization of this symposium possible.

Liège, February 1988 Charles Gerday

Contents

Contributors

You will find the addresses at the beginning of the respective contribution

Introductory Lecture

Inositol Lipids, G-Proteins and Signal Transduction

S. COCKCROFT[1]

1 Introduction

Activation of cell surface receptors generates intracellular signals which are the main mediators of cellular responses. The number of different agonists available to interact with their target receptors are enormous compared to the number of intracellular signals. Cyclic AMP is one such intracellular signal, whose synthesis (catalyzed by adenylate cyclase) is well understood. The adenylate cyclase signalling system consists of three kinds of components: a receptor, a guanine nucleotide regulatory protein (G-protein) and the catalytic unit, adenylate cyclase. Receptors either stimulate or inhibit the formation of cAMP by communicating with a pair of homologous G-proteins, Gs which mediates stimulation of adenylate cyclase and Gi which is responsible for its inhibition (Rodbell 1980; Gilman 1984). The G proteins that regulate adenylate cyclase belong to a larger family of homologous G proteins that includes transducin, a molecule that couples light-sensitive rhodopsin to cyclic cGMP phosphodiesterase (Stryer 1983) and Go (Worley et al. 1986), a molecule present in large amounts in brain whose function is not known.

Another transmembrane signalling system that may resemble the adenylate cyclase system utilizes an enzyme, polyphosphoinositide phosphodiesterase (PPI-pde, also known as phospholipase C) which hydrolyzes the polyphosphoinositides (PPI) to generate water-soluble phosphorylated derivatives of inositol and diacylglycerol. Second messenger roles for one of the inositol phosphates, inositol (1,4,5) triphosphate (IP_3) and diacylglycerol have been recently recognized (Nishizuka 1984; Berridge and Irvine 1984). Like cAMP, diacylglycerol activates a specific protein kinase (commonly known as protein kinase C), whilst IP_3 is responsible for mobilising Ca^{2+} from intracellular stores. Both these second messengers are a product of a single phospholipid, phosphatidylinositol bisphosphate (PIP_2). Diacylglycerol can also be generated from phosphatidylinositol (PI) and phosphatidylinositol phosphate (PIP) building into the system, a mechanism whereby both products need not be formed in equimolar amounts. This article discusses how receptors, that are responsible for hydrolysing these inositol lipids, activate the enzyme, PPI-pde.

1 Department of Experimental Pathology, School of Medicine, University College London, University Street, London WC1E 6JJ, England

Ch. Gerday, R. Gilles, L. Bolis (Eds.)
Calcium and Calcium Binding Proteins

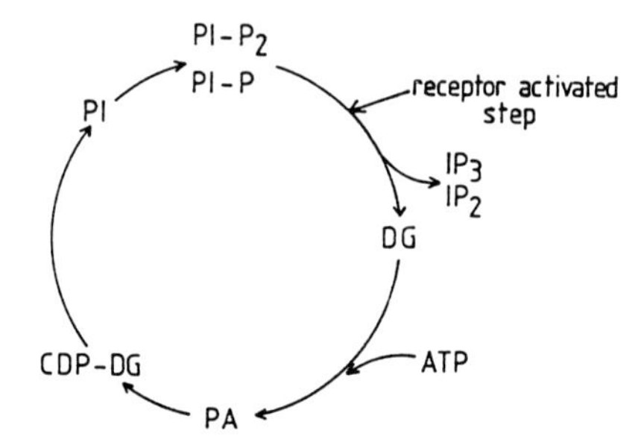

Fig. 1. The inositol lipid cycle. The first event that occurs on cell activation is the hydrolysis of PIP_2 (and possible PIP) to generate inositol trisphosphate and diacylglycerol (DG). DG is phosphorylated by Mg.ATP to make phosphatidic acid (PA) which is then converted to PI via CDP-DG. The inositol lipid kinases then phosphorylate PI sequentially to regenerate the polyphosphoinositides (PIP and PIP_2)

1.1 Inositol Lipids

Inositol lipids constitute about 5–10% of the total cellular lipids and are present at both the plasma membrane and intracellular membranes. These lipids have a characteristic fatty acid composition in that stearic acid ($C_{18:0}$) is esterified at position one and arachidonic acid ($C_{20:4}$) at position two of the glycero-backbone. PI is the predominant inositol lipid and from this is derived its phosphorylated derivatives. PI is phosphorylated by a specific kinase utilising Mg.ATP to form PI-4-P and this enzyme is localised in various regions of the cell including secretory granules, plasma membranes, golgi and lysosomes (Cockcroft et al. 1985b). PIP is a substrate for a second kinase which appears to be exclusively localised at the plasma membrane to form PI-4,5-P_2 (Cockcroft et al. 1985b).

In 1953, it was shown that acetylcholine stimulated the turnover rate of PI in the pancreas (Hokin and Hokin 1953) and it has taken nearly three decades to elucidate the reactions involved in the stimulated inositide metabolism. Figure 1 illustrates the cycle of reactions that take place. Before 1981, it was thought that PI was the initial substrate to be hydrolysed during receptor stimulation (Michell 1975) but it is now accepted that PIP_2 (and possibly PIP) is the initial target for the enzyme (Michell et al. 1981). Hence PI consumption in part is due to its being used as a substrate to generate the phosphorylated derivatives of PI. However the bulk of the PI that disappears from many cells (e.g. platelets and neutrophils) may be hydrolysed directly by reactions not involved in transmembrane signalling (Cockcroft et al. 1985a; Majerus et al. 1985). PI may be removed by phospholipase A_2 for the liberation of arachidonic acid.

1.2 Identification of PPI-pde that is Coupled to Receptors

Two enzymes have been identified that can hydrolyse the inositol lipids. The cytosolic enzyme was discovered by Dawson in 1959 (Dawson 1959) and was initially thought to attack PI only. Recognition of PPIs as the target of receptor activation soon led to the demonstration that the cytosolic enzyme can hydrolyse all three inositol lipids (Rittenhouse 1983; Wilson et al. 1984). A membrane-bound enzyme was discovered in red cells in 1978, but since this enzyme was specific for the PPIs, it was not considered as a potential candidate as a receptor-controlled enzyme at that time. Absence of Ca^{2+}-mobilising receptors on red cells also discouraged studies on this enzyme. It is now clear that the membrane-bound and the cytosolic forms of the enzyme are present in all cells (where it has been sought) and here we consider which of the two enzymes are the potential targets for receptors to couple to.

Cytosolic Phosphodiesterase Active Against PI, PIP and PIP_2. The cytosolic enzyme is characterised by its inability to hydrolyse its substrate when presented in a membrane (Hirasawa et al. 1981; Shukla 1982; Hofmann and Majerus 1982). Studies with this enzyme have mainly been carried out with either pure substrates or substrate presented in a mixture of lipids but not intact membranes. Thus it has been suggested that the receptor causes a perturbation in the membrane which then allows the enzyme to hydrolyse the inositol lipids (Irvine et al. 1984). All three inositol lipids can be hydrolysed by this enzyme (Rittenhouse 1983; Wilson et al. 1984) but preference for PIP_2 can be observed if the state of presentation of the substrate and the ionic environment of the enzyme are manipulated (Irvine et al. 1984; Wilson et al. 1984). Ca^{2+} is also required for the hydrolysis of all three inositol lipids and manipulation of the experimental conditions can modulate the Ca^{2+} requirement from a μmolar range to a mmolar range (Irvine et al. 1984; Wilson et al. 1984).

One characteristic of the cytosolic enzyme is that the water-soluble products formed are not only the linear inositol phosphates but also the cyclic forms, since the enzyme can also function as a transferase by using part of the substrate molecule as an acceptor (Wilson et al. 1985). This transferase reaction is analogous to the enzymatic reaction by which cAMP is formed from ATP by adenylate cyclase. The cyclic derivative of IP_3 [$cI(I{:}2,4,5)P_3$] is also active in mobilising intracellular Ca^{2+} and secretion in platelets with a potency similar to $I(1,4,5)P_3$ (Wilson et al. 1985). Moreover, $cI(1{:}2,4,5)P_3$ was more potent that $I(1,4,5)P_3$ in producing a change in membrane conductance similar to that evoked by light when injected into Limulus photoreceptor cells (Wilson et al. 1985). Inositol cyclic trisphosphate has been reported to be formed in thrombin-stimulated platelets (Ishii et al. 1986).

A Membrane-Bound PPI-pde. A membrane-bound enzyme against PIP and PIP_2 has also been identified in some cells and this enzyme is characterised by its ability to hydrolyse the endogenous substrate present in the membranes. The enzyme is inactive under physiological conditions, that is conditions where the level of Ca^{2+} is maintained in the 100 nmolar range. Raising the level of Ca^{2+} to μ-mmolar levels can stimulate the enzyme to attack PIP and PIP_2 but not PI. The presence of Mg^{2+} and increasing ionic strength raises the Ca^{2+} requirement for enzyme activation. Such a membrane-bound enzyme was initially discovered in red blood cells (Allan and Michell 1978; Downes and Michell 1982) and was subsequently identified in other cell types as well, including nerve ending membranes (Van Rooijen et al. 1983) and neutrophils (Cockcroft et al. 1984).

From the calcium requirement of the enzyme it was suggested that Ca^{2+} could not be the physiological activator in receptor-stimulated cells (Downes and Michell 1982). Of the two enzymes identified, it was considered that the membrane-bound pde, rather than the cytosolic species, be the enzyme initially activated by the occupied receptor. Since the agonist, fMetLeuPhe, had no direct effect on the membrane-bound PPI-pde (Cockcroft et al. 1984), we considered the possibility that like the adenylate cyclase system, a guanine nucleotide regulatory protein may be involved.

2 Guanine Nucleotide Regulation of PPI-pde

Direct demonstration that guanine nucleotides regulated the activation of PPI-pde came from studies with neutrophils and blowfly salivary glands (Cockcroft and Gomperts 1985; Litosch et al. 1985). However, there were many indications in the literature which hinted that a G-protein may be involved in the regulation of Ca^{2+}-mobilising receptors. It had been shown that guanine nucleotides reduced the affinity of the agonist for some Ca^{2+}-mobilising receptors, analogous to an effect commonly observed with receptors coupled to adenylate cyclase (Birnbaumer et al. 1985). As long ago as 1974; Glossman et al. had shown that binding of angiotensin II to its receptor was reduced in the presence of GTP or GppNHp (Glossman et al. 1974). Other Ca^{2+}-mobilising receptors where guanine nucleotides have been shown to modulate agonist binding to the receptors include muscarinic receptors in heart (Berrie et al. 1979) and in 1321 N1 astrocytoma cells (Evans et al. 1985), fMetLeuPhe receptors in neutrophils (Koo et al. 1983), α-adrenergic receptors (Goodhardt et al. 1982) and thyrotropin-releasing hormone (TRH) receptors in GH_3 cells (Hinkle and Kinsella 1984). Some of the early studies were open to a different interpretation in that they often failed to categorically distinguish receptors that coupled to adenylate cyclase system or inositide metabolism. This was necessary, since many agonists can interact with more than one kind of receptor coupling to a different signal transduction mechanism.

More recent work which also pointed in this direction included studies with secretory cells that showed that exocytosis could be stimulated by GTP and its non-hydrolysable analogues. Incorporating guanine nucleotide analogues into mast cells allowed these cells to respond to external Ca^{2+} (Gomperts 1983). Studies with permeabilised platelets showed that guanine nucleotides enhanced the calcium sensitivity of the secretory mechanism, apparently by generating diacylglycerol (Haslam and Davidson 1984a,b).

The PPI-pde in plasma membranes of neutrophils respond to the guanine nucleotides analogues, GTPγS, GppNHp and $GppCH_2p$ in the presence of 100 nM Ca^{2+} and high ionic strength, conditions that prevail in unstimulated cells. We called the putative G-protein, G_p (Cockcroft and Gomperts 1985). Hydrolysis of PPI is accompanied by a quantitative generation of IP_2 and IP_3 establishing a cleavage of the phospholipase C-type. This result has been confirmed in a variety of membrane preparations and in permeabilised cells (see Table 1). Although the mechanism of activation of PPI-pde by guanine nucleotides is not known, it is likely that it bears a resemblence to the other G-proteins. Thus agonists can also activate the PPI-pde in the presence of GTP.

2.1 Coupling of Receptors to PPI-pde by a G-Protein, G_p

Litosch et al. (1985) and Smith et al. (1985) first demonstrated that the receptor could also couple to the PPI-pde in the presence of GTP. Table 1 shows a list of cells where it has now been shown that agonist in the presence of GTP or its analogues alone can stimulate the activity of the PPI-pde. The coupling factor is present in the membrane and can be removed by 0.5% cholate, conditions which can extract Gs from membranes (Martin et al. 1986b).

Table 1. Systems in which it has been shown that PPI-pde can be activated by GTP (or its analogues) alone or in combination with an agonist

Cell-type	Agonist	Reference
Neutrophils	fMetLeuPhe	Smith et al. (1986) Cockcroft and Gomperts (1985)
GH_3 cells	TRH	Martin et al. (1986a,b)
Pancreas	Carulein, carbachol	Merritt et al. (1986)
Blowfly salivary	5-Hydroxytryptamine	Litosch et al. (1985)
Cerebral cortex	GTPγS	Gonzales and Crews (1985)
Smooth muscle	Acetylcholine	Sasaguri et al. (1986a,b)
Platelets	Thrombin	Brass et al. (1986)
Liver	Vasopressin epinephrine angiotensin	Uhring et al. (1986)
Thyroid	GTPγS	Burch et al. (1986) Field et al. (1987)
Mast cells	GTPγS	Cockcroft et al. (1987)
Astrocytoma cells	Carbachol	Hepler and Harden (1986)
Fish olfactory cilia	L-Alanine	Huque and Bruch (1986)
WRK-1 cells	Vasopressin	Guillon et al. (1986a)
Sycamore cells (Acer pseudoplatanus)	GTPγS	Dillenschneider et al. (1986)

Receptor-mediated formation of IP_3 in membranes is rapid and can be blocked by GDPβS (Litosch et al. 1985; Martin et al. 1986b; Uhing et al. 1986). Mg.ATP, by its ability to maintain the inositol lipids in the phosphorylated state, can enhance the response and also maintain the response for a longer period (Cockcroft, unpublished; Martin et al. 1986a,b). Most studies have been made using the endogenous substrates but addition of exogenous substrate has also been done (e.g. Sasaguri et al. 1986a; Litosch and Fain 1985). The mechanism by which the exogenously added substrate is utilised by the enzyme is not known but it will be interesting to establish whether the lipid has to intercalate with the lipid bilayer or whether the enzyme can utilise the substrate present as a monomer or in micelles. For the purification of the PPI-pde, it will be important to establish assay methods which are not dependent on the presence of the endogenous substrate.

2.2 Regulation of the Cytosol Enzyme by GTPγS

Two types of studies have been carried out to investigate G-protein regulation of the PPI-pde. Activation has been studied either in membranes (examples are neutrophils, liver) or in permeabilised cells (examples are pancreas, GH_3 cells). Whilst it is clear from the studies with membranes that the entire machinery to stimulate receptor-mediated hydrolysis of PIP_2 remains intact in the membrane, the studies with permeabilised cells cannot exclude a role for the cytosolic enzyme.

Preliminary evidence suggests that the soluble enzyme can also be regulated by guanine nucleotides (Baldassare and Fisher 1986). Cytosol prepared from platelets was found to potentiate the hydrolysis of PPI due to GTPγS but not due to thrombin in platelet plasma membranes. The authors ascribe the potentiation to some component(s) present in the cytosol. To demonstrate that the component may be the soluble enzyme, they showed that PIP_2 incorporated into lipid vesicles could be hydrolysed on the addition of cytosol and this effect was enhanced by addition of GTPγS but not GDPβS. The authors of this study concluded that a G-protein could regulate the cytosolic PPI-pde. Banno et al. (1986) came to a rather different conclusion when they studied the effect of GTP, GDP and GTPγS on a partially purified cytosolic phospholipase C (pde) from platelets. They concluded that GTP and GDP (but not GTPγS) have a direct effect on the enzyme. Judgement should be withheld about the role of the cytosolic enzyme until the purified enzyme can be shown to be regulated by G-proteins. The second problem that has to be considered is whether the receptor can also couple to the cytosolic enzyme.

2.3 Identification of the Products of Inositide Hydrolysis

Studies from intact cells have shown that on receptor stimulation $I(1,3,4)P_3$ and $I(1,3,4,5)P_4$ is generated in addition to $I(1,4,5)P_3$ (Hawkins et al. 1986). In contrast, experiments with membrane preparations only stimulate the formation of $I(1,4,5)P_3$ (Cockcroft, unpublished; Martin et al. 1986b). Of all the isomers formed, $I(1,4,5)P_3$ has been shown to mobilise Ca^{2+} from intracellular stores. The function of $I(1,3,4,5)P_4$ is thought to regulate the entry of Ca^{2+} ions into cells from the external medium (Michell 1986; Irvine and Moor 1986). Experiments done on membrane preparations support the conclusion that the $I(1,3,4)P_3$ is indirectly generated from $I(1,4,5)P_3$ probably by phosphorylation of $I(1,4,5)P_3$ to $I(1,3,4,5)P_4$ by a cytosolic enzyme followed by a specific removal of the 5′-phosphate (Irvine et al. 1986; Hawkins et al. 1986).

The other product of PPI-pde hydrolysis is DG and this has been monitored by following its conversion to ^{32}P-labelled PA by γ-labelled ATP in fMetLeuPhe stimulated neutrophils (Cockcroft, unpublished) and in thrombin-stimulated platelets (Brass et al. 1986). Increases in labelled diacylglycerol have also been monitored in platelets (Haslam and Davidson 1984a) and in GH_3 cells (Martin et al. 1986b).

3 G-Protein Activation

Gs, Gi and transducin have been purified and their structures are well known. All three of these G-proteins are heterotrimers consisting of αβγ subunits. Whilst the βγ subunits are identical in all three G-proteins, the α-subunit is different. The α-subunit contains the binding site for the guanine nucleotide and it also possesses GTPase activity. GTPγS activates the G-protein by displacing the bound GDP and taking its place leading to subunit dissociation of the α-subunit from the βγ-subunits. Mg^{2+} is essential for subunit dissociation. The Mg^{2+} requirement is high for Gs compared to Gi (Birnbaumer et al. 1985). The free α-GTPγS complex is then responsible for stimulating the catalytic unit. Because GTPγS cannot be hydrolysed, the α-subunit is permanently activated. In the presence of the ligand and GTP, the ligand catalyses the exchange of GDP to GTP, and the free α-GTP complex remains active as long as the GTP is not hydrolysed by the intrinsic GTPase. Hydrolysis terminates the activity of the free α-GTP complex.

3.1 Fluoride Activation of PPI-pde

If the G-protein responsible for hydrolysing inositol lipids bears any structural resemblance to the other well-defined G-proteins, then one would predict that fluoride is a universal activator of PPI-pde. The basis of fluoride activation of G-proteins in general is well understood in that its actions resemble those of the non-hydrolysable GTP analogues. Aluminium ion is required as a co-ion for fluoride to work, the active species being $[ALF_4]^-$. This can interact with bound GDP on the α-subunit of the G-protein and leads to subunit dissociation liberating the activated free α-subunit GDP-$[ALF_4]^-$ complex which can now interact with the catalytic unit to cause persistant stimulation. Thus fluoride can activate Gs, Gi and transducin.

Fluoride has been shown to stimulate a variety of cell functions which cannot be explained on the basis of its ability to interact with the adenylate cyclase system. Fluoride can stimulate histamine secretion from mast cells (Patkar et al. 1978), respiratory burst in neutrophils (Curnutte et al. 1979), secretion in platelets (Poll et al. 1986), activation of phosphorylase in hepatocytes (Blackmore et al. 1985) and excitation of *Limulus* photoreceptors (Corson et al. 1983). All these systems are characterised by their ability to respond to a rise in cytosol $[Ca^{2+}]$ and indeed it has been demonstrated that fluoride can cause a rise in cytosol Ca^{2+} in platelets (Poll et al. 1986; Brass et al. 1986), hepatocytes (Blackmore et al. 1985) and neutrophils (Strnad and Wong 1985).

Since a rise in cytosol Ca^{2+} can be caused by IP_3, it was therefore not surprising to observe that fluoride can also stimulate the formation of IP_3 in both intact hepatocytes (Blackmore et al. 1985) and neutrophils (Strnad et al. 1986) but also in isolated membranes from liver (Cockcroft and Taylor 1987), WRK-1 cells (a mammary tumour cell-line) (Guillon et al. 1986b) and GH_3 cells (a pituatory cell-line) (Martin et al. 1986b). Fluoride, like GTPγS, activates the PPI-pde in a concentration-dependent manner (1–20 mM). The time-course of activation is slower than that of GTPγS, but like GTPγS it can be inhibited by GDPβS (Cockcroft and Taylor 1987).

All this suggests that the G-protein that regulates the PPI-pde must bear structural resemblance to the other well-characterised G-proteins, i.e. Gp is also made up of αβγ-subunits.

3.2 Termination of Activated Gp by an Intrinsic GTPase

Since termination of adenylate cyclase involves GTP hydrolysis, a similar mechanism of G-protein inactivation may also occur with Ca^{2+}-mobilising receptors. A GTPase activity is stimulated in membrane preparations from neutrophils by fMetLeuPhe (Hyslop et al. 1984), from liver by vasopressin (Fain et al. 1985), and from GH_3 cells with TRH (Wojcikiewicz et al. 1986) and in all these cases the evidence points to a GTPase activity intrinsic to the putative Gp.

3.3 Ca^{2+} Requirement of PPI-pde

The PPI-pde is nominally inactive when the Ca^{2+} level is fixed in the 100 nmolar range, conditions that pertain in the unstimulated cell. Under these conditions, the PPI-pde is inactive but it can be activated by guanine nucleotide analogues or GTP plus agonist (Cockcroft 1986; Martin et al. 1986a; Uhing et al. 1986). The presence of Ca^{2+} is obligatory since reduction of Ca^{2+} levels from 100 nM to 1 nM and below inhibits the activity of PPI-pde stimulated by guanine nucleotide analogues (Cockcroft 1986; Martin et al. 1986a). Thus although Ca^{2+} is essential, its requirement is met by the presence of resting levels of Ca^{2+} which is present in all cells. Secondly, the rise in cytosol Ca^{2+} may further activate the PPI-pde to potentiate the response perhaps by recruitment of the cytosolic enzyme as discussed below.

In contrast to experiments in neutrophils (Cockcroft 1986) and smooth muscle (Sasaguri et al. 1986a), it has been reported that Ca^{2+} in the range of 0.1–10 μM alone can activate the PPI-pde in cell-free systems. Examples include pancreas (Taylor et al. 1986), islets (Best 1986), sea urchin eggs (Whitaker and Aitchison 1985), GH_3 cells (Martin et al. 1986a) and hepatocytes (Uhing et al. 1986). Since Ca^{2+} increases from 0.1 μM to over 1 μM in hormonally stimulated cells, it is possible that Ca^{2+} may have an effect on stimulating the PPI-pde. In hepatocytes the Ca^{2+} requirement could be shifted to higher concentrations by increasing the concentration of Mg^{2+}. The presence of Mg.ATP also affected the Ca^{2+} requirement at least in GH_3 cells. In the absence of Mg.ATP there was a high Ca^{2+} requirement for PPI-pde activation compared to the requirement in its presence. Since the studies in neutrophils and smooth muscle were conducted in the absence of Mg.ATP, it may explain why Ca^{2+} in the range of 0.1–10 μM does not activate the PPI-pde. It is clear that, depending on the experimental conditions, Ca^{2+} in the range that occurs in stimulated cells can activate the PPI-pde. Indeed, raising the level of cytosol Ca^{2+} in intact cells (by mechanisms including Ca^{2+}-ionophores and membrane depolarization by high K^+) has been shown to activate inositide metabolism in pancreas (Taylor et al. 1986), islets (Best 1986), neutrophils (Cockcroft et al. 1981; Meade et al. 1986).

The PPI-pde can be fully activated by Ca^{2+} in the millimolar range in cell types where it has been studied. This effect of Ca^{2+} is analogous to the forskolin activation of adenylate cyclase. What is not known is how Ca^{2+} stimulates the activity of the enzyme. The PPIs are negatively charged phospholipids with an ability to bind divalent cations including Ca^{2+} (Buckley and Hawthorne 1972). Thus it is possible that Ca^{2+} binding to the PPIs makes the substrate susceptible to hydrolysis. It should be noted that although Ca^{2+} and Sr^{2+} are effective, Mg^{2+} is not (Cockcroft et al. 1984; Cockcroft 1986). On the other hand, Ca^{2+} may interact with the enzyme and thus stimulate it. Since neither of these effects are mutually exclusive, it is possible that Ca^{2+} activation results from both substrate presentation and enzyme activation. Activation by Ca^{2+} cannot be inhibited by GDPβS implying that the guanine regulatory protein is not involved (Cockcroft 1986).

In Fig. 2, the site of the Ca^{2+} requirement is placed at the enzyme. It can be envisaged that the free α_p-GTP complex increases the affinity for Ca^{2+} making Ca^{2+} the activator of the enzyme. On the other hand, the Ca^{2+} requirement may lie at the level of the G-protein itself. It is interesting to note that (in contrast to the well-defined requirement of Mg^{2+} in causing subunit dissociation of Gi and Gs) Mg^{2+} is not obligatory for GTPγS regulation of PPI-pde (Cockcroft and Gomperts 1985; Martin et al. 1986a). It is conceivable that Ca^{2+} is necessary to fulfill this role.

3.4 Neomycin as an Inhibitor of PPI-pde

Neomycin is an aminoglycoside antibiotic that can bind tightly to PPI (Schacht 1978) and so prevent its hydrolysis (Schacht 1976; Downes and Michell 1981). It most probably interacts with these anionic lipids due to the cationic nature of neomycin. Thus neomycin will also interact with other negatively charged molecules such as ATP and IP_3 (Prentki et al. 1985). The Ca^{2+}-stimulated hydrolysis of PPI can be blocked by neomycin in erythrocytes (Downes and Michell 1981) and nerve ending membranes (Van Rooijen and Agranoff 1985). This effect of neomycin has been used to study the relevance of inositol lipid metabolism in signal transduction mechanisms in a variety of cells. Neomycin inhibits GTPγS-dependent histamine secretion (Cockcroft and Gomperts 1985) and inositol phosphate formation (Cockcroft et al. 1987) from mast cells, GTPγS induced contraction in skeletal muscle (Verghara et al. 1985), cholecystokinin-induced inositol phosphates formation and Ca^{2+} release (Streb et al. 1985), thrombin stimulated inositol phosphate formation and cell proliferation in fibroblasts, Ca^{2+}-stimulated exocytosis in sea urchin egg plasma membrane (Whitaker and Aitchison 1985) and GTPγS stimulated inositol phosphate formation in FRTL-5 thyroid cells (Burch et al. 1986) supporting the notion that hydrolysis of PIP_2 is essential for cell activation. In all of these studies neomycin (because of its impermeant nature) was introduced into the cell either by membrane permeabilisation or was employed in cell-free systems.

Siess and Lapetina (1986) have also reported that neomycin can also block the inositol phosphate formation, secretion and aggregation in platelets responding to thrombin but not to other platelet agonists (vasopressin, platelet activating factor, collagen or prostaglandin endoperoxide analogues). They concluded from their studies that neomycin has multiple effects on platelets that are unrelated to a specific inhibition of inositol lipid degradation. It was clear from their studies that neomycin was affecting the binding of thrombin to the receptor rather than acting as a inhibitor of the PPI-pde. Neomycin was applied to intact platelets in their work hence it is not surprising that they could not monitor inhibition of PPI-pde.

Although neomycin is a "dirty" tool, nonetheless it can still be used in dissecting the role of PPI-pde activation in certain situations. However, caution has to be applied before accepting such results.

4 Pertussis and Cholera Toxins as Modifiers of G-Proteins

Two bacterial toxins, one from *Vibrio* cholerae and the other from *Bordetella* pertussis have become useful probes in distinguishing the G-proteins of the adenylate cyclase system. Each toxin is composed of A (active) and B (binding) protomer. The B subunit facilitates the entry of the A subunit inside the cells where it can catalyse the transfer of the ADP-ribosyl portion from NADPH, onto an amino acid residue. Cholera toxin ADP-ribosylates the α_s-subunit leading to inhibition of the intrinsic GTPase activity and hence persistant activation of adenyl cyclase. In contrast, ADP ribosylation of α_i by pertussis toxin leads to a decrease in the receptor stimulation of Gi-dependent hydrolysis of GTP. It only blocks the effect of GTP to activate Gi without affecting the activity of its non-hydrolysable analogues (Hildebrant et al. 1982). G_o and transducin are also targets for pertussis toxin and transducin is a target for cholera toxin. Is Gp also a target for these toxins?

No general consensus emerges, as it does with Gi and Gs. Pertussis toxin can inhibit cellular responses, including the activation of PPI-pde in many cell types but this does not appear to be universal (see Table 2). Pertussis toxin ADP-ribosylates a 41 kDa protein that appears to be distinct from Gi in cells such as neutrophils and NG08-15 cells suggesting that the pertussis substrate in those cells that are coupled to PPI-pde activation occur through a distinct protein that is neither Gi or Go (Milligan et al. 1986; Gierchik et al. 1986). Cholera toxin has been shown to inhibit the PPI-pde in only one system. More studies are needed before a clear picture may emerge from the present chaos. However, in the HL60 cells, a neutrophil-like cell line, pertussis toxin has been exploited to show added G_o or Gi can restore fMetLeuPhe stimulation of PPI-pde when native Gp is inactivated (Kikuchi et al. 1986). Although this does not allow one to conclude the actual identity of the putative G-protein that couples the PPI-pde, it does strongly suggest that Gp must belong to that same family of G-proteins as Gs, Gi and Go.

It is interesting to note that Gs is known to exist in two different forms which differ in the sequence of the amino acids. This arises not because Gs is a product of two separate genes but due to a differential splicing of the mRNA. It is not known what controls the production of the two transcripts and it is equally possible that Gp may also be a product of a single gene which is differentially processed in various cell-types. This may explain the differential effects of pertussis toxin in different cell-types.

If it is found that Gp also contains identical $\beta\gamma$-subunits to the other G-proteins, then it can be envisaged that liberation of free $\beta\gamma$ subunits from any of the G-proteins will affect the basal turnover rate of the other signalling systems. This would allow a cell to integrate its response to a hormonal stimulation by affecting not only the signalling system that it is coupled to but also the parallel signalling systems that are present in those cells. It is interesting to note that many cell-types possess two receptor subtypes which are coupled to two parallel signalling systems such as adenylate cyclase and PPI-pde which respond to one agonist. Examples include adrenaline on parotid, 5-hydroxytryptamine on blowfly salivary glands, TSH on thyroid (Field et al. 1987), secretin in pancreas (Trimble et al. 1986), glucagon on liver (1986) and adreno-

Table 2. Pertussis or cholera toxin sensitivity of those receptor systems which operate through stimulated PIP_2 hydrolysis

Inhibited by pertussis toxin	
Mast cells	Nakamura and Ui (1984, 1985)
Neutrophils	Okajima and Ui (1984)
HL60	Kikuchi et al. (1986), Krause et al. (1985)
Macrophages	Holian (1986)
Lymphocytes	Spangrude et al. (1985)
Fibroblasts	Paris and Pouyssegur (1986)
Renal mesangial cells	Pfeilschifter and Bauer (1986)
Platelets	Brass et al. (1986)
Aortic vascular smooth muscle cells	Kanaide et al. (1986)
Liver (EGF receptor only)	Johnson et al. (1986)
Not affected by pertussis toxin	
Brown adipocytes	Schimmel and Eliott (1986)
Astrocytoma cells	Masters et al. (1985), Hepler and Harden (1986)
Heart cells	Masters et al. (1985)
GH_3 cells	Martin et al. (1986a,b)
Panreas	Merritt et al. (1986)
Liver	Lynch et al. (1986)
FRTL-5 (thyroid) cells	Burch et al. (1986)
Coronary artery-smooth muscle	Sasaguri et al. (1986b)
Adrenal glomerulosa cells	Kojima et al. (1986)
Cerebral striata	Kelly et al. (1985)
Inhibited by cholera toxin	
T-cell-line (Jurkat)	Imboden et al. (1986)
Not affected by cholera toxin	
GH_3 cells	Martin et al. (1986a,b)
Liver	Uhing et al. (1986)
Pancreas	Merritt et al. (1986)

corticotrophin (ACTH) on adrenal cells (Farese et al. 1986). Receptor activation of cells by an agonist can no longer be analysed simply on the basis of affecting one signalling system.

5 Conclusions

We now have a working model of how receptors that couple to PPI-pde do so via a putative G-protein, Gp (Fig. 2). Analogy is drawn from the adenylate cyclase system but there are important differences between the two. The major difference lies in the

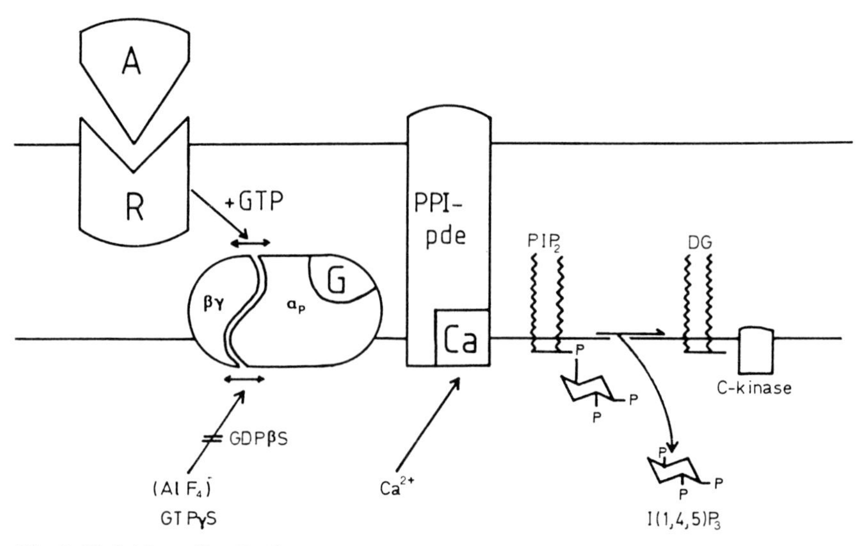

Fig. 2. Model for cell activation

fact that adenylate cyclase utilises a substrate that is not rate-limiting (i.e. ATP) compared to PPI-pde which utilises a lipid, PIP_2 to generate its second messengers. Substrate regeneration by the inositol lipid kinases may themselves be subject to regulation.

Figure 2 envisages a G-protein made up of a heterotrimer, $\alpha\beta\gamma$-subunits. The α-subunit binds the guanine nucleotide and in the basal state it is occupied with GDP. This subunit also possess an intrinsic GTPase activity. When the receptor is occupied with an appropriate ligand, it catalyses the exchange of GDP to GTP on the α-subunit of the heterotrimer. This leads to subunit dissociation releasing free α_p-GTP complex which can activate the PPI-pde providing that nmolar levels of Ca^{2+} is available. The site of Ca^{2+} activation is not known but it could be either the G protein or the enzyme. If it is the G-protein then it becomes analogous to the obligatory Mg^{2+} requirement for adenylate cyclase.

Termination of the signal occurs when the intrinsic GTPase on the α_p-subunit hydrolyses the GTP to GDP. This inactivates the α_p subunit which is now available to re-associate with $\beta\gamma$-subunits. GTP analogues such as GppNHp, GTPγS can displace the GDP in the absence of the ligand and so cause subunit dissociation. Fluoride complexed with aluminium ions can interact with GDP mimicking GTP and thereby causes subunit dissociation. GDPβS can inhibit agonist plus GTP, GTP analogues and fluoride activation most likely by GDPβS by occupying the site on the α-subunit and so preventing its dissociation. Ca^{2+} can also directly activate the PPI-pde by interacting with a site on the enzyme which is not inhibited by GDPβS.

Since Gp has not been purified this model should be considered as a working hypothesis. The next goal is to purify the individual components of the PPI-pde signalling system in order to understand fully how it functions.

Acknowledgements. SC is a Lister Institute Fellow and would like to thank the MRC for their support.

References

Allan D, Michell RH (1978) A calcium activated polyphosphoinositide phosphodiesterase in the plasma membrane of human and rabbit erythrocytes. Biochim Biophys Acta 508:277–286

Baldassare JJ, Fisher GJ (1986) Regulation of membrane-associated and cytosolic phospholipase C activities in human platelets by guanosine triphosphate. J Biol Chem 261:11942–11944

Banno Y, Nakashima S, Tohmatsu T, Nozawa Y, Lapetina EG (1986) GTP and GDP will stimulate platelet cytosolic phospholipase C independently of Ca^{2+}. Biochim Biophys Res Commun 140: 728–734

Berridge MJ, Irvine RF (1984) Inositol trisphosphate, a novel second messenger in cellular signal transduction. Nature (Lond) 312:315–320

Berrie CP, Birdsall NJM, Burgen ASV, Hume EC (1979) Guanine nucleotides modulate muscarinic receptor binding in the heart. Biochem Biophys Res Commun 87:1000–1005

Best L (1986) A role for calcium in the breakdown of inositol phospholipids in intact and digitonin-permeabilized pancreatic islets. Biochem J 238:773–779

Birnbaumer L, Codina J, Mattera R, Cerione RA, Hildebrandt JD, Dunyer T, Rojas FJ, Caron MG, Lefkowitz RJ, Iyengar R (1985) Regulation of hormone receptors and adenyl cyclases by guanine nucleotide binding N proteins. Rec Prog Horm Res 41:41–94

Blackmore PF, Bocckino SB, Waynick LE, Exton JE (1985) Role of guanine nucleotide binding protein in the hydrolysis of hepatocyte phosphatidylinositol 4,5-bisphosphate by calcium-mobilizing hormones and the control of cell calcium. Studies utilizing aluminium fluoride. J Biol Chem 260:14477–14483

Brass LF, Laposta M, Banga HS, Rittenhouse SE (1986) Regulation of the phosphoinositide hydrolysis pathway in thrombin-stimulated platelets by a pertussis toxin-sensitive guanine nucleotide-binding protein. J Biol Chem 261:16838–16847

Buckley JT, Hawthorne JN (1972) Erythrocyte membrane polyphosphoinositide metabolism and the regulation of calcium binding. J Biol Chem 247:7218–7223

Burch RM, Luini A, Axelrod J (1986) Phospholipase A_2 and phospholipase C are activated by distinct GTP-binding proteins in response to α_1-adrenergic stimulation in FRTL5 thyroid cells. Proc Natl Acad Sci USA 83:7201–7205

Carney DH, Scott OL, Gordon EA, La Belle ES (1985) Phosphoinositides in mitogenesis: neomycin inhibits thrombin stimulated phosphoinositide turnover and initiation of cell proliferation. Cell 42:479–488

Cockcroft S (1986) The dependence on Ca^{2+} of the guanine nucleotide-activated polyphosphoinositide phosphodiesterase in neutrophil plasma membranes. Biochem J 240:503–507

Cockcroft S, Gomperts BD (1985) Role of guanine nucleotide binding protein in the activation of polyphosphoinositide phosphodiesterase. Nature (Lond) 314:534–536

Cockcroft S, Taylor JA (1987) Fluoroaluminates mimic GTPγS in activating the polyphosphoinositide phosphodiesterase of hepatocyte membranes: Role for the guanine nucleotide binding protein, G_P, in signal transduction. Biochem J 241:409–414

Cockcroft S, Bennett JP, Gomperts BD (1981) The dependence on Ca^{2+} of phosphatidylinositol breakdown and enzyme secretion in rabbit neutrophils stimulated by formylmethionylleucylphenylalanine or ionomycin. Biochem J 200:501–508

Cockcroft S, Baldwin JM, Allan D (1984) The Ca^{2+}-activated polyphosphoinositide phosphodiesterase of human and rabbit neutrophil membranes. Biochem J 221:477–482

Cockcroft S, Barrowman MM, Gomperts BD (1985a) Breakdown and synthesis of polyphosphoinositides in fMetLeuPhe-stimulated neutrophils. FEBS Lett 181:259–263

Cockcroft S, Taylor JA, Judah JD (1985b) Subcellular localisation of inositol lipid kinases in rat liver. Biochim Biophys Acta 845:163–170

Cockcroft S, Howell TW, Gomperts BD (1987) Two G-proteins act in series to control stimulus-secretion coupling in mast cells: G_P regulates transmembrane signalling and G_E regulates exocytosis. J Cell Biol (in press)

Corson DW, Fein A, Walthall W (1983) Chemical excitation of Limulus photoreceptors. 11. Vanadate, GTPγS, and fluoride prolong excitation evoked by dim flashes of light. J Gen Physiol 82:659–677

Curnutte JT, Babior BM, Karnovsky ML (1979) Fluoride-mediated activation of the respiratory burst in human neutrophils: a reversible process. J Clin Invest 63:637–647

Dawson RMC (1959) Studies on the enzymic hydrolysis of monophosphoinositide by phospholipase preparations from *P. Notatum* and ox pancreas. Biochim Biophys Acta 33:68–77

Dillenschneider M, Hetherington A, Graziana G, Alibert G, Berta P, Haiech J, Ranjeva R (1986) The formation of inositol phosphate derivatives by isolated membranes from Acer pseudoplatanus is stimulated by guanine nucleotides. FEBS Lett 208:413–417

Downes CP, Michell RH (1981) The polyphosphoinositide phosphodiesterase of erythrocyte membranes. Biochem J 198:133–140

Downes CP, Michell RH (1982) The control by Ca^{2+} of the phosphoinositide phosphodiesterase and the Ca^{2+}-pump ATPase in human erythrocytes. Biochem J 202:53–58

Evans T, Hepler JR, Masters S, Brown JH, Harden T (1985) Guanine nucleotide regulation of agonist binding to muscarinic cholinergic receptors. Relation to efficacy of agonists for stimulation of phosphoinositide breakdown and Ca^{2+}. Biochem J 232:751–757

Fain JN, Brindley DN, Pittner RA, Hawthorne JN (1985) Stimulation of specific GTPase activity in by vasopressin isolated membranes from cultured rat hepatocytes. FEBS Lett 192:251–254

Farese RV, Rosic N, Babischkin J, Farese MG, Foster R, Davis JS (1986) Dual activation of the inositol-triphosphate-calcium and cyclic nucleotide intracellular signalling systems by adrenocorticotropin in rat adrenal cells. Biochem Biophys Res Commun 135:742–748

Field JB, Ealey PA, Marshall NJ, Cockcroft S (1987) Thyroid stimulating hormone stimulates increases in inositol phosphates as well as cyclic AMP in the thyroid: Evidence for different inositol lipid pools. Biochem J 247:519–524

Gierchik P, Falloon J, Milligan G, Pines M, Gallin JI, Spiegel A (1986) Immunochemical evidence for a novel pertussis toxin substrate in human neutrophils. J Biol Chem 261:8058–8062

Gilman AG (1984) G proteins and dual control of adenylate cyclase. Cell 36:577–579

Glossman H, Baukal A, Catt KJ (1974) Angiotensin II receptors in bovine adrenal cortex. Modification of angiotensin II by guanyl nucleotides. J Biol Chem 249:664–666

Gomperts BD (1983) Involvement of guanine nucleotide-binding protein in the gating of Ca^{2+} by receptors. Nature (Lond) 306:64–66

Gonzales RA, Crews FT (1985) Guanine nucleotides stimulate production of inositol trisphosphate in rat cortical membranes. Biochem J 232:799–804

Goodhardt M, Ferry N, Geynet P, Hanoune J (1982) Hepatic alpha-1 adrenergic receptors show agonist-specific regulation by guanine nucleotides. Loss of nucleotide effect after adrenalectomy. J Biol Chem 257:11577–11583

Guillon G, Balestre M-N, Mouillac B, Devilliers G (1986a) Activation of membrane phospholipase C by vasopressin: A requirement for guanyl nucleotides FEBS Lett 196:155–159

Guillon G, Mouillac B, Balestre M-N (1986b) Activation of polyphosphoinositide phospholipase C by fluoride in WRK1 cell membranes. FEBS Lett 204:183–188

Haslam R, Davidson M (1984a) Receptor-induced diacylglycerol formation in permeabilised platelets: possible role for a GTP-binding protein. J Receptor Res 4:605–629

Haslam R, Davidson M (1984b) Guanine nucleotides decrease the free $[Ca^{2+}]$ required for secretion of serotonin from permeabilized blood platelets: Evidence of a role for a GTP-binding-protein in platelet activation. FEBS Lett 174:90–95

Hawkins PT, Stephens L, Downes CP (1986) Rapid formation of inositol 1,3,4,5-tetrakisphosphate and inositol 1,3,4-trisphosphate in rat parotid glands may both result indirectly from receptor-stimulated release of inositol 1,4,5-trisphosphate from phosphatidylinositol 4,5-bisphosphate. Biochem J 238:507–516

Hepler JR, Harden TK (1986) Guanine nucleotide-dependent pertussis-toxin0insensitive stimulation on inositol phosphate formation by carbachol in a membrane preparation from human astrocytoma cells. Biochem J 239:141–146

Hildebrandt JD, Sekura RD, Codina J, Iyengar R, Manclark CR, Birnbaumer L (1982) Stimulation and inhibition of adenylyl cyclases mediated by distinct regulatory proteins. Nature (Lond) 302:706–709

Hinkle PM, Kinsella PA (1984) Regulation of thyrotropin-releasing hormone binding by monovalent cations and guanyl nucleotides. J Biol Chem 259:3445–3449

Hiraswa K, Irvine RF, Dawson RMC (1981) The hydrolysis of phosphatidylinositol monolayers at an air/water interface by the calcium-ion-dependent phosphatidylinositol phosphodiesterase of pig brain. Biochem J 193:607–614

Hofmann SL, Majerus PW (1982) Modulation of phosphatidylinositol-specific phospholipase C activity by phospholipid interactions, diglycerides, and calcium ions. J Biol Chem 257:14359–14364

Hokin MR, Hokin LE (1953) Enzyme secretion and the incorporation of ^{32}P into phospholipides of pancreas slices. J Biol Chem 203:967–977

Holian A (1986) Leukotriene B_4 stimulation of phosphatidylinositol turnover in macrophages and inhibition by pertussis toxin. FEBS Lett 201:15–19

Huque T, Bruch RC (1986) Odorant- and guanine nucleotide-stimulated phosphoinositide turnover in olfactory cilia. Biochem Biophys Res Commun 137:36–42

Hyslop P, Oades Z, Jesaitis A, Painter R, Cochrane C, Sklar L (1984) Evidence for N-formyl chemotactic peptide-stimulated GTPase activity in human neutrophil homogenates. FEBS Lett 166:165–169

Imboden JB, Shoback DM, Pattison G, Stobo JD (1986) Cholera toxin inhibits the T-cell antigen receptor-mediated increases in inositol trisphosphate and cytoplasmic free calcium. Proc Natl Acad Sci USA 83:5673–5677

Irvine RF, Moor RM (1986) Micro injection of inositol 1,3,4,5-tetrakisphosphate activates sea urchineggs by a mechanism dependent on external Ca^{2+}. Biochem J 240:917–920

Irvine RF, Letcher AJ, Dawson RMC (1984) Phosphatidylinositol-4,5-bisphosphate phosphodiesterase and phosphomonoesterase activities of rat brain. Some properties and possible control mechanisms. Biochem J 218:177–185

Irvine RF, Letcher AJ, Heslop JP, Berridge MJ (1986) The inositol tris/tetrakisphosphate pathway-demonstration of Ins(1,4,5)P_3 3-kinase activity in animal tissues. Nature (Lond) 320:631–634

Ishii H, Connolly TM, Bross TE, Majerus PW (1986) Inositol cyclic trisphosphate [inositol 1,2-(cyclic)-4,5-trisophosphate] is formed upon thrombin stimulation of human platelets. Proc Natl Acad Sci USA 83:6397–6401

Johnson RM, Connelly PA, Sisk RB, Pobiner BF, Hewlett EL, Garrison JC (1986) Pertussis toxin or phorbol 12-myristate 13-acetate can distinguish between epidermal growth factor- and angiotensin-stimulated signals in hepatocytes. J Biol Chem 83:2032–2036

Kanaide H, Matsumoto T, Nakamura M (1986) Inhibition of calcium transients in cultured vascular smooth muscle cells by pertussis toxin. Biochem Biophys Res Commun 140:195–203

Kelly E, Rooney TA, Nahorski SR (1985) Pertussis toxin separates two muscarinic receptor-effector mechanisms in the striatum. Eur J Pharmacol 119:129–130

Kikuchi A, Kozawa O, Kaibuchi K, Katada T, Ui M, Takai Y (1986) Direct evidence for involvement of a guanine nucleotide-binding protein in chemotactic peptide-stimulated formation of inositol bisphosphate and trisphosphate in differentiated human leukemic (HL-60) cells. J Biol Chem 261:11558–11562

Kojima I, Shibata H, Ogata E (1986) Pertussis toxin blocks angiotensin II-induced calcium influx but not inositol trisphosphate production in adrenal glomerulosa cell. FEBS Lett 204:347–351

Koo C, Lefkowitz R, Snyderman R (1983) Guanine nucleotides modulate the binding affinity of the oligopeptide chemoattractant receptor on human polymorphonuclear leukocytes. J Clin Invest 72:748–753

Krause K-H, Schlegel W, Wollheim CB, Anderson T, Wadlvogel FA, Lew DP (1985) Chemotactic peptide activation of human neutrophils and HL-60 cells. Pertussis toxin reveals correlation between inositol trisphosphate generation, calcium ion transients, and cellular activation. J Clin Invest 76:1348–1354

Litosch I, Fain JN (1985) 5-methyltryptamine stimulates phospholipase C-mediated breakdown of exogenous phosphoinositides by blowfly solivary gland membranes. J Biol Chem 260: 16052–16055

Litosch I, Wallis C, Fain J (1985) 5-Hydroxytryptamine stimulates inositol phosphate production in a cell-free system from blowfly salivary glands. Evidence for a role of GTP in coupling receptor activation to phosphoinositide breakdown. J Biol Chem 260:5464–5471

Lynch CJ, Prpic V, Blackmore PF, Exton JH (1986) Effect of islet-activating pertussis toxin on the binding characteristics of Ca^{2+}-mobilizing hormones and on agonist activation of phosphorylase in hepatocytes. Mol Pharmacol 29:196–203

Majerus P, Wilson D, Connolly T, Bross T, Neufeld E (1985) Phosphoinositide turnover provides a link in stimulus response coupling. Trends in Biochem Sci 10:168–171

Martin TFJ, Lucas DO, Majjalieh SM, Kowalchyk JA (1986a) Thyrotropin-releasing hormone activates a Ca^{2+}-dependent polyphosphoinositide phosphodiesterase in permeable GH_3 cells. GTPγS potentiation by a cholera and pertussis toxin-insensitive mechanism. J Biol Chem 261: 2918–2927

Martin TJF, Majjalieh SM, Lucas DO, Kowalchyk JA (1986b) Thyrotropin-releasing hormone stimulation of polyphosphoinositide hydrolysis in GH_3 cell membranes is GTP dependent but insensitive to cholera or pertussis toxin. J Biol Chem 261:10041–10049

Masters SB, Martin MW, Harden TK, Brown JH (1985) Pertussis toxin does not inhibit muscarinic-receptor-mediated phosphoinositide hydrolysis or calcium mobilisation. Biochem J 227:933–937

Meade CJ, Turner GA, Bateman PE (1986) The role of polyphosphoinositides and their breakdown products in A23187-induced release of arachidonic acid from rabbit polymorphonuclear leukocytes. Biochem J 238:425–436

Merritt JE, Taylor CW, Rubin RP, Putney JW (1986) Evidence suggesting that a novel guanine nucleotide regulatory protein couples receptors to phospholipase C in exocrine pancreas. Biochem J 236:337–343

Michell RH (1975) Inositol phospholipids in cell surface receptor function. Biochim Biophys Acta 415:81–147

Michell RH (1986) A second messenger function for inositol tetrakisphosphate. Nature (Lond) 324:613

Michell RH, Kirk CJ, Jones LM, Downes CP, Creba JA (1981) The stimulation of inositol lipid metabolism that accompanies calcium mobilisation in stimulated cells: defined characteristics and unanswered questions. Phil Trans R Soc B 296:123–138

Milligan G, Gierschik P, Spiegel AM, Klee WA (1986) The GTP-binding regulatory proteins of neuroblastoma x glioma, NG08-15, and glioma, C6, cells: Immunochemical evidence of a pertussis toxin substrate that is neither N_i nor N_O. FEBS Lett 195:225–230

Nakamura T, Ui M (1984) Islet activating factor, pertussis toxin inhibits Ca^{2+}-induced and guanine nucleotide dependent releases of histamine and arachidonic acid from rat mast cells. FEBS Lett 173:414–418

Nakamura T, Ui M (1985) Simultaneous inhibitions of inositol phospholipid breakdown, arachidonic acid release, and histamine secretion in mast cells by islet-activating protein, pertussis toxin: A possible involvement of the toxin-specific substrate in the Ca^{2+}-mobilizing receptor-mediated biosignaling system. J Biol Chem 260:3584–3593

Nishizuka Y (1984) The role of protein kinase C in cell surface signal transduction and tumour promotion. Nature (Lond) 308:693–698

Okajima F, Ui M (1984) ADP-ribosylation of the specific membrane protein by islet-activating protein, Pertussis toxin, associated with inhibition of a chemotactic peptide-induced arachidonate release in neutrophils: A possible role of the toxin substrate in Ca^{2+}-mobilizing biosignaling. J Biol Chem 259:13863–13871

Paris S, Puyssegur J (1986) Pertussis toxin inhibits thrombin-induced activation of phosphoinositide hydrolysis and Na^+/H^+ exchange in hamster fibroblasts. EMBO J 5:55–60

Patkar S, Kazimierczak W, Diamant B (1978) Histamine release by calcium from sodium fluoride-activated rat mast cells. Further evidence for a secretory process. Int Arch Allergy Appl Immunol 57:146–157

Pfeilschifter J, Bauer C (1986) Pertussis toxin abolishes angiotensin II induced phosphoinositide hydrolysis and prostaglandin synthesis in rat renal mesangial cells. Biochem J 236:289–294

Poll C, Kyrle P, Westwick J (1986) Activation of protein kinase C inhibits sodium fluoride-induced elevation of human platelet cytosolic free calcium and thromboxane B_2 generation. Biochem Biophys Res Commun 136:381–389

Prentki M, Deeney JT, Matschinsky FM, Joseph SK (1985) Neomycin: a specific drug to study the inositol-phospholipid signalling system. FEBS Lett 197:285–288

Rittenhouse SE (1983) Human platelets contain phospholipase C that hydrolyses polyphosphoinositides. Proc Natl Acad Sci USA 80:5417–5420

Rodbell M (1980) The role of hormone receptors and GTP-regulatory proteins in membrane transduction. Nature (Lond) 284:17–22

Sasaguri T, Hirata M, Kuriyama H (1986a) Dependence on Ca^{2+} of the activities of phosphatidylinositol 4,5-bisphosphate phosphodiesterase and inositol 1,4,5-trisphosphate phosphatase in smooth muscles of the porcine coronary artery. Biochem J 231:497–503

Sasaguri T, Hirata M, Itoh T, Koga T, Kuriyama H (1986b) Guanine nucleotide binding protein involved in muscarinic responses in the pig coronary artery is insensitive to islet-activating protein. Biochem J 239:567–574

Schacht J (1976) Inhibition by neomycin of polyphosphoinositide turnover in subcellular fractions of guinea pig cerebral cortex in vitro. J Neurochem 27:1119–1124

Schacht J (1978) Purification of polyphosphoinositides by chromatography on immobilized neomycin. J Lipid Res 1063:1067

Schimmel R, Elliott M (1986) Pertussis toxin does not prevent alpha-adrenergic-stimulated breakdown of phosphoinositides or respiration in brown adipocytes. Biochem Biophys Res Comm 135:823–829

Shukla S (1982) Phosphatidylinositol specific phospholipase C. Life Sci 30:1323–1335

Siess W, Lapetina EG (1986) Neomycin inhibits inositol phosphate formation stimulated by thrombin but not other agonists. FEBS Lett 207:53–57

Smith CD, Lane BC, Kusaka I, Verghese MW, Snyderman R (1985) Chemoattractant receptor induced hydrolysis of phosphatidylinositol 4,5-bisphosphate in human polymorphonuclear leukocytes membranes: Requirement for a guanine nucleotide regulatory protein. J Biol Chem 260:5875–5878

Smith CD, Cox CC, Snyderman R (1986) Receptor-coupled activation of phosphoinositide specific phospholipase C by an N protein. Science 232:97–100

Spangrude GJ, Sacchi F, Hill HR, van Epps DE, Daynes RA (1985) Inhibition of lymphocyte and neutrophil chemotaxis by pertussis toxin. J IMmunol 135:4135–4143

Streb H, Heslop JP, Irvine RF, Schulz I, Berridge MJ (1985) Relationship between secretogogue-induced Ca^{2+} release and inositol polyphosphate production in permeabilized pancreatic acinar cells. J Biol Chem 260:7309–7315

Strnad CF, Wong K (1985) Calcium mobilization in fluoride activated human neutrophils. Biochem Biophys Res Commun 133:161–167

Strnad CF, Parente JE, Wong K (1986) Use of fluoride ion as a probe for the guanine nucleotide-binding protein involved in the phosphoinositide-dependent neutrophil transduction pathway. FEBS Lett 206:21–24

Stryer L (1983) Transducin and the cyclic GMP phosphodiesterase: Amplifier proteins in vision. Cold Spring Harbor Symposia on Quantitative Biology 48:841–852

Taylor CW, Merritt JE, Putney JJW, Rubin RP (1986) Effects of Ca^{2+} on phosphoinositide breakdown in exocrine pancreas. Biochem J 238:765–772

Trimble ER, Bruzzone R, Biden TJ (1986) Secretin induces rapid increases in inositol trisphosphate, cytosolic Ca^{2+} and diacylglycerol as well as cyclic AMP in rat pancreatic acini. Biochem J 239:257–261

Uhing RJ, Prpic V, Jiang H, Exton JH (1986) Hormone-stimulated polyphosphoinositide breakdown in rat liver plasma membranes. Roles of guanine nucleotides and calcium. J Biol Chem 261:2140–2146

Van Rooijen L, Agranoff B (1985) Inhibition of polyphosphoinositide phosphodiesterase by aminoglycoside antibiotics. Neurochem Res 10:1019–1024

Van Rooijen L, Seguin E, Agranoff B (1983) Phosphodiesteratic breakdown of endogenous polyphosphoinositides in nerve endings. Biochem Biophys Res Commun 112:919–926

Vergara J, Tsien RY, Delay M (1985) Inositol 1,4,5-trisphosphate: A possible chemical link in excitation-contraction coupling in muscle. Proc Nat Acad Sci USA 82:6352–6356

Wakelam MJO, Murphy GJ, Hruby VJ, Houslay MD (1986) Activation of two signal-transduction systems in hepatocytes by glucagon. Nature (Lond) 323:68–71

Whitaker M, Aitchinson M (1985) Calcium-dependent polyphosphoinositide hydrolysis is associated with exocytosis in vitro. FEBS Lett 182:119–124

Wilson DB, Bross TE, Hofmann SL, Majerus PW (1984) Hydrolysis of polyphosphoinositides by purified sheep seminal vesicle phospholipase C enzymes. J Biol Chem 259:11817–11824

Wilson D, Connolly T, Bross T, Majerus P, Sherman W, Tyler A, Rubin L, Brown J (1985) Isolation and characterisation of the inositol cyclic phosphate products of polyphosphoinositide cleavage by phospholipase C. Physiological effects in permeabilised platelets and limulus photoreceptor cells. J Biol Chem 260:13496–13501

Wojcikiewicz RJH, Kent PA, Fain JN (1986) Evidence that thyrotropin-releasing hormone-induced increases in GTPase activity and phosphoinositide metabolism in GH_3 cells are mediated by a guanine nucleotide-binding protein other than G_s or G_i. Biochim Biophys Res Commun 138: 1383–1389

Worley PF, Baraban JM, Van Dop C, Neer EJ (1986) Go, a guanine nucleotide-binding protein: Immunohistochemical localization in rat brain resembles distribution of second messenger systems. Proc Natl Acad Sci USA 83:4561–4565

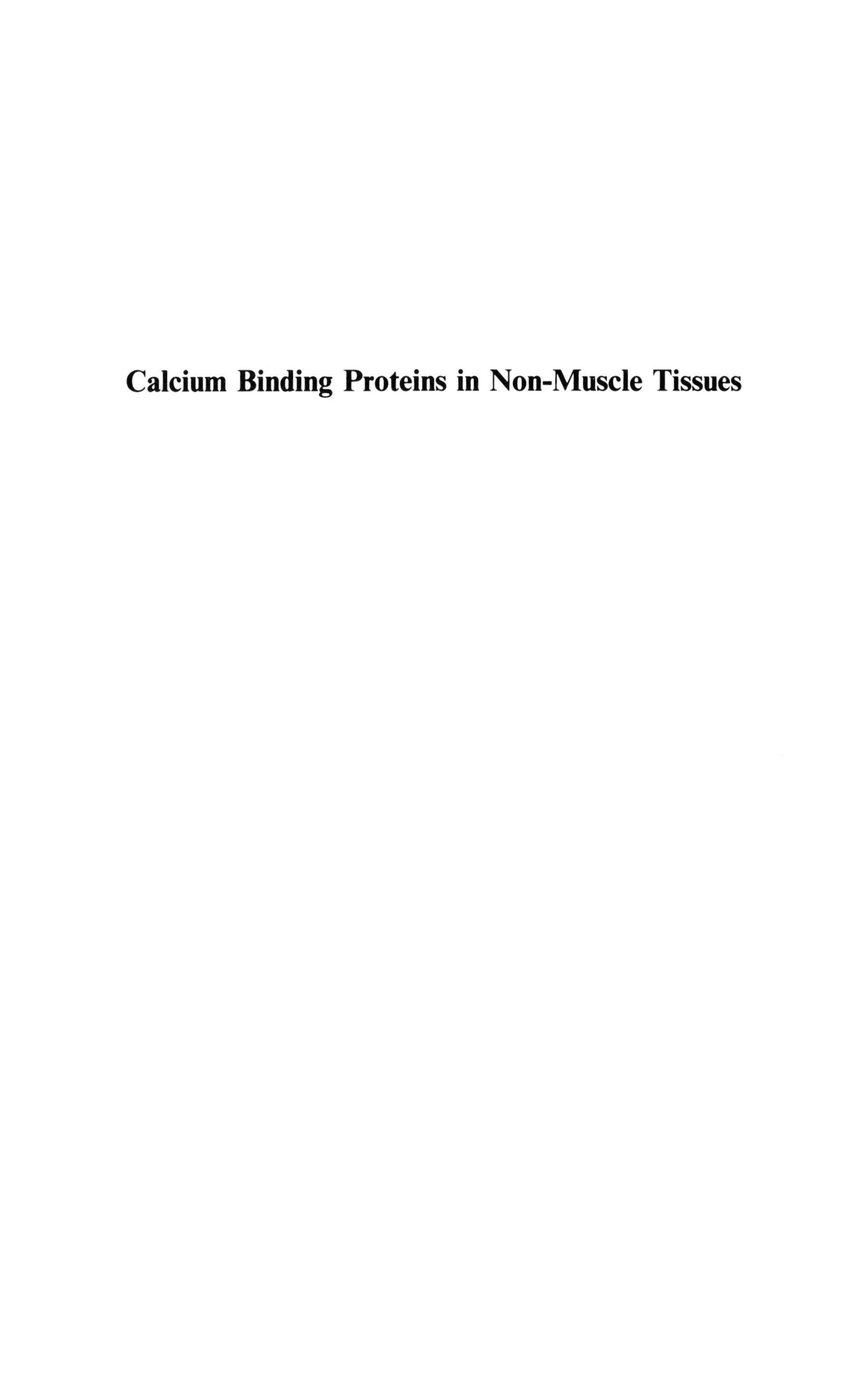

Calcium Binding Proteins in Non-Muscle Tissues

Soluble Calcium Binding Proteins in Vertebrate and Invertebrate Muscles

CH. GERDAY[1]

1 Introduction

The beginning of the story of EF hand calcium modulated proteins dates back to 1965, when three isotypes of a soluble calcium binding protein were isolated from carp skeletal muscle (Hamoir and Konosu 1965; Konosu et al. 1965). These proteins were found to be a constant feature of fish muscle and the general name parvalbumin was given. It was believed for a long time that these proteins were specific to the white muscles of lower vertebrates, fish and amphibians, in which their concentration is often close to 1 mM.

However, Lehky et al. (1974) and Péchère (1974) succeeded in isolating parvalbumin from rabbit skeletal muscle, they were also extracted from human muscle (Lehky et al. 1974), chicken muscle (Blum et al. 1977; Heizmann et al. 1977; Strehler et al. 1977), rat (Rinaldi et al. 1982) and mouse muscles (Heizman 1984). In search of parvalbumins in invertebrate muscles, Benzonana et al. (1974) discovered another type of soluble calcium binding proteins (SCBP) in crustacean muscle. They were also identified in other invertebrates, such as amphioxus (Kohler et al. 1978), sandworm (Cox and Stein 1981; Gerday et al. 1981), scallop (Collins et al. 1983) and earthworm (Gerday et al. 1986b). Like parvalbumins, they are acidic proteins, and their concentration in muscle is of the same order of magnitude as that of parvalbumin in vertebrate muscle. For reviews concerning parvalbumin and SCBP see (Gerday 1982; Demaille 1982; Wnuk et al. 1982; Heimann 1984).

2 Purification

Parvalbumin is evenly distributed in the cellular space, whereas it has been reported that SCBP could be apparently localized at the level of the isotropic band, its distribution following that of actin (Benzonana et al. 1977). Both proteins are, however, easily extracted by homogenization of the muscle tissue in low ionic strength buffers or even in water. Fractionation of the extract can be made either with acetone (Bhushana Rao et al. 1969), ammonium sulfate (Gosselin-Rey and Gerday 1977) or heat treatment (Strehler et al. 1977). Molecular sieve and DEAE cellulose chromatography are

1 Laboratoire de Biochimie Musculaire, Institut de Chimie B6, Sart Tilman, 4000 Liege, Belgium

Ch. Gerday, R. Gilles, L. Bolis (Eds.)
Calcium and Calcium Binding Proteins

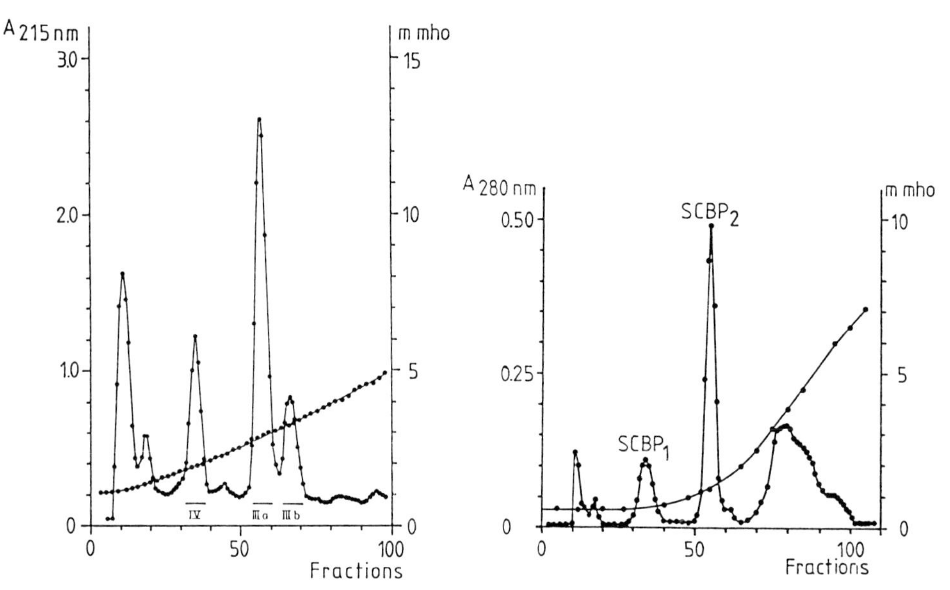

Fig. 1. **A** Separation of the parvalbumin isotypes (IV, IIIa and IIIb) from *Leptodactylus insularis* muscle on a DEAE cellulose column (1.5 × 30 cm) equilibrated in 15 mM piperazine-HCl pH 6 buffer. Gradient 200–200 ml, 0.15 M NaCl. **B** Separation of the SCBP isotypes ($SCBP_1$ and $SCBP_2$) from *Lumbricus terrestris* muscles on a DEAE cellulose column (2.5 × 30 cm). The column was eluted first with 100 ml of a 15 mM HCl-Tris, 2 mM 2-mercaptoethanol buffer pH 7.8 and then with a linear gradient 250–250 ml 0.2 M NaCl

usually the additional methods used to complete the purification of these proteins. As an example, the Fig. 1A and B shows the chromatographic profiles obtained on DEAE cellulose illustrating the isolation of the three parvalbumin isotypes from tropical frog *Leptodactylus insularis* and of the two SCBP isotypes from the obliquely striated muscle of the earthworm, *Lumbricus terrestris.*

3 General Properties

Parvalbumins show a remarkable constancy in their physico-chemical properties. Their molecular weight is close to 12,000, they have isoelectric points ranging from 3.9 to 6.6, high diffusion constant and a high phenylalanine to tyrosine and tryptophan ratio so that the fine absorption bands of phenylalanine residues are always visible on the UV spectrum. They have more than 50% of their amino acid residues organized into helical regions. They bind two calcium with high affinity around 10^8 M^{-1} and the binding sites can also accommodate magnesium with a lower affinity $K_{ass} = 10^4$ M^{-1}. These sites can therefore be compared to the two Ca^{2+}-Mg^{2+} sites of troponin-C. They are evenly distributed in the sarcoplasm and are, an exception for mammalian muscle, present in the cellular space under the form of several isotypes up to five in lungfish skeletal muscle (Gerday et al. 1979b). SCBP from invertebrate muscles are also acidic proteins, they have a molecular weight close to 20,000 but appear in crustacean muscle under the form of homo and hetero dimers. They bind one to three calcium per 20,000 M.W. with a high affinity $K_{ass} = 10^8 M^{-1}$ and also Mg^{2+} with a lower affinity $K_{ass} = 10^4 M^{-1}$. The polypeptide chain is also largely organized in helical regions whereas the UV absorption spectra are rather classical with a maximum at 280 nm. Polymorphism is absent or reduced to two isotypes very similar to each other contrarily to the parvalbumin isotypes. Their concentration in fast skeletal muscle is higher than 100 μmol/kg wet weight.

4 Structural Characteristics

The amino acid sequences of parvalbumins determined so far are shown in Fig. 2, the parvalbumin isolated from the very fast muscle of the swimbladder of the toadfish being the last we have recently added to the list. Large variations between the amino acid sequences of different parvalbumins are often encountered even between isotypes found in the same muscle cell. They are, for instance, 49 amino acids replacements including additions or deletions, when the primary structures of parvalbumins Pike II and Pike III are compared. They apparently belong to two different evolutionary lineages (Gerday 1976). 24 amino acid residues are common to parvalbumins of ectothermic animal origin and these residues are identically positioned in the two mammalian parvalbumin sequences known so far. They include residues 51, 53, 55, 57 and 62 along with residues 90, 92, 94, 96 and 101, which are implicated in the coordination of the two calcium bound to the protein. Typical characteristic features of mammalian parvalbumins appear to be the triplet of lysine residues at position 36–38 and the sequence 83–89 KTLMAAG. The three-dimensional structure of parvalbumin was first determined using carp III parvalbumin (Kretsinger and Nockolds 1973) and also recently a terbium derivative of the main parvalbumin isolated from the swimbladder of the toadfish (Wery et al. 1985). The α-carbon backbones are very similar, the largest discrepancy being found at the level of the NH_2-terminal end of the chain, a region, however, of low electron density and of delicate interpretation. There is no similarity in the packing of carp and toadfish parvalbumin terbium derivative. In both cases, however, the polypeptide chain is made of three homologous regions A-B, C-D and E-F, each representing a domain defined as an "EF hand" domain (Kretsinger 1976, 1980) consisting of a loop of 11 residues flanked on either sides by two helices (A, B, C, D and E, F) each of about ten residues. In parvalbumin the domain A-B has lost the ability to bind calcium.

The three-dimensional structure of the soluble calcium binding proteins from invertebrate muscles has not yet been solved, but five amino acid sequences of SCBP polypeptide chains isolated from shrimp α and β chain (Takagi and Konishi 1984a,b), scallop (Takagi et al. 1984), sandworm (Kobayashi et al. 1984) and amphioxus (Takagi et al. 1986) are known. These amino acid sequences are shown in Fig. 3, the residues have been aligned in order to maximize the isology between the polypeptide chains. One can see that the primary structure of these proteins is not very conservative, since only nine residues are common to the five polypeptide chains. These chains are also

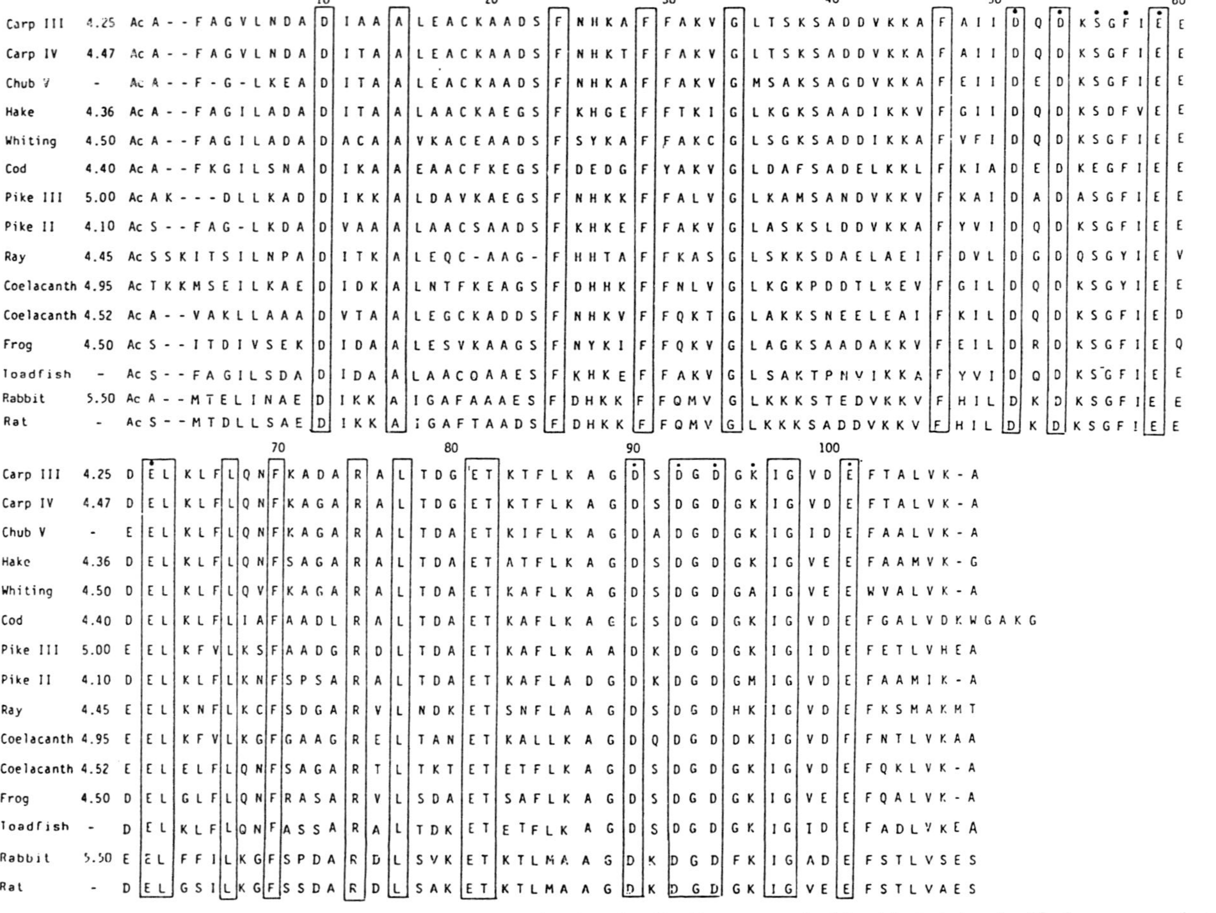

Fig. 2. Amino acid sequences of the parvalbumin determined so far. Residues *topped with a black dot and a black square* are implicated in the coordination of calcium in the CD and EF site respectively. The sequences were taken from the following references: carp **pI** 4.25 (Coffee and Bradshaw 1973); carp **pI** 4.47 (Coffee et al. 1974); chub V (Gerday et al. 1978); hake **pI** 4.36 (Capony and Pechère 1973); whiting **pI** 4.5 (Joassin and Gerday 1977); Cod **pI** 4.4 (Elsayed and Bennich 1975); pike III **pI** 5.0 (Frankenne et al. 1973; Gerday, unpublished); pike II **pI** 4.1 (Gerday 1976); ray **pI** 4.45 (Thatcher and Pechère 1977); coelacanth **pI** 4.95 (Pechère et al. 1978); coelacanth **pI** 4.52 (Jauregui-Adell and Pechère 1978); frog **pI** 4.5 (Capony et al. 1975); toadfish (Gerday et al., unpublished), rabbit (Capony et al. 1976); rat (Berchtold et al. 1982)

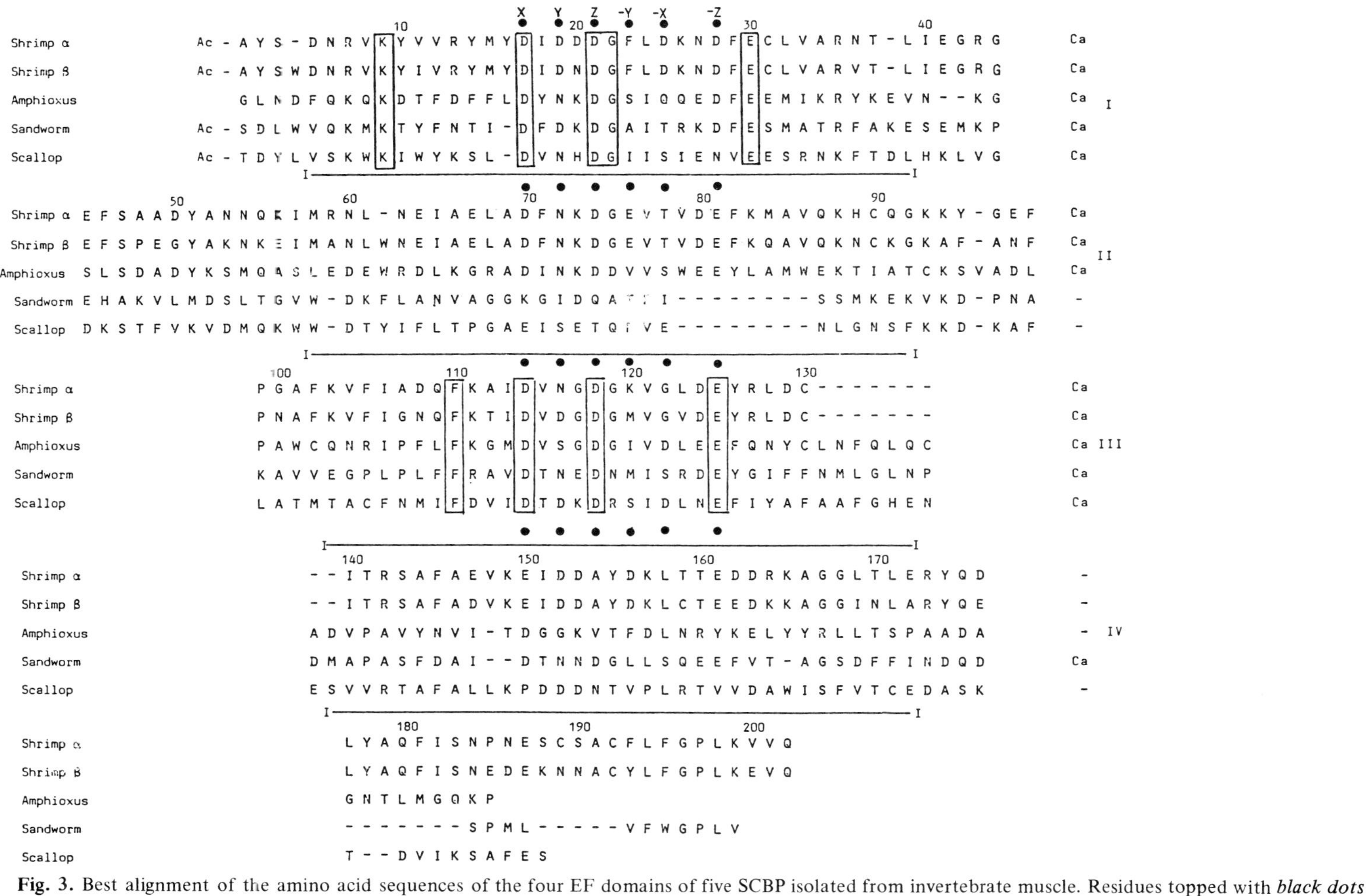

Fig. 3. Best alignment of the amino acid sequences of the four EF domains of five SCBP isolated from invertebrate muscle. Residues topped with *black dots* are theoretically those implicated in calcium coordination. Amino acid sequences have been taken from the following references: shrimp α chain (Takagi and Konishi 1984a); shrimp β chain (Takagi and Konishi 1984b); amphioxus (Takagi et al. 1986); sandworm (Kobayashi et al. 1984); scallop (Takagi et al. 1984)

made of four "EF hand" domains, some of them having lost their ability to bind calcium as a consequence of point mutations having affected residues normally implicated in calcium coordination. Isology between parvalbumin and SCBP can be hardly detected exception being at the level of the presumed positions of residues implicated in calcium binding. One can indeed easily discern in SCBP, by analogy with the calcium binding sites of parvalbumin, the sites which have kept the ability to bind Ca^{2+}: domain I and III for all SCBP, domain II in α and β chain of shrimp and in amphioxus and domain IV in sandworm SCBP. In domain II of sandworm and scallop SCBP there is a deletion at position-Z, a replacement of the Asp residue, usually occupying position X, by a Lys and a Glu residue respectively and similar changes at the level of position Z. In domain IV only the sandworm SCBP still binds calcium, in all other, point mutations have mainly affected position Z.

The homology between parvalbumin and SCBP has also been confirmed by peptide mapping (Gerday et al. 1981).

5 Structural Relations with Other Calcium-Binding Proteins

This point has been exhaustively discussed in the original papers of Collins et al. (1973), Weeds and McLachlan (1974), Vanaman et al. (1977), Goodman and Pechère (1977), Goodman et al. (1979) and in the review articles mentioned above. To summarize, it has been proposed by Kretsinger (1976, 1980) that the calcium modulated proteins, the biological function of which is switched on and off as a function of calcium binding, are made by the assembly of domains homologous to the EF hand domain of parvalbumin. Inversely, any calcium binding protein containing at least one EF hand domain is a calcium modulated protein. The comparison of the amino acid sequences of several calcium binding proteins has allowed the detection, mainly at the level of the presumably calcium binding sites, of a high degree of sequence isology with the amino acid sequence of the calcium binding site of parvalbumin. This is the case for troponin-C, myosin light chains, calmodulin, calbindin-D, oncomodulin, S-100 proteins and SCBP. It has been proposed (Weeds and McLachlan 1974) that these proteins derived from a four-domain ancestor having four calcium binding sites. The recent elucidation of the three-dimensional structure of vitamin-D-dependent calcium binding protein (Calbindin-D) by Szebenyi et al. (1981), Szebenyi and Moffat (1986), troponin-C (Herzberg and James 1985; Sundaralingam et al. 1985) and calmodulin (Babu et al. 1985) has confirmed that these proteins effectively contain, as postulated by R.H. Kretsinger, domains homologous to the EF hand calcium binding domain of parvalbumin. Another calcium binding protein can now be added to the list, the B-subunit of calcineurin, a calmodulin binding protein containing four EF hand domains, present in mammalian brain (Aitken et al. 1984). There is also a strong probability, although the amino acid sequence of this protein is not known that CBP-18 a 18,000 M.W. protein containing a high affinity binding site (Manalan and Klee 1984) also belongs to the EF hand family. Table 1 summarizes some of the characteristics of the different EF hand proteins mentioned above.

Table 1. EF-hand calcium binding proteins

Protein	M.W. (±)	EF domains	Ca^{2+} binding	Main source
Parvalbumin	12,000	3	2	Muscle
Troponin-C	18,000	4	4	Muscle
Calmodulin	17,000	4	4	Ubiquitous
P-light chains (myosin)	20,000	4	1	Muscle
Alkali light chains (myosin)	20,000	4	0	Muscle
Calbindin (Vit D)	10,000	2	2	Mammal intestine
Calbindin (Vit D)	28,000	4–6	2	Avian intestine
S-100	2 × 10,500	2	2	Brain
Oncomodulin	12,000	2	2	Tumor
Calcineurin-B subunit	19,000	4	4	Brain
CBP-18	18,000	?	1	Brain
SCBP	20,000 or dimers	4	1–3	Invertebrate muscle

6 Distribution in Muscle Tissue

6.1 Skeletal Muscle

The distribution of parvalbumins in various muscles has been evaluated qualitatively and quantitatively using polyacrylamide gel electrophoresis, enzyme-linked immunoassays (ELISA) and HPLC techniques. Polyacrylamide gel electrophoresis of low ionic strengh extracts of nine muscles of the lungfish (Gerday et al. 1979b) and 18 muscles of the carp including a pool of six muscles involved in eye movements (Hamoir et al. 1981) show that the parvalbumin pattern varies from one muscle to the other not only at the level of the overall parvalbumin concentration, but also at the level of the relative distribution of isotypes; i.e., each muscle exhibits a specific parvalbumin pattern. In general, the amount of parvalbumin in fish muscles seems to be inversely related to their myoglobin content. An apparent exception mentioned by Hamoir et al. (1981) is, for instance, the eye muscles which despite a low myoglobin content shows a parvalbumin pattern resembling that of a red muscle, i.e., low content in parvalbumin mainly under the form of isotype I. This could be due, however, to the presence of an important proportion of slow tonic fibers usually found in the orbital layer of the extra-occular muscles (Heizmann 1984).

In the case of fish, slow twitch fibers are used for slow movements and sustained swimming at low speed. To generate the power required to accomplish fast movements, fish recruit muscle fibers of a faster type, usually white fibers, which are also fast twitch glycolytic fibers. This would mean that, in fish, parvalbumin is mainly associated with fast twitch glycolytic fibers showing essentially anaerobic metabolism. The relation between parvalbumin content and speed of contraction was also illustrated by Hamoir et al. (1980), who showed that the parvalbumin concentration in the superfast muscle of toadfish swimbladder was nearly threefold higher than that found in the parietal muscle of the same species.

Table 2. Parvalbumin content in various mammalian muscles. (After Heizmann et al. 1982)

Animal	Muscle	Fiber type composition %	Half relaxation time ms	Parvalbumin g/kg
Mouse	Gast	100 IIB	–	4.9
	EDL	100 IIA + IIB	7.1	4.4
	Soleus	50 I + 50 IIA	22.5	0.01
Rat	Gast	100 IIB	–	3.3
	EDL	5 I, 95 (IIA + IIB)	8	2.4
	EOM	20–30 I, 80–70 (IIA+IIB)	6.3	1
	Soleus	85 I, 15 IIA	49	0.004
Man	Vastus	50 I, 50 (IIA + IIB)	–	0.001
	Triceps	50 I, 50 (IIA + IIB)	>50	0.001

Abbreviations: Gast, gastrocnemius; EDL, extensor digitorum longus; EOM, extraocular muscle.

Clearly also invertebrate SCBP is associated with fast contracting muscle; Cox et al. (1976) have indeed shown that the concentration of SCBP in crayfish tail muscle is about three times higher than that found in the slow muscle of the claw. Immunohistochemical techniques (Celio and Heizmann 1982) have also demonstrated that in rat skeletal muscle fiber, parvalbumin is exclusively found in IIB fibers which are fast twitch glycolytic fibers.

More quantitative studies of the distribution of parvalbumin have been carried out by Heizmann et al. (1982) on various muscles of mammalian origin, including man. The authors were looking especially for the possible relation which could exist between the physiological properties of the muscle, such as contraction and relaxation velocities, fiber type composition and parvalbumin content. A summary of their results is given in Table 2. One can see that parvalbumin is mainly associated with IIB fibers as in fish muscle. IIA fibers (fast twitch oxidative fibers) and type I fibers (slow fibers) are almost completely devoid of parvalbumin. Clearly also, the parvalbumin concentration is inversely related to the half-relaxation time of the muscle.

6.2 Smooth and Cardiac Muscle

The search for parvalbumins in smooth and cardiac muscles of numerous animals has been carried out using several techniques, mainly polyacrylamide gel electrophoresis, radioelectrophoresis and immunochemical techniques (Pechère et al. 1973; Gosselin-Rey 1974; Baron et al. 1975; Heizmann et al. 1977; Le Peuch et al. 1978; Gerday et al. 1979; Hamoir et al. 1980). All these investigations have led to the conclusion that smooth muscles do not contain any parvalbumin. The presence of parvalbumin in cardiac muscle is certainly more controversial. In mammals, radio electrophoresis (Le Peuch et al. 1978) has apparently revealed the presence of parvalbumin (19 μmol/kg) in shrew heart muscle, which is characterized by a very fast beating rate (1000 beats/min). Heart chicken also seems to contain significant amount of parvalbumin (Heizmann et al. 1977) as well as fish heart (Gerday et al. 1979b) and especially ventricular muscle of myoglobin-hemoglobin free antarctic fish showing a clear-cut parvalbumin electrophoretic pattern different, however, from that of trunk muscle (Hamoir and Gérardin-Otthiers 1980).

7 Parvalbumin and Animal Development

The parvalbumin concentration in skeletal muscles has been followed during the prenatal and postnatal development of chicken (Le Peuch et al. 1979), rat (Celio and Heizmann 1982; Berchtold and Means 1985) and rabbit (Leberer and Pette 1986) using radioimmunoassay, ELISA and RNA blot hybridization techniques. In chicken, it was shown that parvalbumin appears in skeletal muscle around the time of hatching, together with the sarcoplasmic reticulum ATPase. In rat gastrocnemius, parvalbumin m-RNA increases constantly up to 20-fold from day 5 to day 20, contrary to what happens in brain, where the parvalbumin m-RNA content is comparatively constant during development. The increase in parvalbumin m-RNA is apparently correlated to the differentiation of immature muscles fibers into fast twitch glycolytic fiber. This is corroborated by the study of Leberer and Pette (1986), who have followed the parvalbumin content in various fast and slow twitch muscles during the embryonic and postnatal development of the rabbit. They have shown that parvalbumin synthesis is switched on almost immediately after birth and that the amount of this protein, in presumptive fast muscle, increases progressively to reach a maximum about 90 days after birth. This also occurs synchronously with the onset of high frequency neural activity and the concomitant loss of polyneural innervation.

8 Biological Function

The fact that soluble calcium binding proteins are associated with fast relaxation time suggests that these proteins could be involved in the relaxation process of fast twitch fibers. Various experiments have been carried out in order to define the role of these proteins in the contraction cycle. We (Gerday and Gillis 1976; Gillis and Gerday 1977) and other (Blum et al. 1977; Pechère et al. 1977) have demonstrated that calcium-free parvalbumin was able to inactivate the ATPase activity of myofibrils and that in turn sarcoplasmic reticulum could decalcify calcium-loaded parvalbumin. Piront et al. (1980) have also obtained similar results using myofibrils, SCBP and sarcoplasmic reticulum from the tail muscle of the crayfish. From these experiments, it was inferred that the soluble calcium binding proteins could act as a shuttle mechanism transporting calcium from the myofibrils to the sarcoplasmic reticulum during the relaxation phase. These hypothesis has received support from computer simulation studies (Gillis et al. 1982), although similar analysis (Robertson et al. 1981) has led to the opposite conclusion, that parvalbumins cannot be involved in muscle relaxation. In this latter study, however, the authors have chosen a parvalbumin concentration equal to that of troponin 70 μmol/kg amount which is six times lower than that found in rat gastrocnemius, an homogeneous IIb fiber muscle (Heizmann et al. 1982) and 5 to 20 times lower than the concentration found in homogeneous fast twitch glycolytic muscles of fish and amphibians. Although quite realistic for rabbit whole muscle the concentration of 70 μmol/kg does not represent the real parvalbumin concentration found in IIb fiber, but rather an average value which strongly depends on the relative proportion of IIA, IIB and I type fibers in the muscle (Heizmann et al. 1982). The idea of the existence of parvalbumin-calcium transients well after the relaxation phase agrees well with the finding that when the muscle tension has gone, a large part of the liberated calcium from myofibrils is still found in the cytoplasm obviously bound to the parvalbumin and not located into the sarcoplasmic reticulum (Somlyo et al. 1981; Gillis 1985). The formation of a parvalbumin-calcium complex during relaxation also explains the time course of the labile heat, first described by Aubert (1956), resulting from the exothermic binding process of calcium to parvalbumin (Peckham and Woledge 1986).

The presumptive participation of parvalbumin and more generally of the soluble calcium binding proteins in muscle relaxation is also supported by cross-innervation and denervation experiments carried out on various rabbit muscles (Leberer and Pette 1986). It was shown that denervation of fast twitch muscle induces a dramatic decrease of the parvalbumin content, which reaches the level of a slow muscle (soleus) within 75 days. Low frequency chronic stimulation of a fast twitch muscle also results in an even more pronounced effect, since after 3 weeks the parvalbumin content falls within the concentration range found in slow twitch soleus.

It has also been suggested (Gillis 1980) that soluble calcium binding proteins could be of particular importance in ectothermic animals which have no internal regulation of temperature. One can suppose indeed that the temperature dependence of the SR pumping rate could be substantially greater than that of calcium binding to parvalbumin or SCBP. Indeed, in the case of frog (*Rana temporaria*), the Q10 of the calcium removal from the cytoplasm by the sarcoplasmic reticulum, within the range of 0–25°C, is around 3. So ectothermic animals which have the ability to maintain their muscle physiologically active at low temperature must have compensating mechanisms which could be an increase in their parvalbumin content. In collaboration with S. Taylor from the Mayo Clinic in Rochester (Minnesota), we have tested the Gillis hypothesis by comparing first the parvalbumin content in three groups of muscles (foreleg, abdominal and thigh) of two populations of frogs (*Rana temporaria*) respectively kept a 18 and 2°C for a period of 6 months. The parvalbumin content was measured in whole extracts by three different techniques:

1. Measurement of the calcium content in exhaustively dialyzed whole extracts of muscles and in the parvalbumin-fraction obtained by chromatography of an aliquot of the extract on an ACA54 column equilibrated in 50 mM NH_4HCO_3.
2. ELISA sandwich test using antibodies raised against parvalbumin IVa and IVb and a 50:50 (w/w) mixture of the purified antibodies coupled to peroxidase.
3. HPLC chromatography of aliquots of heated extracts (5 min at 60°C) on a reverse phase column 25 × 0.4 cm (Nucleosil 300-7C_8 – Macherey Nagel – Germany). In the conditions used, the parvalbumin IVa and IVb, present in frog muscles, are well separated from each other and from the other components of the extract

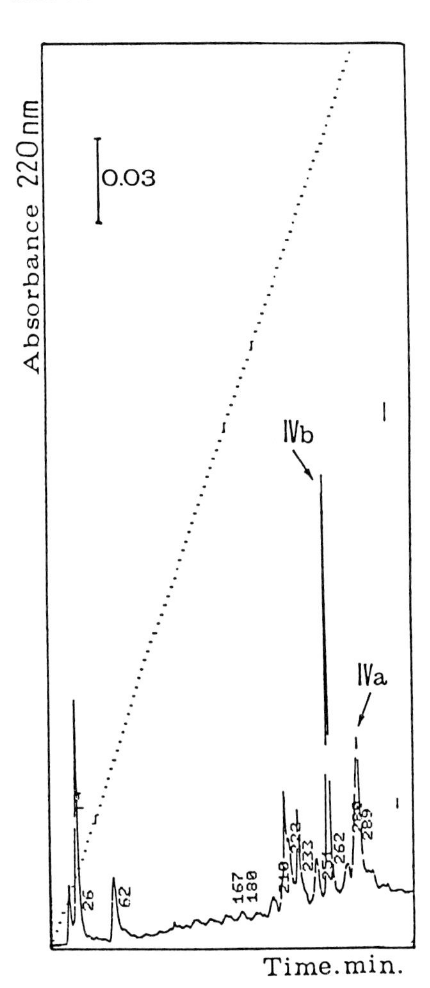

Fig. 4. High performance liquid chromatography of 50 μl of a heated whole extract of an equivalent of 23 mg of foreleg fresh muscle of *R. temporaria*. Nucleosil 300-7RP column (0.4 × 12.5 cm). Buffer A: 0.1% H_3PO_4 10 mM $NaClO_4$. Buffer B: 60% V/V acetonitrile in the buffer. Linear gradient 0–100% B in 35 min

(Fig. 4). This has been checked by PAGE of the fractions corresponding to the parvalbumin isotypes, collected from the HPLC column.

Another test was to compare the parvalbumin concentration in muscle of an amphibian from temperate regions (*Rana temporaria*) with that of an amphibian from tropical regions (*Leptodactylus insularis*) which never experiences temperature below 18°C. The parvalbumin was measured by Ca^{2+} titration on exhaustively dialyzed whole extracts and by HPLC chromatography of the heat stable extract. One has to note that muscle fibers isolated from *Rana temporaria* and *Leptodactylus insularis* manifest a completely different behavior on exposure to cold temperature. Indeed, the tropical muscle fibers are unable to produce tension upon exposure to 4°C unlike muscle fibers from *R. temporaria* which display a potentiation in their force producing ability at the colder temperature.

Table 3 summarizes the results obtained with "cold" and "warm" adapted *Rana temporaria.* The analyzed material was also sampled as a function of sex but no differ-

Table 3. Parvalbumin concentration in three groups of muscles (thigh, foreleg and abdominal) from cold- and warm-adapted *Rana temporaria*

	Parvalbumin in mmol/kg muscle[a] as determined by different techniques					
Muscle group	Ca^{2+} determination[b]		ELISA		HPLC	
	18°C	2°C	18°C	2°C	18°C	2°C
Thigh	0.228	0.287	0.258	0.253	0.206	0.224
Foreleg	0.240	0.284	0.303	0.359	0.204	0.352
Abdominal	0.225	0.252	0.218	0.317	0.220	0.364

[a] The value are averages of at least three determinations on pooled muscle from several specimens.
[b] The value are average of the data obtained on whole extracts and after chromatography on ACA54 column.

ence was observed between male and female specimens. The different techniques used each have their advantages and disadvantages: the calcium measurement on the dialyzed extract or on the parvalbumin fraction isolated from molecular sieve chromatography (ACA54) gives reproducible results, but is not sensitive enough to evaluate the parvalbumin concentration in single muscle fibers. The ELISA test shows a variation coefficient of about 10%. Using a 10 μg/ml coating concentration of a mixture of affinity purified anti-parvalbumin immunoglobulin the sandwich test gives a linear answer over a parvalbumin range of 25 to 250 ng/ml. This allows a reliable estimation of parvalbumin in single muscle fiber. If the method is easy to use when only one parvalbumin isotype is present, matters become more complicated if the muscle fiber contains several isotypes because they often show, as in the case of *Rana temporaria* parvalbumins, immunological reactions of partial identity. A complete analysis of a heated muscle extract by HPLC is achieved within 40 min and the technique allows the detection at 220 nm of 1 μg of isotype IVB in 50 μl of the buffer. This corresponds to a parvalbumin concentration of about 3 μmol/kg fresh muscle, which is among the lowest values of parvalbumin concentration encountered in skeletal muscle. Difficulties occur however with some parvalbumins, such as *R. temporaria* parvalbumin IVA which are easily denatured under a large variety of conditions and give, under purified form, unsymetrical peak on the HPLC column due to a physical denaturation of the protein. The situation is apparently better in extracts which, of course, support a minimum of manipulations. The high sensitivity of some parvalbumin to denaturation seems always to be restricted to parvalbumins containing several cysteine residues such as *R. temporaria* component IVA (Gosselin-Rey and Gerday 1977) or whiting parvalbumin (Joassin and Gerday 1977). This precludes, of course, the precise estimation of parvalbumin concentration by the HPLC method, the validity of which is limited to stable parvalbumins.

From Table 3 and taking into account the relative precision of the methods discussed above one can say that there is no significant difference in the muscle parvalbumin content of the two groups of frogs, adapted or not to cold temperature, nor in the muscle of the tropical amphibian *L. insularis*. In this latter case, the parvalbumin concentration (three isotoypes present) varies from 0.2–0.3 mmol/kg fresh muscle depending on the specimen investigated. So it seems clear that these soluble calcium-binding proteins do not compensate for the loss of efficiency of the calcium pumping of the sarcoplasmic reticulum at low temperature.

Another intriguing point is the fact that parvalbumins can bind nucleotides with affinities enabling the formation of protein-nucleotide complexes at physiological concentrations (Permyakov et al. 1982). The data indicate that during the resting state parvalbumin-ATP complexes are formed with the metal-free or magnesium-loaded form of the protein, whereas during relaxation the formation of parvalbumin nucleotide complexes occurs mainly under the form of ADP complex with calcium-loaded parvalbumin. These changes in affinity for either ATP and ADP modulated by the metal state of parvalbumin suggest a functional importance of nucleotides binding to parvalbumin during the contraction-relaxation cycle but this has still to be clarified.

9 Conclusion

This chapter has summarized what is presently known about the physicochemical characteristics of muscle soluble calcium binding proteins and discussed their possible implication in the relaxation phase of muscle contraction. There is little doubt that they act as a soluble relaxing factor in fast twitch muscle fibers which use anaerobic substrates degradation as their main source of energy (IIB fiber) and, second condition, which are characterized by short half-relaxation time, smaller than 20 ms.

In these fibers it is also clear that the concentration of parvalbumin is inversely related to the half relaxation time. Fast twitch fibers of the other type, known as intermediate fibers (IIA), making use of the catalytic capacity of oxidative enzymes to produce the power required for sustained performance do not apparently contain parvalbumin nor do the slow twitch fibers. The reason is that fatigue-resistant fibers generally relax more slowly than IIB fiber and there is probably no correlation between the parvalbumin concentration in a muscle and its metabolic strategy, the unique parameter conditioning the amount of parvalbumin being the relaxation time. It remains, however, that in the case of high frequency repetitive stimulation, the limiting factor is still the calcium pumping rate of the sarcoplasmic reticulum which has the function of rendering the parvalbumin free of calcium during the period of recovery. Indeed, long-lasting calcium transients have been recorded well after the period of relaxation (Baylor et al. 1982); they correspond to the gradual release of calcium by parvalbumin. If this period is shortened too much, the proportion, after relaxation, of calcium-parvalbumin complexes, will gradually increase and will eventually become the unique form present. If this happens, the relaxation time will become independent of the presence of the soluble calcium binding proteins.

Acknowledgments. The author wishes to thank the Fonds de la Recherche Scientifique et Médicale, Belgium (programme no 3.451385) for financial support and N. Gerardin-Otthiers for expert technical assistance.

References

Aitken A, Klee CB, Cohen P (1984) The structure of the B subunit of calcineurin. Eur J Biochem 139:663–671

Aubert X (1956) Le couplage énergétique de la contraction musculaire. Thèse d'agrégation de l'Enseignement Supérieur. Edition Arscia, Bruxelles

Babu YY, Sack JS, Greenbough TJ, Bugg CE, Means AR, Cook WJ (1985) Three-dimensional structure of calmodulin. Nature (Lond) 316:37–40

Baron G, Demaille J, Dutruge E (1975) The distribution of parvalbumins in muscle and in other tissues. FEBS Lett 56:156–160

Baylor SM, Chandler WK, Marshall MW (1982) Use of metallochromic dyes to measure changes in myoplasmic calcium during activity in frog skeletal muscle fibres. J Physiol 331:139–177

Benzonana G, Cox J, Kohler L, Stein EA (1974) Caractérisation d'une nouvelle métallo-protéine calcique du myogène de certains crustacés. CR Acad Sci Paris 279:1491–1493

Benzonana G, Wnuk W, Cox JA, Gabbiani G (1977) Cellular distribution of sarcoplasmic calcium-binding proteins by immunofluorescence. Histochemistry 51:335–347

Berchtold MW, Means AR (1985) The Ca^{2+}-binding protein parvalbumin: molecular cloning and developmental regulation of mRNA abundance. Proc Natl Acad Sci USA 82:1414–1418

Berchtold MW, Heizmann CW, Wilson KJ (1982) Primary structure of parvalbumin from rat skeletal muscle. Eur J Biochem 127:381–389

Bhushana Rao KSP, Focant B, Gerday Ch, Hamoir G (1969) Low molecular weight albumins of cod white muscle (*Gadus callarias*). Comp Biochem Physiol 30:33–48

Blum HE, Lehky P, Kohler L, Stein EA, Fischer E (1977) Comparative properties of vertebrate parvalbumins. J Biol Chem 252:2834–2838

Capony JP, Pechère JF (1973) The primary structure of the major parvalbumin from hake muscle. Eur J Biochem 32:88–108

Capony JP, Demaille J, Pina C, Pechère JF (1975) The amino-acid sequence of the most acidic major parvalbumin from frog muscle. Eur J Biochem 56:215–227

Capony JP, Pina C, Pechère JF (1976) Parvalbumin from rabbit muscle. Isolation and primary structure. Eur J Biochem 70:123–135

Celio MR, Heizmann CW (1982) Calcium-binding protein parvalbumin is associated with fast contracting muscle fibres. Nature (Lond) 297:504–506

Coffee CJ, Bradshaw RA (1973) Carp muscle calcium binding protein. Characterization of the tryptic peptides and the complete aminoacid sequence of component B. J Biol Chem 248: 3305–3312

Coffee CJ, Bradshaw RA, Kretzinger RH (1974) The coordination of calcium ions by carp muscle calcium binding proteins A, B and C. In: Friedman M (ed) Protein metal interactions. Plenum Publishing Corporation, New York, pp 211–233

Collins JA, Potter JD, Horn MJ, Wilshire G, Jackman N (1973) The amino-acid sequence of rabbit skeletal muscle troponin C: gene replication and homology with calcium binding proteins from carp and hake muscle. FEBS Lett 36:268–272

Collins JH, Johnson JD, Szent-Györgyi AG (1983) Purification and characterization of a scallop sarcoplasmic calcium-binding protein. Biochemistry 22:341–345

Cox JA, Stein EA (1981) Characterization of a new sarcoplasmic calcium-binding protein with magnesium induced cooperativity in the binding of calcium. Biochemistry 20:5430–5436

Cox JA, Wnuk W, Stein EA (1976) Isolation and properties of a sarcoplasmic calcium-binding protein from crayfish. Biochemistry 15:2613–2618

Demaille JG (1982) Calmodulin and calcium-binding proteins. In: Cheung WY (ed) Evolutionary diversification of structure and function in calcium and cell function. Vol 2. Academic Press, New York, pp 111–144

Elsayed SM, Bennich H (1975) The primary structure of allergen M from cod. Scand J Immunol 4:203–208

Frankenne F, Joassin L, Gerday Ch (1973) The amino acid sequence of the Pike (*Esox lucius*) parvalbumin III. FEBS Lett 35:145–147

Gerday Ch (1976) The primary structure of the parvalbumin II of Pike (*Esox lucius*). Eur J Biochem 70:305–318

Gerday Ch (1982) Soluble calcium-binding proteins from fish and invertebrate muscle. Mol Physiol 2:63–87

Gerday Ch, Gillis JM (1976) The possible role of parvalbumin in the control of contraction. J Physiol 258:96–97P

Gerday Ch, Collin S, Piront A (1978) The amino acid sequence of the parvalbumin V of chub (*Leuciscus cephalus* L). Comp Biochem Physiol 61B:451–457

Gerday Ch, Collin S, Piront A (1979a) Phylogenetic relationship between cyprinidae parvalbumins II. The amino acid sequence of the parvalbumin V (*Leuciscus cephalus*). Comp Biochem Physiol 61B:451–457

Gerday Ch, Joris B, Gérardin-Otthiers N, Collin S, Hamoir G (1979b) Parvalbumins from the lungfish (*Protopterus dolloi*). Biochimie 61:589–599

Gerday Ch, Collin S, Gerardin-Otthiers N (1981) The soluble calcium binding protein from sandworm (*Nereis virens*) muscle. J Muscle Res Cell Motil 2:225–238

Gerday Ch, Collin S, Gérardin-Otthiers N (1986a) The amino acid sequence of the parvalbumin of the swimbladder muscle of the toadfish (*Opsanus tau*). (in preparation)

Gerday Ch, D'Haese J, Huch R (1986b) A soluble calcium-binding protein from the obliquely striated muscle of the earthworm *Lumbricus terrestris*. J Muscle Res Cell Motil 7:64

Gillis JM (1980) The biological significance of muscle parvalbumins. In: Siegel FL, Carafoli E, Kretsinger RH, MacLennan DH, Wasserman RH (eds) Calcium-binding proteins. Structure and function. Elsevier, North-Holland, New York, pp 309–311

Gillis JM (1985) Relaxation of vertebrate skeletal muscle. A synthesis of the biochemical and physiological approaches. Biochim Biophys Acta 811:97–145

Gillis JM, Gerday Ch (1977) Calcium movements between myofibrils; parvalbumins and sarcoplasmic reticulum in muscle. In: Wasserman RH, Corradino RA, Carafoli E, Kretsinger RH, MacLennan DH, Siegel FL (eds) Elsevier, North-Holland, New York, pp 193–196

Gillis JM, Thomason D, Lefevre J, Kretsinger RH (1982) Parvalbumin and muscle relaxation: a computer stimulation study. J Muscle Res Cell Motil 3:377–398

Goodman M, Pechère JF (1977) The evolution of muscle parvalbumins investigated by the maximum parsimony method. J Mol Evol 9:131–158

Goodman M, Pechère JF, Haiech J, Demaille J (1979) Evolutionary diversification of structure and function in the family of intracellular calcium binding proteins. J Mol Evol 13:331–352

Gosselin-Rey C (1974) Fish parvalbumin. In: Drabikowski W, Strzelecka-Golaszewska H, Carafoli E (eds) Immunochemical reactivity and biological distribution in calcium binding proteins. Polish Scientific Publishers, Warszawa, pp 679–701

Gosselin-Rey C, Gerday Ch (1977) Parvalbumin from frog skeletal muscle (*Rana temporaria*). Biochim Biophys Acta 492:53–63

Hamoir G, Gérardin-Ottiers N (1980) Differentiation of the sarcoplasmic proteins of white yellowish and cardiac muscles of an antarctic hemoglobin-free fish *Champsocephalus gunnari*. Comp Biochem Physiol 65B:199–206

Hamoir G, Honosu S (1965) Carp myogens of white and red muscles. General composition and isolation of low molecular weight components of abnormal amino-acid composition. Biochem J 96:85–97

Hamoir G, Gérardin-Otthiers N, Focant B (1980) Protein differentiation of the superfast swimbladder muscle of the toadfish *Opsanus tau*. J Mol Biol 143:155–160

Hamoir G, Gérardin-Otthiers N, Grodent V, Vandewalle P (1981) Sarcoplasmic differentiation of head muscles of the carp *Cyprinus carpio* (Pisces, Cypriniforme). Mol Physiol 1:45–58

Heizmann CW (1984) Parvalbumin, an intracellular calcium binding protein, distribution, properties and possible roles in mammalian cells. Experientia 40:910–921

Heizmann CW, Häuptle M, Eppenberger HM (1977) The purification, characterization and localization of a parvalbumin like protein from chicken-leg muscle. Eur J Biochem 80:433–441

Heizmann CW, Berchtold MW, Rowlerson AM (1982) Correlation of parvalbumin concentration with relaxation speed in mammalian muscles. Proc Natl Acad Sci 79:7243–7247

Herzberg O, James MNG (1985) Structure of the calcium regulatory muscle protein troponin-C at 2.8A resolution. Nature (Lond) 313:653–659

Joassin L, Gerday Ch (1977) The amino acid sequence of the major parvalbumin of the whiting (*Gadus merlangus*). Comp Biochem Physiol 57B:159–161

Jauregui-Adell J, Pechère JF (1978) Parvalbumins from coelacanth muscle. Amino acid sequence of the major component. Biochem Biophys Acta 536:275–282

Kobayashi T, Takasaki Y, Takagi T, Konishi K (1984) The amino acid sequence of sarcoplasmic calcium-binding protein obtained from sandworm (*Perinereis vancaurica tetra denta*). Eur J Biochem 144:401–408

Kohler LG, Cox JA, Stein EA (1978) Sarcoplasmic calcium binding proteins in protochordate and cyclostome muscle. Mol Cell Biochem 20:85–93

Konosu S, Hamoir G, Pechère JF (1965) Carp myogens of white and red muscles. Properties and amino acid composition of the main low molecular weight components of white muscle. Biochem J 96:98–112

Kretsinger RH (1976) Calcium binding proteins. Ann Rev Biochem 45:239–266

Kretsinger RH, Nockolds CE (1973) Carp muscle calcium-binding protein. Structure determination and general description. J Biol Chem 248:3319–3326

Leberer E, Pette D (1986) Neural regulation of parvalbumin expression in mammalian skeletal muscle. Biochem J 235:67–73

Lehky P, Blum H, Stein EA, Fisher EH (1974) Isolation and characterization of parvalbumins from the skeletal muscle of higher vertebrates. J Biol Chem 249:4332–4334

Le Peuch CJ, Demaille J, Pechère JF (1978) Radioelectrophoresis: a specific microassay for parvalbumins. Application to muscle biopsies from man and other vertebrates. Biochim Biophys Acta 537:153–159

Le Peuch CJ, Ferraz C, Walsch MP, Demaille JG, Fischer EH (1979) Calcium and cyclic nucleotide dependent regulatory mechanisms during development of chicken embryo skeletal muscle. Biochemistry 24:5267–5273

Manalan AS, Klee CB (1984) Purification and characterization of a novel Ca^{2+}-binding protein (CBP-18) from bovine brain. J Biol Chem 259:2047–2050

Pechère JF (1974) Isolement d'une parvalbumine du muscle de lapin. CR Acad Sci Paris 278: 2577–2579

Pechère JF, Capony JP, Demaille J (1973) Evolutionary aspects of the structure of muscles parvalbumins. Syst Zool 22:533–548

Pechère JF, Derancourt J, Haiech J (1977) The participation of parvalbumins in the activation-relaxation cycle of vertebrate fast skeletal muscle. FEBS Lett 75:111–114

Pechère JF, Rochat H, Ferraz C (1978) Parvalbumins from coelacanth muscle. Amino-acid sequence of the major component. Biochim Biophys Acta 536:275–282

Peckham M, Woledge RC (1986) Labile heat and changes in rate of relaxation of frog muscles. J Physiol 374:123–135

Permyakov EA, Kalinichenko CP, Yarmolenko VV, Burstein EA, Gerday Ch (1982) Binding of nucleotides to parvalbumins. Biochem Biophys Res Commun 105:1059–1065

Piront A, Lebacq J, Gillis JM (1980) Physiological properties of calcium binding proteins from crayfish muscle. J Muscle Res Cell Motil 1:468–469

Rinaldi ML, Haiech J, Pavlovitch J, Rizk M, Ferras C, Derancourt J, Demaille JG (1982) Isolation and characterization of a rat skin parvalbumin like calcium-binding protein. Biochem 21: 4805–4810

Robertson SP, Johnson JD, Potter JD (1981) The time-course of Ca^{2+} exchange with calmodulin, troponin, parvalbumin and myosin in response to transient increases in Ca^{2+}. Biophys J 34: 559–569

Somlyo AV, Gonzales-Serratos H, Schuman H, McCleilan G, Somlyo AP (1981) Calcium release and ionic changes in the sarcoplasmic reticulum. J Cell Biol 90:577–594

Strehler EE, Eppenberger HM, Heizmann CW (1977) Isolation and characterization of parvalbumin from chicken leg muscle. FEBS Lett 78:127–133

Sundaralingam M, Bergstrom R, Strasburg G, Rao SP, Roychowdihury P (1985) Molecular structure of troponin-C from chicken skeletal muscle at 3-angstrom resolution. Science 227:945–948

Szebenyi DME, Moffat K (1986) The refined structure of vitamin D-dependent calcium-binding protein from bovine intestine. J Biol Chem 261:8761–8777

Szebenyi DM, Obendorf SK, Moffat K (1981) Structure of vitamin D dependent calcium-binding protein from bovine intestine. Nature (Lond) 294:327–332

Takagi T, Konishi K (1984a) Amino acid sequence of α chain of sarcoplasmic calcium binding protein obtained from shrimp tail muscle. J Biochem 95:1603–1615

Takagi T, Konishi K (1984b) Amino acid sequence of the β chain of sarcoplasmic calcium binding protein (SCP) obtained from shrimp tail muscle. J Biochem 96:59–67

Takagi T, Kobayashi T, Konishi K (1984) Amino-acid sequence of sarcoplasmic calcium binding protein from scallop (*Patinopecten Yessoensis*) adductor striated muscle. Biochim Biophys Acta 787:252–257

Takagi T, Konishi K, Cox JA (1986) The amino acid sequence of two sarcoplasmic calcium-binding proteins from the protochordate amphioxus. Biochemistry 25:3585–3592

Tatcher DR, Pechère JF (1977) The amino-acid sequence of the major parvalbumin from Thornback Ray muscle. Eur J Biochem 75:121–132

Vanaman TC, Sharief F, Watterson DM (1977) Structural homology between brain modulator and muscle TNC_S. In: Wasserman RH, Corradino RA, Carafoli E, Kretsinger RH, MacLennan DH, Siegel FL (eds) Calciumbinding proteins and calcium function. Elsevier, North-Holland, New York, pp 107–116

Weeds AG, McLachlan AD (1974) Structural homology of myosin light chains, troponin C and carp calcium binding protein. Nature (Lond) 252:646–649

Wery JP, Dideberg O, Charlier P, Gerday Ch (1985) Crystallization and structure at 3.2 A resolution of a terbium parvalbumin. FEBS Lett 182:103–106

Wnuk W, Cox JA, Stein EA (1982) Parvalbumin and other soluble high affinity calcium binding proteins from muscle. In: Cheung WY (ed) Calcium and cell function. Vol 2. Academic Press, New York, pp 243–278

The Rat Parvalbumin Gene

M. W. Berchtold [1]

1 Introduction

Parvalbumin belongs to the family of high affinity Ca^{2+}-binding proteins (Heizmann and Berchtold 1987). In mammals, parvalbumin is synthesized in high amounts in fast contracting/relaxing muscles in a development-dependent manner (Celio and Heizmann 1982; Berchtold and Means 1985). Parvalbumin is also expressed in nonmuscle tissues such as brain (Celio and Heizmann 1981; Berchtold et al. 1985) and certain endocrine tissues (Berchtold et al. 1984; Endo et al. 1985).

Parvalbumin cDNA was cloned from a rat gastrocnemius muscle library using a synthetic 17-mer oligonucleotide as a probe (Berchtold and Means 1985). Full length parvalbumin cDNA contains two putative polyadenylation signals; both are probably used, because all parvalbumin-positive tissues contain distinct parvalbumin mRNA species of 700 and 1100 bp respectively (Epstein et al. 1986).

Parvalbumin cDNA was used to isolate the parvalbumin gene in order to analyze its structure, to study its regulation, and to probe the potential evolutionary relationship among it and genes that specify other members of the family of high affinity Ca^{2+}-binding proteins.

2 Results and Discussion

Parvalbumin cDNA was used to isolate six independent genomic clones (Fig. 1). Two of the latter clones (33, 201) were analyzed by Southern blotting using fragments from full-length parvalbumin cDNA, and intron/exon boundaries were sequenced by M13 nonrandom sequencing strategies. The parvalbumin gene contained five exons interrupted by introns; one intron had an exceptional length of 9 Kb. The entire transcription unit comprised 15.5 Kb.

When the gene organization of rat parvalbumin, chicken and *Drosophila* calmodulin (Epstein et al. 1987; Beckingham et al. 1987), mouse myosin light chain 3f (Robert et al. 1984) and sea urchin Spec 1 (Hardin et al. 1985) were compared, several similarities were apparent (Fig. 2). Several splice site positions were highly conserved

1 Institut für Pharmakologie und Biochemie, Universität Zürich-Irchel, Winterthurerstraße 190, CH–8057 Zürich, Switzerland

Ch. Gerday, R. Gilles, L. Bolis (Eds.)
Calcium and Calcium Binding Proteins

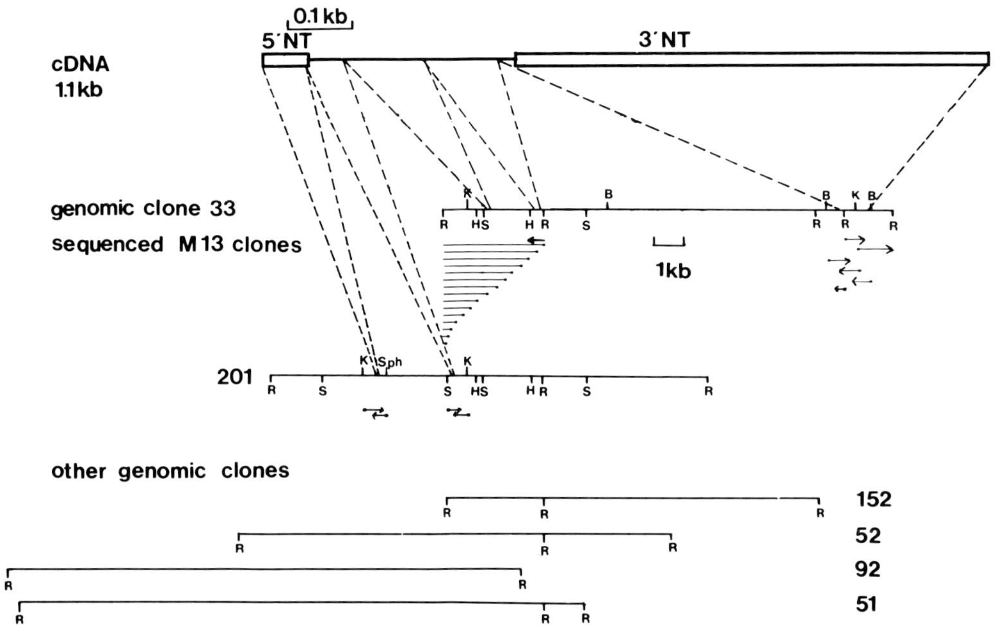

Fig. 1. Alignment of six rat genomic parvalbumin clones (33, 201, 152, 52, 92, 51) and mapping of clones 201 and 33. The different parvalbumin clones from a λ Charon 4A library were aligned with respect to their *Eco*RI restriction sites. *Horizontal lines* show the relative position of these overlapping clones. Southern blots were performed using full length cDNA fragments for hybridization. Clones 33 and 201 were analyzed in more detail. Restriction enzymes compatible for M13 mp 18 and 19 cloning were used to map clones 33 and 201. Restriction enzyme abbreviations are: *B Bam*HI; *R Eco*RI; *H Hind*III; *K Kpn*I; *S Sst*I; *Sph Sph*I. *Dashed lines* indicate the position of exons with respect to the full length cDNA. *Arrows* show the direction of sequencing and length of sequenced DNA fragments

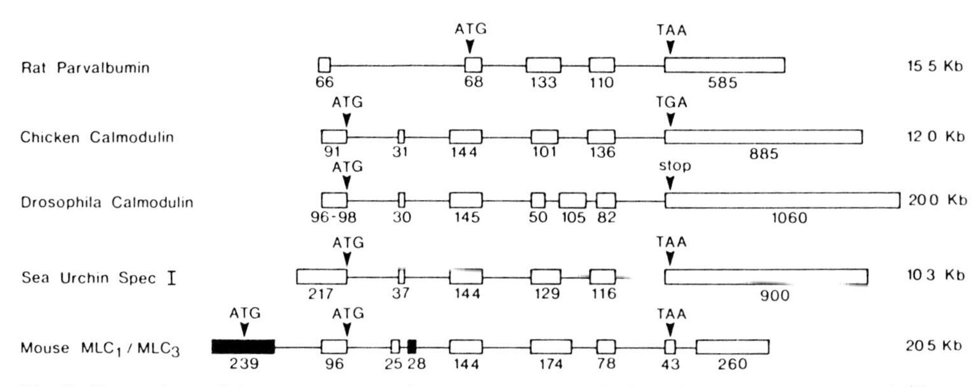

Fig. 2. Comparison of the gene structure of rat parvalbumin with that of other genes encoding Ca^{2+}-binding proteins. The structural organization of the rat parvalbumin gene was compared to that of the genes encoding chicken and *Drosophila* calmodulin, mouse myosin light chain 3f, and sea urchin Spec I protein. Alignments are based on protein sequence homologies. Numbers below exon boxes indicate their sizes. Introns are not drawn to scale. The approximate length of each transcription unit is marked on the right side of each gene

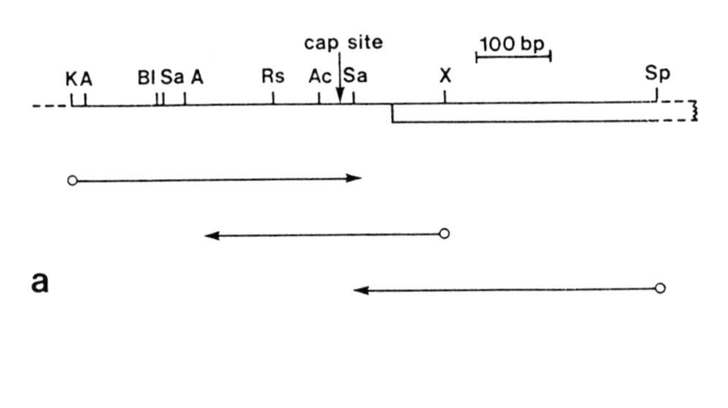


```
     -370        -360        -350        -340        -330
 AAAGACAC  AGTATGTGTA  GGGGACCTGA  CCTAGTCCCT  TTACCCCAAG

     -320        -310        -300        -290        -280
CAACCTCTGT  TCCTTTCCTT  CCCTCTTCCT  TTCCTTGCAA  GTTTAGAACA

     -270        -260        -250        -240        -230
AAGTGTGGCT  GCCACCCAAG  GTGATCATGG  AAGGCTTCAT  GCAAGAGGTG

     -220        -210        -200        -190        -180
TGCCCTGCTT  GGACCTTATT  GCTGGGTGCT  GTAGGGAGGT  AGAGGCCAAG

     -170        -160        -150        -140        -130
GATTATTAAT  AGGCCTGGCT  CAATAGGCCT  GGATGGGGGT  GAATGTGATA

     -120        -110        -100         -90         -80
CAGACACCAG  CACCGTGGCT  GGGGAGTACC  TGACACCGGA  AGGGGAGGGG

      -70         -60         -50         -40         -30
GCCAGGGGCT  GGGGGAGCGC  CACCTCCCAA  AATAGCCAAG  GCTGTAGACT

      -20         -10          +1
ATTTAAGTGA  CTGTCCCACC  ACAAGTCTCA  TTTCC
```

b

Fig. 3a,b. Restriction map and sequence of the parvalbumin promoter. **a** A restriction map of part of clone 201 containing the parvalbumin promoter was established using the following restriction enzymes: *Acc*I (*Ac*), *Ava*II (*A*), *Bcl*I (*BI*), *Rsa*1 (*Rs*), *Sau* 3A (*Sa*), *Kpn*I (*K*), *Sph*I (*Sp*) and *Xba*I (*X*). *Arrows* indicate direction and length of sequenced DNA stretches. ***Open box*** represents the 5′ end of the first intron in the parvalbumin gene. **b** The parvalbumin promoter was sequenced up to −377 bp from the cap site. The following fragments were cloned into M13 mp 18 and 19 and sequenced using the M13 dideoxy chain termination method: *KpnI* – *Sph*I from both ends; *Xba*I – *Kpn*I fragment from the *Xba*I site, and *Kpn*I – *Eco*RI fragment from the *Kpn*I end. The cap site was assigned position (+1) on the assumption that the 5′ noncoding sequence is 72 bp long. A "TATA" box (−24) and two "CAAT" boxes homologous sequences (−47, −156) are shown by open boxes.

Underlined sequences signal either homologies to promoter regions of other genes or direct repeats: Homologies; −106 to −75 to chicken myosin light chain 3f (Nabeshima et al. 1984); −88 to −65 to chicken calmodulin (Epstein et al. 1987), and −77 to −58 to the histocompatibility gene H-2K^{b} (Kimura et al. 1986). Direct repeats are located at positions −158 to −148 and −172 to −162; −304 to −294 and −320 to −310

with respect to amino acid sequences, for example, all genes examined contained a splice site 26 bp upstream from the translation stop codon. The latter observation strongly supports the view that the genes specifying the above Ca^{2+}-binding proteins comprise a gene superfamily. Because most splice sites are located within the Ca^{2+}-binding loops at different positions, introns were probably not present in the putative ancestral gene encoding a single Ca^{2+}-domain protein. However, introns existed most likely before the present genes evolved and were probably not involved in the duplication events. The parvalbumin gene displays an intron 8 bp upstream from the ATG translation start codon. Since the latter is an unusual location not seen in other genes for Ca^{2+}-binding proteins, the parvalbumin gene may have evolved from a four domain ancestor by an unique, alternative splicing event leading to a deletion in the 5′ region of the gene. 377 bp of the putative promoter region of the rat parvalbumin gene were sequenced (Fig. 3). A "TATA" box was located 24 bp upstream from the cap site and two "CAAT" boxes were found at positions −47 and −156. The parvalbumin promoter region was structurally similar to promoters of genes encoding the Ca^{2+}-binding proteins myosin light chain 3f (Nabeshima et al. 1984) and chicken calmodulin (Epstein et al. 1987). Two doublets of 11 bp repeats were found in the rat parvalbumin 5′ untranscribed gene region. Structural requirements for the tissue specific and development-dependent regulation of the parvalbumin gene are under investigation.

Acknowledgment. This work was supported by Grant 3.634-0.87 from the Swiss National Foundation.

References

Beckingham K, Doyle KE, Maune FJ (1987) The calmodulin gene of Drosophila melanogaster. Methods Enzymol 139:230–247

Berchtold MW, Means AR (1985) The Ca^{2+}-binding protein parvalbumin: molecular cloning and developmental regulation of mRNA abundance. Proc Natl Acad Sci USA 82:1414–1418

Berchtold MW, Celio MR, Heizmann CW (1984) Parvalbumin in non-muscle tissues of the rat. Quantitation and immunohistochemical localization. J Biol Chem 259:5189–5196

Berchtold MW, Celio MR, Heizmann CW (1985) Parvalbumin in human brain. J Neurochem 45: 235–239

Celio MR, Heizmann CW (1981) Calcium binding protein parvalbumin as a neuronal marker. Nature (Lond) 293:300–302

Celio MR, Heizmann CW (1982) Calcium binding protein parvalbumin is associated with fast contracting muscle fibres. Nature (Lond) 297:504–506

Endo T, Takazawa K, Onaya T (1985) Parvalbumin exists in rat endocrine glands. Endocrinology 117:527–531

Epstein P, Means AR, Berchtold MW (1986) Isolation and characterization of a rat parvalbumin gene and full length cDNA. J Biol Chem 261:5886–5891

Epstein P, Simmen RCM, Tanaka T, Means AR (1987) Isolation and structural analysis of the chromosomal gene for chicken calmodulin. Methods Enzymol 139:217–229

Hardin SH, Carpenter CD, Hardin PE, Bruskin AM, Klein WH (1985) Structure of the Spec 1 gene encoding a major calcium-binding protein in the embryonic ectoderm of the sea urchin, Stongylocentrotus purpuratus. J Mol Biol 186:243–255

Heizmann CW, Berchtold MW (1987) Expression of parvalbumin and other Ca^{2+}-binding proteins in normal and tumor cells: a topical review. Cell Calcium 8:1–41

Kimura A, Israel A, Le Bail O, Kourilsky P (1986) Detailed analysis of the mouse H-2K^{b} promoter: enhancer-like sequences and their role in the regulation of class I gene expression. Cell 44:261–272

Nabeshima Y, Fuji-Kuriyama Y, Muramatsu M, Ogata K (1984) Alternative transcription and two modes of splicing result in two myosin light chains from one gene. Nature (Lond) 308:333–338

Robert B, Daubas P, Akimenki MA, Cohen A, Garner I, Guenet JL, Buckingham M (1984) A single locus in the mouse encodes both myosin light chains 1 and 3, a second locus corresponds to a related pseudogene. Cell 39:129–140

Calcium Binding to Troponin C and the Regulation of Muscle Contraction: a Comparative Approach

W. WNUK [1]

1 Introduction

Muscle contraction consists of the cyclic attachment and detachment of the heads of myosin in the thick filament to actin in the thin filament. The attachment is followed by a change in the angle of myosin-actin attachment, so that the thick and thin filaments slide past each other and contractile force is generated. The energy for this process is supplied by ATP and is released by the interaction of actin with myosin, which activates the ATPase activity of myosin. The regulation of the actin-myosin-ATP interaction has been studied by analyzing the actin-activated myosin ATPase activity, which is the in vitro correlate of muscle contraction (Taylor 1979; Adelstein and Eisenberg 1980).

The calcium ion is now generally recognized as being involved in the regulation of muscle contraction. Resting muscle has a free [Ca^{2+}] of about 10^{-7} M. A rise in [Ca^{2+}] to about 10^{-5} M initiates contraction. In all types of muscle, Ca^{2+} binding to a sensor of calcium transients triggers a series of events leading to contraction. However, the manner in which Ca^{2+} acts to regulate contraction varies for different types of muscle. To date, three major types of regulation have been described. Each of them involves a distinct sensor molecule, the onset of contracting being triggered by binding of Ca^{2+} to one of the following proteins: (1) troponin C (TnC), the Ca^{2+}-binding component of troponin (Tn), in the actin-linked regulation found in vertebrate striated muscle (Ebashi and Endo 1968); (2) the regulatory light chain of myosin, in the myosin-linked regulation found in molluscs (Szent-Györgyi et al. 1973); (3) calmodulin, which in turn activates the myosin light chain kinase in the myosin-linked regulation via phosphorylation, found in vertebrate smooth muscle (Dabrowska et al. 1978). Many cells contain dual and perhaps multiple regulatory systems. For example, the thin filament-linked system coexists with the myosin-linked system in many invertebrate muscles (Lehman and Szent-Györgyi 1975). In addition, the same regulatory system may play a different role in different muscles. Thus all the necessary components of a myosin phosphorylation system are present in vertebrate striated (skeletal, cardiac) muscle but phosphorylation appears to play only a modulating role. This chapter deals only with comparative aspects of the structure and function of TnC in the actin-associated regulatory system. The reader is referred to recent reviews relevant to the actin-linked regulation (Potter and Johnson 1982; Zot and Potter 1984; Leavis and Gergely

1 Division de Gastroentérologie et Nutrition, Hôpital Cantonal Universitaire, Geneva, Switzerland

Ch. Gerday, R. Gilles, L. Bolis (Eds.)
Calcium and Calcium Binding Proteins

1984) and to the myosin-regulated systems (Kendrick-Jones and Scholey 1981; Hartshorne and Siemankowski 1981).

As a result of the key biological function of TnC, there has been an enormous effort directed toward a detailed understanding of how calcium binding to TnC can regulate muscle contraction. Consequently, much has been learned about the structure of TnC from fast skeletal muscle, including its three-dimensional organization. However, little is known still about a molecular basis for the understanding of how the binding of Ca^{2+} to TnC acts as a conformational switch and how this conformational signal is transmitted to other thin filament proteins. One approach that can be used to learn more about TnC is to study this protein from widely divergent species to see how natural selection preserved structural requirements in distinct regions of the TnC molecule. It is also of interest to investigate the actomyosin ATPase controlled by the regulatory system containing TnC from species more primitive than vertebrates, in the hope that some basic features of the regulation appears.

2 Thin Filament-Linked Regulation of Muscle Contraction

In striated muscle, the thin filament is composed of filamentous actin, tropomyosin (Tm) and Tn in a 7:1:1 molar ratio (Ebashi et al. 1969; Potter 1974). Polymeric actin forms a double-stranded helix in which its two parallel strands create a pitched groove (Hanson and Lowy 1962). Tm is composed of two helical subunits that coil about each other (Smillie 1979). The elongated Tm molecules form head-to-tail filaments that lie in the groove of the actin filament, with each Tm molecule interacting with seven actin monomers on each strand. Tn, which is closely associated with Tm, is localized in the thin filament at regular intervals (Hanson 1968; Ebashi et al. 1969). Tm is present in all types of muscle, so it is Tn, and the interaction of one of its subunits, TnC, with Ca^{2+}, that differentiates the actin-linked regulation from the other types of regulation. Addition of the Tm-Tn complex to the mixture of purified actin and myosin from vertebrate striated muscle suppresses the actin-activated ATPase activity of myosin in the absence of Ca^{2+}. A rise in $[Ca^{2+}]$ to 10^{-5} M relieves this inhibition. Thus, in striated muscle there is a Ca^{2+}-independent actin-activated myosin ATPase activity which is repressed by the binding of Tm-Tn to actin and is derepressed by Ca^{2+}-binding to TnC.

On the basis of X-ray diffraction and electron microscope investigations, the steric blocking model has been proposed as a mechanism for the calcium-regulated control of striated muscle contraction (Haselgrove 1972; Huxley 1972; Wakabayashi et al. 1975). This all-or-none model suggests that, when TnC is Ca^{2+}-free, Tn keeps Tm in the relaxed position where the Tm molecules lie slightly toward the periphery of the thin filament, thus sterically preventing the binding of the myosin heads to actin. Following the binding of Ca^{2+} to TnC, Tn allows Tm to move to a more central position in the groove of the double helix of actin, where Tm does not block the binding of myosin cross-bridges. Recent data suggest that the Tn-regulated control may be better viewed as an allosteric system (Adelstein and Eisenberg 1980; Chantler 1982). Both biochemical and structural evidence indicates various cooperative phenomena among Tm units and the associated actin monomers. The most radical challenge comes from recent experiments suggesting that the essence of regulation lies not in the control of the binding of myosin to actin but rather in changes in certain crucial kinetic transitions of actin-bound myosin (Chalovich and Eisenberg 1982).

Tn consists of three subunits, TnC, troponin I (TnI) and troponin T (TnT), that self-assemble into a 1:1:1 molar complex (Potter 1974). The molecular weights determined for subunits from rabbit fast skeletal muscle are: TnC, 17,965 (Collins et al. 1977); TnI, 20,864 (Wilkinson and Grand 1975); TnT, 30,503 (Pearlstone et al. 1976). TnT, the Tm binding subunit, serves as a link to attach the Tn complex to the thin filament via Tm. TnI, the inhibitory subunit, by itself prevents ATPase activity in a system containing Tm and actomyosin. When TnC and TnI are added to the Tm-regulated actomyosin, the ATPase activity is no longer inhibited but remains insensitive to Ca^{2+}. The Ca^{2+}-dependent regulation is possible only when the all three components of Tn are present in the system (Greaser and Gergely 1971; Eisenberg and Kielley 1974).

Within the Tn complex, each of its three subunits interacts with the other two (Potter and Johnson 1982; Leavis and Gergely 1984). Strong interactions exist between TnC and TnI and between TnI and TnT. As for connections between TnC and TnT, the latter are still under debate and their physiological significance remains to be resolved. In the thin filament, when calcium is absent, not only TnT is associated with Tm, but also TnI binds to both actin and Tm. In the presence of Ca^{2+}, virtually all these interactions are modified. Bonds involving TnC with TnI and perhaps with TnT are strengthened, while those of TnI to actin and Tm are abolished, and those of TnT to Tm are weakened. The interactions between Tm and actin are also changed. These observations are consistent with a model in which the binding of Ca^{2+} to TnC causes TnI to dissociate from sites provided by Tm and actin, whereas TnT continues anchoring the TnC-TnI complex to Tm (Potter and Johnson 1982). Consequently, the Ca^{2+}-induced conformational change in the disposition of the Tm-Tn complex may either relieve the inhibition of binding of myosin to actin, whether or not a simple steric blocking view is correct, or may suppress the inhibition of the rate-limiting step in the actomyosin ATPase cycle.

As mentioned in Section 1, many invertebrate muscles contain both actin-linked and myosin-linked regulatory systems. Yet, in fast striated muscle of some invertebrates such as arthropods (Kerrick and Bolles 1981; Watanabe et al. 1982), Ca^{2+} binding to TnC seems to be the primary trigger for contraction. The presence of Tn was described in striated muscle of arthropods (Bullard et al. 1973; Benzonana et al. 1974; Regenstein and Szent-Györgyi 1975; Lehman 1975) and molluscs (Goldberg and Lehman 1978), in obliquely "striated" muscle of molluscs (Konno 1978) and in smooth muscle of protochordates (Endo and Obinata 1981). Similarly to its vertebrate counterpart, Tn isolated from invertebrates appears to consist of three components. The latter differ, however, from the corresponding subunits of vertebrate Tn in molecular weight and amino acid composition. The smallest (Mr = 16,000 to 18,000) and medium (Mr = 23,000 to 29,000) subunits are analogous in function to vertebrate TnC and TnI, respectively (Regenstein and Szent-Györgyi 1975; Lehman 1975; Lehman et al. 1980; Tsuchiya et al. 1982; Wnuk et al. 1984). As for the TnT-like protein (Mr = 45,000 to 55,000), its involvement in Tn function has not yet been studied in detail (Lehman 1982).

3 Chemistry of Troponin C

3.1 Primary Structure

TnC is thought to have evolved from an ancestor composed of four Ca^{2+}-binding domains, common to many intracellular calcium-binding proteins. For our understanding of the history of these proteins, perhaps most important were the observations of Collins et al. (1973) and Weeds and McLachlan (1974), who recognized four homologous regions in the sequence of TnC and myosin alkali light chain, which were also homologous to similar regions previously described in soluble sarcoplasmic Ca^{2+}-binding proteins, the parvalbumins (Wnuk et al. 1982). The putative ancestral protein arose from two successive tandem duplications of the precursor gene coding for a single-domain polypeptide of about 40 amino acids, consisting of a calcium-coordinating loop flanked by two helices. Extant genealogical data suggest that a major divergence separated the lineages of calmodulin, TnC and the alkali light chain of myosin in one branch from the lineages of parvalbumin and the regulatory light chain of myosin in the other branch (Goodman et al. 1979).

Among the Ca^{2+}-binding proteins sequenced thus far, the primary structure of calmodulin is most remarkably conserved throughout evolution (Klee and Vanaman 1982) and is thought to resemble most closely that of the ancestral protein. As calmodulin regulates a broad spectrum of intracellular enzymes and other target proteins known to be Ca^{2+} sensitive, the stringent structural constraints upon calmodulin may be necessary for its many functions, each of which may impose its own structural requirement. By way of contrast, the sequences and Ca^{2+}-binding properties of the TnCs reported so far show significant variations even when different muscles of the same species are compared. Moreover, while no polymorphic forms have been reported within the same muscle in the case of vertebrate TnC, its invertebrate counterparts from some phyla appear as multiple forms. For example, the fast striated muscle from lobster and crayfish contains as many as three (Lehman and Ferrell 1980) and four (Wnuk et al. 1986) TnC isotypes, respectively. As the regulatory function of TnC resides only in the Ca^{2+}-dependent interactions with TnI and perhaps with TnT in the thin filament, there is no apparent reason for an invariance of amino acids which are not decisive for the specific role of TnC.

At this time, nine TnC sequences are already known. Among vertebrate TnCs, the primary structure of TnC from the fast skeletal muscle of rabbit (Collins et al. 1973, 1977), human (Romero-Herrera et al. 1976), chicken (Wilkinson 1976) and frog (Van Eerd et al. 1978) has been determined, as has that of TnC from the cardiac muscle of beef (Van Eerd and Takahashi 1976) and rabbit (Wilkinson 1980). TnC from slow skeletal muscle is identical with its cardiac counterpart in the same vertebrate species (Wilkinson 1980). As for invertebrate TnCs, the sequence of TnC from the smooth muscle of protochordate ascidian has been published (Takagi and Konishi 1983) and that of two major TnC isoforms from the striated muscle of crayfish has been recently established (Takagi, pers. commun.). These complete amino acid sequences contain 150 (crayfish TnC isotype β and γ) to 162 residues (TnC from the fast skeletal muscle of chicken and frog). Figure 1 shows the aligned sequences of TnC

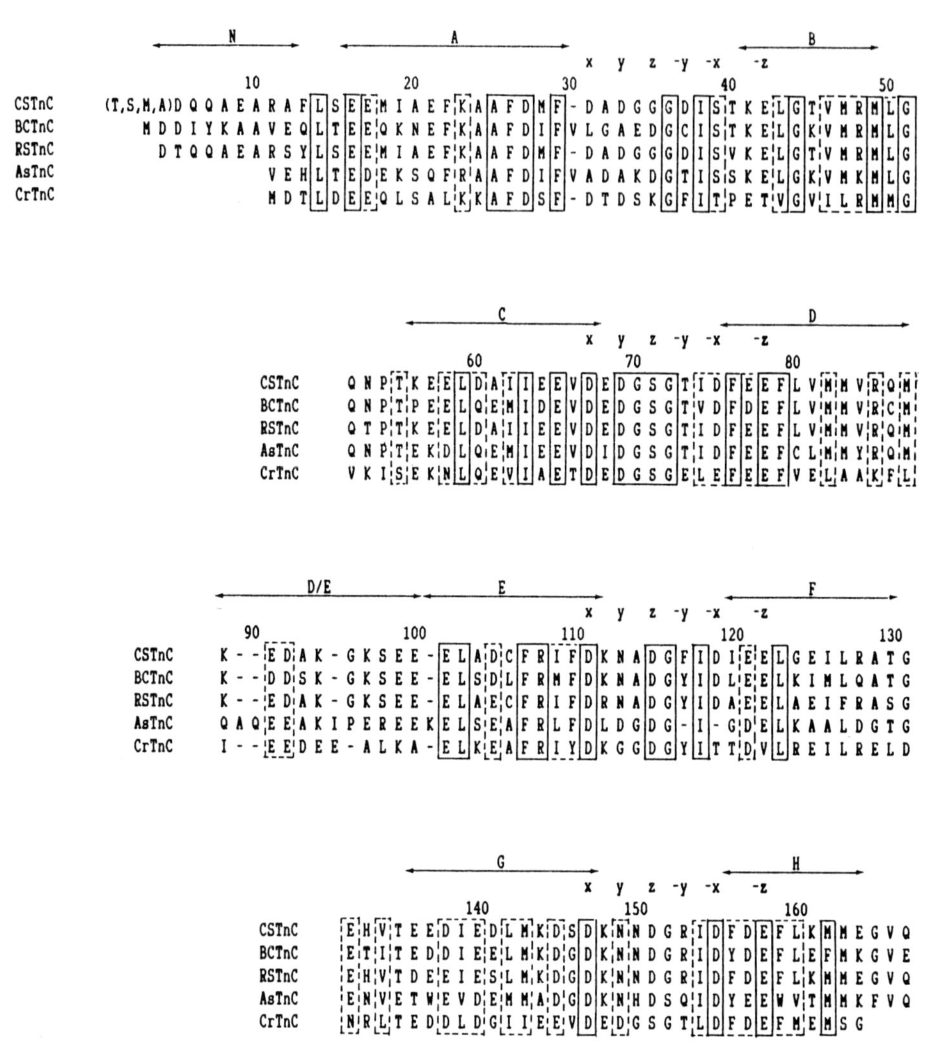

Fig. 1. Alignment of the amino acid sequences of TnC from chicken fast skeletal muscle (CSTnC; Wilkinson 1976), bovine cardiac muscle (BCTnC; Van Eerd and Takahashi 1976), rabbit fast skeletal muscle (RSTnC; Collins et al. 1977), ascidian smooth muscle (AsTnC; Takagi and Konishi 1983) and crayfish striated muscle (CrTnC, isotype γ; Takagi, pers. commun.). Amino acid residues, written using the one-letter code, are presented in four rows the alignment of which corresponds to the four homologous domains of the proteins (Collins et al. 1973). On the basis of the three-dimensional structure of TnC from avian (turkey and chicken) fast skeletal muscle (Herzberg and James 1985a; Sundaralingam et al. 1985a), helices (⟷) are denoted N, A to C, D-E, and F to H, from the amino to the carboxyl termini; the calcium-coordinating positions are labeled from x to –z. *Solid line boxes* indicate positions at which the 9 known TnC sequences are identical; *dashed line boxes* show positions at which all nine proteins are assumed to be related by functionally conserved replacements. Gaps (—) have been introduced for the purpose of alignment

Table 1. Homology matrice for the amino acid sequences of TnCs and calmodulin[a]

	CSTnC	RSTnC	BCTnC	AsTnC	BBCam
RSTnC	142[i] 7[c]				
BCTnC	110[i] 18[c]	102[i] 18[c]			
AsTnC	80[i] 24[c]	80[i] 20[c]	89[i] 26[c]		
BBCam	75[i] 29[c]	77[i] 25[c]	76[i] 27[c]	75[i] 25[c]	
CrTnC	58[i] 32[c]	56[i] 33[c]	55[i] 35[c]	50[i] 35[c]	57[i] 38[c]

[a] Pairwise comparisons were made using the alignment of the sequences of TnCs shown in Fig. 1, in which the sequence of calmodulin (Watterson et al. 1980) was also included.
Abbreviations: CSTnC and RSTnC, TnC from the fast skeletal muscle of chicken and rabbit, respectively; BCTnC, TnC from muscle; BBCam, bovine brain calmodulin; CrTnC, TnC isotype γ from crayfish striated muscle.
[i] The number of positions with identical residues.
[c] The number of positions with functionally conserved residues.

from five various species. The calcium-coordinating positions and helical segments are also indicated, as found in the crystal structure of avian skeletal TnC by Herzberg and James (1985a) and Sundaralingam et al. (1985a).

The alignment of all the known TnC sequences reveals many isologies in primary structure. Their N-terminal residue is blocked by an acetyl group. Nearly half of the primary structure is either invariable or conservative. Of the 167 apparent total positions compared in Fig. 1, identical residues are found at 35 positions, while those in 38 positions can be considered functionally conserved. If the sequence of bovine brain calmodulin (Watterson et al. 1980) is included in this alignment (not shown), similar numbers of identical (28) and conserved (39) positions are obtained. The overall three-dimensional structure of calmodulin (Babu et al. 1985) is similar to that of avian TnC. Thus the invariable and conservative positions shown in Fig. 1 seem to constitute the common framework for structures which are composed of two pairs of the helix-loop-helix motifs, connected by additional helical segment (see Sect. 3.2). Therefore it may be predicted that all the TnC assume an overall structure similar to that of avian skeletal TnC and calmodulin.

Quite strikingly, pairwise comparisons show that the level of homology between vertebrate TnCs and some invertebrate TnCs may be lower than that between vertebrate TnCs and calmodulin (Table 1). For example, the TnC from rabbit fast skeletal muscle and the crayfish TnC isotype γ are identical at 56 positions and conserved at 33 others, while 77 positions are invariable and 25 conservative between the rabbit TnC and calmodulin. Nevertheless, the crayfish TnC isotypes restore Ca^{2+} sensitivity to skinned rabbit adductor (fast-twitch) fibers, devoid of endogenous TnC by extraction with an EDTA solution, whereas calmodulin is unable to activate the EDTA-extracted skeletal muscle skinned cells (Hoar et al. 1985). Neither the rabbit TnC (Cox et al. 1981) nor the crayfish TnC isotypes (Wnuk, unpublished observations) can substitute for calmodulin in activating cyclic-nucleotide phosphodiesterase and myosin light chain kinase. The TnCs appear to be highly specialized derivatives of the more generalized calcium receptor, calmodulin. The latter protein possesses numerous regulatory activities not shared by muscle TnCs, but is not capable of substituting for TnC in the native thin filament. The interaction sites on TnC and its target proteins, TnI and TnT, seem to be well conserved throughout evolution, despite the substantial variations in the sequence of TnC, as well as in that of TnI (Wilkinson and Grand 1978) and TnT (see Sect. 2). Hence, examination of the sequence variations in distinct regions of the TnC molecule may shed some light on the structural constraints inherent in the need to conserve the function of TnC.

3.2 Three-Dimensional Structure

In the crystal structure of TnC from avian (turkey and chicken) fast skeletal muscle (Herzberg and James 1985a; Sundaralingam et al. 1985a), 108 of the 162 residues form eight α-helical regions: N, A to C, D-E, F to H (Fig. 1). There are seven loops between the eight helices. The dodecapeptide loops between helices A and B, C and D-E, D-E and F, and between G and H, each possesses suitable positioned Ca^{2+}-coordinating residues for binding this metal (Fig. 2). By convention, the flanking helices are lettered sequentially A to H. An additional helix (N) is present near the N terminus. The connection between helices D and E is made via a helical linker (D/E), thus forming the long nine-turn helix (D-E). Except for the D-E helix, all the helices consist of about three turns.

The molecule of avian TnC resembles a dumb-bell with two knobs joined by the long D-E helix. Each knob contains a pair of Ca^{2+}-binding domains or sites, viz. four helices and two Ca^{2+}-binding loops. In the N-terminal knob containing the extra helix (N), helix D is buried completely in a hydrophobic core which is formed by the other four helices surrounding the D helix. These five helices tend to be antiparallel; the angle between helices A and B is 145°, whereas that between helices C and D is 150°. In the N-terminal knob, no metals were found in either of its two Ca^{2+}-binding loops (I and II) probably because of the low pH in the crystallization. The D helix of site II runs without interruption into the E helix of the C-terminal knob, thus spanning loop II and loop III. In the center of the D-E helix, three turns are fully exposed to the solvent. In the C-terminal globular part of the TnC molecule, loops III and IV are occupied by Ca^{2+} and both inter-helix angles of the EF and GH helices are 107°.

The refinement of the structure of avian TnC at 2.2 Å resolution (Herzberg and James 1985b) has revealed that its Ca^{2+}-containing loops III and IV adopt conformations very similar to those of the two homologous loops found in the crystal structure of parvalbumin (Moews and Kretsinger 1975). Of the 12 positions in the loop region, the side chains in relative positions 1, 3, 5, 9 and 12 (as numbered in Fig. 2), as well as the main-chain at position 7, are the dentates forming a near-octahedral arrangement

TnC	Loop	1	2	3	4	5	6	7	8	9	10	11	12
		x		y		z		-y		-x			-z
CS	I	D	A	D	G	G	G	D	I	S	T	K	E
RS,HS	I	D	A	D	G	G	G	D	I	S	V	K	E
FS	I	D	T	D	G	G	G	D	I	S	T	K	E
CS,RS,HS,FS	II	D	E	D	G	S	G	T	I	D	F	E	E
BC,RC	II	D	E	D	G	S	G	T	V	D	F	D	E
As	II	D	I	D	G	S	G	T	I	D	F	E	E
Crβ	II	D	E	D	G	S	G	E	I	E	F	E	E
Crγ	II	D	E	D	G	S	G	E	L	E	F	E	E
CS	III	D	K	N	A	D	G	F	I	D	I	E	E
RS	III	D	R	N	A	D	G	Y	I	D	A	E	E
HS	III	D	R	N	A	D	G	Y	I	D	P	E	E
FS	III	D	K	N	A	D	G	Y	I	D	S	E	E
BC	III	D	K	N	A	D	G	Y	I	D	L	E	E
RC	III	D	K	N	A	D	G	Y	I	D	L	D	E
CS,RS,HS	IV	D	K	N	N	D	G	R	I	D	F	D	E
FS	IV	D	K	N	N	D	G	K	I	D	F	D	E
BC,RC	IV	D	K	N	N	D	G	R	I	D	Y	D	E
As	IV	D	K	N	H	D	S	Q	I	D	Y	E	E
Crβ	IV	D	E	D	G	S	G	T	I	D	F	M	E
Crγ	IV	D	E	D	G	S	G	T	L	D	F	D	E

Fig. 2. Sequence alignment for Ca^{2+}-binding loops of nine TnCs. Amino acid residues are written using the one-letter code. The loop sequences are numbered 1 to 12 from the N-terminus. The Ca^{2+}-coordinating positions are labelled from x to −z, as determined crystalographically for CS or predicted on the basis of sequence homology. Abbreviations: CS, RS, HS and FS, TnC from the fast skeletal muscle of chicken (Wilkinson 1976), rabbit (Collins et al. 1977), human (Romero-Herrera et al. 1976) and frog (Van Eerd et al. 1978), respectively; BC and RC, TnC from the cardiac muscle of beef (Van Eerd and Takahashi 1976) and rabbit (Wilkinson 1980), respectively; As, TnC from ascidian smooth muscle (Takagi and Konishi 1983); Crβ and Crγ, two major TnC isoforms from crayfish striated muscle (Takagi, personal communication). Loops I in BC, RC, As, Crβ and Crγ, as well as loops III in As, Crβ and Crγ, are not shown, since they have not the suitable positioned Ca^{2+}-coordinating residues due to amino acid replacements and deletions or insertions, and therefore are considered as nonfunctional

around the calcium ion. Residues 1 and 9 occupy the ±x vertices, residues 3 and 7, the ±y vertices, and residues 5 and 12, the ±z vertices. In both loops, the cation dentate coordination sphere consists of four oxygens from four side chains (+x, Asp; +y, Asn; +z, Asp; −z, Glu), one peptide carbonyl oxygen (−y, Phe or Arg) and one water molecule (−x) that mediates an indirect interaction between the cation and an aspartic acid. A glycine at position 6 is required to enable both the side chain at position 5 (+z) and the carbonyl main chain at position 7 (−y) to coordinate to the calcium ion. Loops III and IV are coupled by an antiparallel β-sheet conformation containing two hydrogen bonds between isoleucines that are at position 8 in each of the two loops.

3.3 Variable Regions

Looking at the known TnC sequences (Fig. 1), the amino acid replacements and deletions or insertions do not seem to have occurred randomly. Comparison of the tertiary structure of avian TnC with the primary structure of the other RnCs reveals that most differences are located in three regions. The first region is the N-terminal end of about 10 residues, which is present in vertebrate TnCs and absent in their counterparts from ascidian and crayfish. This region is also missing in calmodulin. It seems therefore likely that the decapeptide, comprising the extra helix (N) found in the N-terminal half of the molecule of avian TnC, has been acquired during vertebrate evolution. Although the N helix provides the additional stability to the arrangement of the N-terminal pair of Ca^{2+}-binding domains, the absence of helix N in invertebrate TnCs indicates that this helix is not essential for TnC function.

The second most variable region corresponds to the exposed middle part of the long D-E helix (linker D/E) found in the structure of avian TnC. In this part of sequence, invertebrate TnCs display extensive changes and ascidian TnC has four additional residues. Nevertheless, most of the residues in all the D/E linkers are of relatively high helix-forming propensity. As argued by Sundaralingam et al. (1985b), the exposed turns of the D/E linker are stabilized by several interhelical salt bridges between charged amino acid side chains. A number of such bridges can be predicted for the D/E linker of TnCs from ascidian and crayfish. The third most variable region corresponds to helix F in the avian TnC molecule. The helical conformation in this region, however, is probably conserved in the other TnCs, as suggested by its higher potential for helix formation than for making other conformations. On the other hand, of the 35 identical positions found in all nine TnCs, only 4 identities occur in domain IV (Fig. 1). The whole domain IV has suffered many substitutions and only half of them can be considered as functionally conserved. Clearly, the evolutionary constraints were less stringent for the C-terminal half of the TnC molecule than for its N-terminal half.

4 Calcium Binding to Troponin C

4.1 Equilibrium Measurements

The equilibrium binding parameters indicate that TnC from rabbit fast skeletal muscle contains two classes of calcium-binding sites (Potter and Gergely 1975). The first category contains two high affinity Ca^{2+} sites with $K_{Ca} = 2.1 \times 10^7\ M^{-1}$. These sites can also bind Mg^{2+} competitively with $K_{Mg} = 2 \times 10^3\ M^{-1}$ (Ca^{2+}-Mg^{2+} sites). The second category of sites contains two low affinity Ca^{2+} sites with $K_{Ca} = 3.2 \times 10^5\ M^{-1}$. At physiological (millimolar) levels of free Mg^{2+}, these sites appear to bind only Ca^{2+} (Ca^{2+}-specific sites). Their Ca^{2+} affinity is, however, reduced at nonphysiological high [Mg^{2+}], presumably by the binding of Mg^{2+} to a distinct class of sites with $K_{Mg} = 2 \times 10^2\ M^{-1}$ (Potter et al. 1981). Ca^{2+}-binding studies on rabbit TnC, using its proteolytic fragments (Leavis et al. 1978) and by modification of carboxylate-containing side chains (Sin et al. 1978), have established that domains I and II include the Ca^{2+}-

Table 2 A. Topography of Ca^{2+}-binding sites on various TnCs[a]

Site	I	II	III	IV
RSTnC	Ca^{2+}-specific	Ca^{2+}-specific	Ca^{2+}-Mg^{2+}	Ca^{2+}-Mg^{2+}
BCTnC	----[n]	Ca^{2+}-specific	Ca^{2+}-Mg^{2+}	Ca^{2+}-Mg^{2+}
CrTnC	----[n]	Ca^{2+}-specific	----[n]	Ca^{2+}-specific
AsTnC	----[n]	+[f]	----[n]	+[f]

[a] *Abbreviations:* RSTnC, TnC from rabbit fast skeletal muscle; BCTnC, TnC from bovine cardiac muscle; AsTnC, TnC from ascidian smooth muscle; CrTnC, TnC isotype γ from crayfish striated muscle.
[f] Functional site; affinity and specifity for Ca^{2+} not determined.
[n] Nonfunctional site.

Table 2 B. Binding constants[b] of Ca^{2+} and Mg^{2+} to various TnCs

	Site	K_{Ca} (M^{-1})	$K_{Ca(Mg)}$ (M^{-1})	K_{Mg} (M^{-1})	References
RSTnC	Ca^{2+}-specific (I, II)	3.2×10^5	1.1×10^5	2.0×10^2	Potter and Gergely (1975)
BCTnC	Ca^{2+}-specific (II)	2.5×10^5	2.5×10^5		Holroyde et al. (1980)
CrTnC	Ca^{2+}-specific (II, IV)	2.1×10^4	4.2×10^4	1.5×10^2	Wnuk et al. (1986)
RSTnC	Ca^{2+}-Mg^{2+} (III, IV)	2.1×10^7	2.8×10^6	2–4×10^3	Potter and Gergely (1975)
BCTnC	Ca^{2+}-Mg^{2+} (III, IV)	1.4×10^7	3.6×10^6	0.7×10^3	Holroyde et al. (1980)

[b] K_{Ca} and K_{Mg}, the affinities for Ca^{2+} and Mg^{2+}, respectively; $K_{Ca(Mg)}$, the affinity for Ca^{2+} in the presence of millimolar concentrations of Mg^{2+}.

specific sites while domains III and IV correspond to the Ca^{2+}-Mg^{2+} sites (Table 2). The above metal-binding properties seem to be common to TnCs from the fast skeletal muscle of most vertebrates, since they have been also found in TnC from the same type of muscle in a lower vertebrate (carp, Wnuk and Stein 1978). This view is also supported by the level of homology between all the four known sequences of TnCs from fast skeletal muscle, particularly in their regions of Ca^{2+}-binding loops (Fig. 2).

Based on the crystalline structure of parvalbumin and the sequence homology between this protein and TnC from both rabbit fast skeletal muscle and bovine heart, Van Eerd and Takahashi (1976) predicted that the cardiac TnC should bind only 3 mol Ca^{2+}/mol. Its loop I is unable to bind calcium, since two key coordinating aspartic acid residues at positions 1 and 3 are substituted for respectively by a leucine and an alanine (Fig. 1) which cannot contribute to Ca^{2+} coordination. In agreement with this prediction, the thorough studies of Holroyde et al. (1980) have provided evidence that bovine cardiac TnC possesses three Ca^{2+}-binding sites with an affinity in the range of physiological free [Ca^{2+}]: two Ca^{2+}-Mg^{2+} sites and one Ca^{2+}-specific site. Their affinities for Ca^{2+} (and Mg^{2+}) are similar to those of the analogous sites of TnC from rabbit fast skeletal muscle (Table 2). Protochordate ascidian TnC binds about 2 mol Ca^{2+}/mol at the free [Ca^{2+}] of 10^{-5} M (Endo and Obinata 1981). The Ca^{2+}- and Mg^{2+}-binding properties of this protein has not yet been studied in detail. Nevertheless, on the basis of the sequence alignment of the regions corresponding to Ca^{2+}-binding loops (Fig. 1), it may predicted that ascidian TnC has lost the ability to bind Ca^{2+} at sites I and III (Takagi and Konishi 1983). In loop I Ca^{2+}-coordinating aspartic acid residues at positions 1 and 3 are substituted for by alanines, whereas loop III has apparently suffered two amino acid deletions at positions 7 and 9.

The crayfish TnC isotype γ contains two low affinity Ca^{2+} sites with $K_{Ca} = 2.1 \times 10^4$ M^{-1} (Wnuk et al. 1986). In the presence of 1 mM Mg^{2+}, these sites bind only Ca^{2+} (Ca^{2+}-specific sites). Their Ca^{2+} affinity is reduced at 10 mM Mg^{2+}, suggesting the binding of Mg^{2+} to this protein with $K_{Mg} = 1.5 \times 10^2$ M^{-1}. Since this effect is noncompetitive, Mg^{2+} does not seem to bind at the Ca^{2+}-specific sites. The crayfish TnC isotype γ has no Ca^{2+}-Mg^{2+} sites with high affinity for Ca^{2+}. Also the crayfish TnC isotype β possesses only the two Ca^{2+}-specific sites. The sequence of the Ca^{2+}-binding regions of crayfish TnC isotype γ (Fig. 1) and β (not shown) indicates that their sites I and III are nonfunctional. In their loops I and III the key Ca^{2+}-coordinating glutamic acid residue at position 12 is substituted for by a threonine and a valine, respectively. The two remaining domains (II and IV) must then include the Ca^{2+}-specific sites. Domain IV in vertebrate TnCs includes the Ca^{2+}-Mg^{2+} mixed site and, thus, the presence of the Ca^{2+}-specific site in the fourth domain of crayfish TnCs was rather unexpected. However, although the affinity and specificity of Ca^{2+} sites cannot easily be predicted on the basis of sequence analysis (Reid and Hodges 1980), the sequence of the functional Ca^{2+}-binding loops of crayfish TnCs is consistent with the calcium-binding data. Not only the amino acid sequence of their loop II but also that of loop IV is remarkably similar to the sequences of the Ca^{2+}-specific loops (I and II) of vertebrate TnCs (Fig. 2). For example, only one conservative replacement at position 8 is found between loop IV in the crayfish TnC isotype γ (Leu) and loop II in bovine cardiac TnC (Val).

In order to obtain a better understanding of how the TnC molecule functions, it is crucial to know which Ca^{2+}-binding sites on TnC are the regulatory ones. Work with rabbit skeletal myofibrils indicated that the range of [Ca^{2+}] at which their ATPase rate is activated is independent of [Mg^{2+}] (Potter and Gergely 1975). Consequently, only the Ca^{2+}-specific sites (I and II) have been proposed to be involved in regulation. The lack of a role for the Ca^{2+}-Mg^{2+} sites (III and IV) in muscle contraction has been challenged by experiments showing a substantial effect of Mg^{2+} on the activation of ATPase for both skeletal and cardiac myofibrils from rabbit (Solaro and Shiner 1976). Moreover, every investigator who has studied the effect of Mg^{2+} on the pCa/force relationship in skinned fibers has observed a significant effect of Mg^{2+} on Ca^{2+} sensitivity (Donaldson and Kerrick 1975; Fabiato and Fabiato 1975; Fuchs and Black 1980). As of yet, no satisfactory explanation for this effect of Mg^{2+} has been presented. On the other hand, a large body of evidence (Potter and Johnson 1982) supports the Ca^{2+}-specific sites alone as the regulatory sites for muscle contraction.

4.2 Kinetic Aspects

In the sarcoplasm, the free [Mg^{2+}] is believed to be in the range of 0.5–5 mM (Brinley et al. 1977; Gupta and Moore 1980). Hence, it is likely that at rest the Ca^{2+}-specific sites are free of divalent metals, whereas the Ca^{2+}-Mg^{2+} sites are occupied by Mg^{2+}. When the free [Ca^{2+}] rises in the sarcoplasm, the Ca^{2+}-specific sites become saturated, and the Ca^{2+}-Mg^{2+} sites exchange their Mg^{2+} for Ca^{2+}, provided that the rate of Mg^{2+} dissociation is fast enough to allow this exchange to take place during the contraction-relaxation cycle. In the case of TnC from rabbit fast skeletal muscle, the rate of Mg^{2+} dissociation (k_{off} = 8/s; Johnson et al. 1979) appears to be far too slow in comparison to the time frame of a single twitch (Robertson et al. 1981). Consequently, repeated stimulations of muscle, e.g., smooth tetanus, are required for the saturation with Ca^{2+} of the Ca^{2+}-Mg^{2+} sites. The off-rate of Ca^{2+} from these sites (k_{off} = 1/s) is also considerably slower than the decay of Ca^{2+} transient in relaxing muscle (50–70 ms; Miledi et al. 1977). In contrast, the off rates of Ca^{2+} from the Ca^{2+}-specific sites of rabbit skeletal TnC (k_{off} = 300/s; Johnson et al. 1979) and Tn (k_{off} = 23/s; Johnson et al. 1981) are fast enough to allow the complete dissociation of Ca^{2+} from these sites during relaxation. To date, these kinetic studies provide the strongest evidence implicating the Ca^{2+}-specific sites (I and II) as the triggers of contraction. As for the Ca^{2+}-Mg^{2+} sites (III and IV), under the transient conditions of a single twitch, their participation must be limited to a nonregulatory role.

4.3 Comparative Considerations

Assuming that the basic mechanism of TnC function does not differ from one TnC molecule to another, the topography of the functional Ca^{2+}-binding sites in various TnCs (Table 2) provides additional information on the location of the regulatory site(s) on the TnC molecule. First, the retention of the functional Ca^{2+}-specific site in domain II throughout evolution indicates that the binding of Ca^{2+} to site II is essential for triggering contraction. Indeed, loop II appears to be most conserved among the four Ca^{2+}-binding loops of TnC (Figs. 1 and 2). Of the 12 positions in loop II, seven are invariant and three functionally conservative. The only variable positions are residues 2 and 7, but the former residue does not contribute to Ca^{2+} coordination and the latter residue chelates the calcium cation via its peptide carbonyl oxygen. Second, the loss of Ca^{2+} binding at sites I (vertebrate cardiac and invertebrate TnCs) and III (invertebrate TnCs) suggests that the binding of Ca^{2+} to these sites does not act as a conformational switch. Helix F in domain III is one of the three most variable regions of TnC (see Sect. 3.3). Sequence variations in this region may be due to the less stringent structural requirements when site III no longer binds Ca^{2+}, as in the case of invertebrate TnCs. In contrast, the N-terminal helix (E) of this site appears to be relatively well preserved. In the structure of avian TnC, the latter helix constitutes the C-terminal part of the long D-E helix. It is thus likely that helix E provides an important element of structural stability to all the TnCs. Similarly, no substantial sequence variations are observed in the flanking helices (A and B) of site I, despite its defunct Ca^{2+}-binding loop in some TnCs. It seems that domain I possesses some particular function in the protein and that this is a reason why the overall configuration of site I is preserved, even when the Ca^{2+} binding ability is lost. Third, domain IV is the location of the Ca^{2+}-Mg^{2+} mixed site in vertebrate TnCs, but includes the Ca^{2+}-specific site in crayfish TnCs. If site IV was the regulatory one in the latter proteins, then the molecular mechanism of crayfish TnC function, involving sites II and IV as the triggers of contraction, would be different from that of vertebrate TnC; however, this seems unlikely. In reconstituted thin filaments, only one of the two Ca^{2+}-specific sites on crayfish TnC displays an affinity within the physiological range of $[Ca^{2+}]$ (see Sects. 6 and 7). It is tempting, therefore, to postulate that it is site II which plays a regulatory role in crayfish TnCs.

The presence on crayfish TnCs of the defunct site III and the Ca^{2+}-specific site in domain IV raises the question of the role played by the C-terminal half of the TnC molecule. A structural role for sites III and IV on vertebrate TnCs has been postulated, since these sites would always contain either Mg^{2+} or Ca^{2+} in vivo, and under low ionic strength conditions, their occupation by either divalent metal is required for attachment of TnC to TnI in intact myofibrils (Zot and Potter 1982). Crayfish TnCs possess also several sites with the same very low affinity ($K = 1 \times 10^3\ M^{-1}$) for Ca^{2+} and Mg^{2+} (Wnuk et al. 1984). If one of these sites was the mutated site in domain III, then one could postulate that its occupation by Mg^{2+} in vivo is required for maintaining the integrity of the Tn complex; however, this turns out not to be the case. Preliminary results of ours indicate that removal of divalent metal ions from whole crayfish myofibrils by extractions with metal chelators does not facilitate a dissociation of TnC from the myofibirils. This suggests that in contrast to vertebrate TnC, crayfish TnCs bind to TnI irrespective of Ca^{2+} or Mg^{2+}. The four-domain ancestor of calcium-binding proteins probably possessed only sites specific for Ca^{2+} as in calmodulin (Goodman et al. 1979). Therefore, it is likely that evolution of domains III and IV in TnC proceeded in two directions, toward either the reduction of Ca^{2+} affinity to the range of high, nonphysiological $[Ca^{2+}]$ and the loss of ion-binding capacity (invertebrates) or the acquisition of high affinity Ca^{2+}-Mg^{2+} sites that are most of the time occupied by Mg^{2+} in vivo (vertebrates). In both cases, these two domains maintain the structure of the Tn complex intact independently of the cytosolic levels of Ca^{2+}, thus allowing the Ca^{2+}-specific site in domain II to function in the regulation of Ca^{2+} activation.

5 Calcium-Induced Conformational Changes of Troponin C

Most information on the Ca^{2+}-dependent conformational changes in TnC (Leavis and Gergely 1984) has been obtained from solution studies on the protein from rabbit fast skeletal muscle. In the TnC free of divalent metal ions, domains I and II appear to be highly organized with preformed helical segments. Such a structure is also observed in the crystal of avian skeletal TnC where no metals were found in these two domains (see Sect. 3.2). The distinctive configuration of the helices produces a clustering of hydrophobic residues, which stabilizes the core of the N-terminal half of the molecule when Ca^{2+} is absent. In contrast, in domains III and IV only two of the four helices (F and G) are coiled in the absence of divalent metals. Ca^{2+} binding to sites III and IV induces the coiling of helices E and H (Nagy and Gergely 1979), which accounts for about 90% of the total increase in secondary structure induced in this molecule by Ca^{2+}. Concomitantly, the hydrophobic inter-helix contacts are formed, thus stabilizing the Ca^{2+}-bound state of the C-terminal half of the molecule (Leavis and Gergely 1984; Herzberg and James 1985a; Sundaralingam et al. 1985a). Mg^{2+} binding to sites III and IV induces a smaller change in secondary structure (Kawasaki and Van Eerd 1972) and several differences in the environment of hydrophobic residues (Levine et al. 1978). Nevertheless, the overall conformation of TnC-Mg_2, which is thought to be the form in which the protein exists in relaxed muscle (see Sects. 4.2 and 4.3), is believed to be similar to that of TnC-Ca_2. Ca^{2+} binding to sites I and II results in the loss of hydrophobic contacts in the N-terminal half of the molecule leading to a tertiary structure change in this region (Evans et al. 1980). These conformational transitions are somehow transmitted to the C-terminal half of the molecule (Grabarek et al. 1986).

With its one Ca^{2+}-specific site (II) and two Ca^{2+}-Mg^{2+} sites (III and IV), vertebrate cardiac TnC displays presumably the Ca^{2+}-dependent conformational states similar to those observed in skeletal TnC (Hincke et al. 1981; Potter and Johnson 1982). As for crayfish TnCs, limited proteolysis of γ-TnC with trypsin appears to be independent of Ca^{2+} (Wnuk; unpublished results). Five points of cleavage were found on this molecule; all exclusively in the C-terminal half of the molecule, two in each of the helices (E and F) flanking the defunct loop III and one between helices F and G. This suggests that even in the presence of Ca^{2+}, at least when the protein is free of other thin-filament components, its C-terminal moiety remains less compact than the Ca^{2+}-bound state of the corresponding region in skeletal TnC. In fact, the C-terminal half of skeletal TnC, mainly domain III, is digested by trypsin only in the absence of Ca^{2+} (Grabarek et al. 1981a). In the presence of Ca^{2+}, the skeletal TnC is cleaved preferentially in its exposed helical linker D/E. On the other hand, the N-terminal half of crayfish γ-TnC is resistant to tryptic proteolysis irrespective of Ca^{2+} and therefore seems to have its hydrophobic core as stable as that of skeletal TnC.

Since Ca^{2+} binding to the regulatory site(s) (I and II, or only II) induces virtually no changes in secondary structure and subtle changes in tertiary structure, the question arises as to what conformational transition of TnC triggers muscle contraction. Based on the crystal structure of turkey TnC, Herzberg and James (1985a) envisaged two of the possible structural changes. First, upon Ca^{2+} binding, the conformational transition in the middle of the long central helix (D-E) would bring about a change in the relative disposition of the two halves of the molecule. This implies a change in relative position of TnI that would remove its inhibitory effect. The D/E helix linker of avian TnC is bent (16°) at the central position occupied by a glycine residue. The latter would serve as a pivot about which the two halves of the molecule could freely rotate. The presence of a proline residue in the center of the helical linker of ascidian TnC suggests even a sharper bend in this molecule. However, the absence of glycine and proline in the corresponding region of crayfish TnCs (Fig. 1) raises doubts as to whether the mechanism of TnC action may involve a major angular displacement of the central helix. Moreover, it is not known whether the structure of TnC with a complete separation between the two halves via the D/E helix is relevant to the structure of TnC in the thin filament.

Alternatively, a Ca^{2+}-induced change in the inter-helix angles in domains I and II might stabilize the interaction of this region with TnI, and/or could be transmitted to the C-terminal moiety of the molecule via a change in conformation of the central helix. In the case of avian TnC, sites I and II lack calcium in the crystal and, significantly, have different inter-helix angles when compared with the Ca^{2+}-containing sites III and IV (see Sect. 3.2). Assuming that the crystal structure represents the molecule in relaxed muscle, the basis for the second hypothesis is that upon Ca^{2+} binding to the triggering sites I and II, the regulatory half of the molecule undergoes a conformational transition to become similar in conformation to the Ca^{2+}-bound state of the C-terminal half of the molecule. The computer modeling indicates that such a conformation for the regulatory moiety of the molecule is energetically possible and conforms to accepted protein structure rules (Herzberg et al. 1986). As expected, the modeled Ca^{2+}-activated N-terminal region is more exposed to solvent and its Ca^{2+}-binding loops are rearranged. The model predicts a striking change on the surface of helix B where a hydrophobic "patch" is formed. Its presence suggests a possible contact surface with other components of Tn. However, as far as interactions of TnI with TnC are concerned, helix B on TnC has not been found to be the site of these interactions.

The binding of a variety of the proteolytic fragments of TnC from rabbit fast skeletal muscle to intact TnI has provided evidence that three different regions of TnC interact with TnI (Grabarek et al. 1981b; Leavis and Gergely 1984). When comparing the primary structure of rabbit TnC with the tertiary structure of avian TnC, it appears that the TnI-binding regions are: helix C, helix E (together with the C-terminal end of linker D/E) and helix G. These helices flank respectively loops II, III and IV on the N-terminal side. All have similar distributions of acidic residues aligned along their surface. These residues may form bonds with the abundant basic groups on the TnI surface. Only helix G binds to TnI whether or not Ca^{2+} is present. Interactions with TnI involving helix E require the binding of Ca^{2+} to loops III and IV. However, these loops would be always occupied by Mg^{2+} or Ca^{2+} in vivo (see Sects. 4.2 and 4.3) and therefore the site involving helix E would always be associated with TnI, irrespective of sarcoplasmic $[Ca^{2+}]$. Interactions with TnI involving helix C require the binding of Ca^{2+} to loops I and II, and therefore helix C is the only site which becomes effective only during Ca^{2+} activation of the muscle. As for interaction between TnC and TnT, the TnT-binding sites exist in both halves of the TnC molecule, but further work will be required to localize them more precisely (Leavis and Gergely 1984).

6 Calcium Binding to Troponin C-Troponin I, Troponin and Troponin-Tropomyosin Complexes

In as much as the Ca^{2+}-induced structural transition of TnC is the first stage of muscle contraction, fundamental to current concepts on this process is the propagation of the Ca^{2+}-initiated conformational signal throughout other thin-filament proteins. The mechanism by which the information transfer occurs is not yet clearly understood. A conceptually simple approach that may give some insight into this problem is to study Ca^{2+} binding to various complexes of TnC with other thin-filament proteins to see how

Table 3. Effect of protein-protein interactions on the Ca^{2+}-binding properties of the regulatory site on crayfish γ-TnC[a]

	K_{Ca} (M^{-1})
γ-TnC[b]	3.1×10^4
γ-TnC-TnI[b]	3.9×10^6
γ-TnC-TnI-TnT[b]	5.4×10^6
Tn-Tm[c]	4.2×10^6
Tn-Tm-actin[c]	2.3×10^5
Tn-Tm-actin-myosin[c] (1 mM MgATP)	2.9×10^5
Tn-Tm-actin-myosin[c] (-ATP)	6.8×10^6

[a] The Ca^{2+}-binding parameters were measured by equilibrium dialysis for crayfish γ-TnC and its TnC-TnI, Tn, Tn-Tm complexes, and by double-isotope technique for reconstituted thin filaments in the absence and presence of myosin. K_{Ca}, the affinity for Ca^{2+}, is an average from several experiments carried out in both the absence and presence (1 mM) of Mg^{2+} except for the thin filaments in the presence of myosin, which were studied only at 1 mM [Mg^{2+}].
[b] Wnuk et al. (1986); Wnuk (unpublished results).
[c] Native Tn, containing all the crayfish TnC isoforms in the original ratio, was used (Wnuk et al. 1984).

interactions between these proteins influence the Ca^{2+}-binding properties of TnC. The equilibria between thin-filament proteins are affected not only by Ca^{2+}, but also, in turn, by free energy coupling; these equilibria must then affect the binding of Ca^{2+} to TnC in the complexes.

Potter and Gergely (1975) first reported that the TnC-TnI complex from rabbit fast skeletal muscle displays a ten-fold increase in the affinities of the Ca^{2+}-specific sites for Ca^{2+} and of the Ca^{2+}-Mg^{2+} sites for both Ca^{2+} and Mg^{2+} as compared with TnC alone. It follows from thermodynamic reasoning that the binding of Ca^{2+} to TnC increases the binding constant of TnI to TnC, as confirmed by quantitative studies of Wang and Cheung (1985). In the Tn (Potter and Gergely 1975) and Tn-Tm (Zot et al. 1983) complexes, the two classes of Ca^{2+} sites have affinities only slightly higher than those in the TnC-TnI complex. This suggests that in the Tn-Tm complex, the effect of Ca^{2+} involving a change in TnC is exerted mostly on TnI. In contrast, interactions involving TnT and Tm appear to be much less affected by Ca^{2+}, at least when the Tn-Tm complex is not attached to the actin filament. This is also the case for the complexes from bovine cardiac muscle with respect to both categories of Ca^{2+} sites (Holroyde et al. 1980). Interestingly, in the presence of 2 mM Mg^{2+}, the affinity of the Ca^{2+}-Mg^{2+} sites for Ca^{2+} on rabbit skeletal TnC is hardly influenced in the TnC-TnI complex, whereas the affinity of the Ca^{2+}-specific sites is increased approximately 10-fold (Potter and Gergely 1975). Hence, while Ca^{2+} or Mg^{2+} binding to the empty Ca^{2+}-Mg^{2+} sites (III and IV) alters and is altered by the interaction of TnC with TnI involving undoubtedly helix E on TnC, the Mg^{2+}-Ca^{2+} exchange at these sites, which presumably takes place in vivo, has no major effect on this interaction. This again suggests a structural role for the Ca^{2+}-Mg^{2+} sites in vertebrate systems.

In the case of crayfish γ-TnC, its interaction with TnI results in a 125-fold increase in the affinity of one of the two Ca^{2+}-specific sites (Table 3), whereas the affinity of the second site is hardly affected. These two sites display essentially the same affinities for Ca^{2+} in the Tn and Tn-Tm complexes as in the TnC-TnI complex. Thus, owing to the interactions of crayfish TnC with neighboring proteins, mainly with TnI, one of the Ca^{2+}-binding sites acquires an affinity in the physiological range of free [Ca^{2+}] (regulatory Ca^{2+}-specific site). The role of the other Ca^{2+}-specific site is not clear. The small change in its Ca^{2+} affinity, induced by protein-protein interactions in the TnC-containing complexes, is reminiscent of the situation prevailing in vertebrate complexes when Mg^{2+} is replaced by Ca^{2+} at the Ca^{2+}-Mg^{2+} sites. This might suggest an auxiliary role for the second site, but its Ca^{2+} affinity ($<10^5$ M^{-1}) is too low to be of physiological significance.

7 Calcium Binding to Thin Filaments and Skinned Fibers

Although the Ca^{2+}-binding properties of TnC in the Tn-Tm complex without actin are of considerable interest, interpretation of the mechanism of the actin-linked regulation requires precise information on the thermodynamics of the Ca^{2+}-dependent interactions within the thin filament. As a contribution toward solving this problem, we have taken advantage of the simplicity of Ca^{2+} binding in crayfish TnCs to study the Ca^{2+}-binding properties of thin filaments that are more difficult to discern in the vertebrate regulatory system because of the presence of multiple and diverse sites which bind Ca^{2+} in the physiological range of its concentrations. As shown in Table 3, incorporation of the crayfish Tn-Tm complex into actin filaments results in a 20-fold decrease in the affinity of the regulatory site on crayfish TnCs (Wnuk and Stein 1980; Wnuk et al. 1984). Similar results have been reported by Zot et al. (1983) for the rabbit skeletal system, where an effect of similar magnitude was found exclusively for the Ca^{2+}-specific sites. This is in agreement with studies showing that the binding of Tn-Tm to actin filaments appears to be weaker in the presence of Ca^{2+} than in its absence (Wegner and Walsh 1981). Indeed, if the binding of Ca^{2+} to TnC decreases the binding constant of Tn-Tm to F-actin, the binding of the Tn-Tm complex to actin filament decreases the binding constant of Ca^{2+} to TnC. Interestingly, in the rabbit thin filament, the reduced affinity approaches the level of that of TnC in its isolated state. Our data and those of others suggest that the effect of actin, mediated through TnI and/or transmitted via Tm to TnT and to TnI, depresses those interactions between TnI and TnC which increase the affinity of the Ca^{2+}-specific site(s) on TnC in the absence of actin. Furthermore, if actin alters the conformation of TnC, as indicated by a decrease in the affinity for Ca^{2+}, then Ca^{2+} binding to the regulatory site(s) must affect the structure of actin filaments. The postulated change in actin structure may be the key event in muscle contraction.

Bremel and Weber (1972) first reported a slight increase in the affinity of skeletal TnC for Ca^{2+}, when the thin filaments and myosin filaments form complexes in the absence of ATP hydrolysis, i.e. under equilibrium conditions. Also, measurements of Ca^{2+} binding to glycerinated rabbit skeletal muscle fibers suggest an increase in the total bound Ca^{2+} in the rigor state as opposed to in the presence of ATP (Fuchs 1977). These results are consistent with studies showing that the cooperative binding of isolated myosin heads (S1) to the rabbit skeletal thin filament is stronger in the presence of Ca^{2+} than in its absence (Trybus and Taylor 1980). Furthermore, the calcium sensitivity in the cooperative binding of S1-ADP to the regulated actin filament (Greene and Eisenberg 1980) implies that the interaction between the thin filament and myosin with bound ADP also induces a higher affinity of TnC for Ca^{2+}. More quantitative information on these effects has been obtained from work with crayfish myofibrils (Wnuk and Stein 1980) and actomyosin, regulated by crayfish Tn-Tm (Wnuk et al. 1984). The binding of myosin to the thin filament in the absence of ATP results in a 30-fold increase in the affinity of the regulatory Ca^{2+}-specific site on TnC (Table 3). The affinity of regulated actomyosin for Ca^{2+} is similar to that of the Tn-Tm complex. Hence, in the rigor state where the Tn-Tm complex is "pushed" into the groove of the actin helix by myosin heads, the effect of actin on the regulatory site is suppressed. Thus, the regulatory site(s) on TnC in the thin filament may exist in at least two affinity states (one low and one high), the high affinity state being induced by interactions of myosin with thin filaments.

The rigor and ADP-bound states can be assumed to correspond only to the final steps of an actomyosin ATPase cycle. During steady state ATP hydrolysis, regulated actomyosin displays Ca^{2+}-binding properties similar to those of thin filaments without myosin (Table 3). However, this does not necessarily mean that the cross-bridge-induced increase in the calcium affinity of the Ca^{2+}-specific site(s) on TnC is not instrumental in the kinetic cycle of ATP hydrolysis (see Sect. 8). The affinity of actin to myosin-nucleotide intermediates of the ATPase cycle is lower than in the rigor state and varies over four orders of magnitude, depending on the state of the nucleotide bound to myosin. Therefore, the effect of myosin intermediates on binding of Ca^{2+} to TnC in the thin filament is difficult to measure directly. It should be also noted that under conditions used in the experiments with actomyosin preparations that are under no tension, the fraction of myosin bridges attached to actin at any instant is presumably too small to alter the overall Ca^{2+} affinity of TnC in the thin filament. The increased calcium binding to myofilaments as a result of cross-bridge interaction has been shown for muscle fibers in the isometric state (Ridgway et al. 1983). Under such physiological conditions, about 20% of myosin heads would be strongly bound to actin, as compared with 100% in rigor (Irving 1985).

8 Relationship of Calcium Binding to Troponin C and the Activation of Actomyosin Adenosine Triphosphatase and Tension Development

The steep pCa/ATPase curve of rabbit skeletal myofibrils has been attributed to the requirement that all four sites (Weber and Murray 1973) or both Ca^{2+}-specific sites (Potter and Gergely 1975) on TnC must be filled by Ca^{2+} for activation to occur. However, more recent experimental data do not support such an explanation. First, as monitored by tension development of rabbit psoas fibers (Brandt et al. 1980) and by ATPase of rabbit skeletal myofibrils (Murray and Weber 1980; Zot and Potter 1982), the responses to Ca^{2+} are much too steep to be explained solely by a requirement for two or even four calcium ions bound on TnC. Second, although indirect measurements of Ca^{2+} binding to rabbit skeletal TnC in the thin filament in the presence of myosin and ATP suggest a cooperativity in Ca^{2+} binding to the Ca^{2+}-specific sites, the ATPase activity of regulated actomyosin rises more steeply with [Ca^{2+}] than does the bound Ca^{2+} at sites I and II (Grabarek et al. 1983). Third, the pCa/ATPase curve of cardiac myofibrils is sloping as sharply as that of skeletal myofibrils, despite the presence of only one regulatory Ca^{2+}-specific site on cardiac TnC (Holroyde et al. 1980). Fourth, actomyosin, regulated by crayfish Tn-Tm complex with only one site binding Ca^{2+} at its physiological levels, displays a steep rise of ATPase activity with [Ca^{2+}] which is similar to that for the rabbit system (Wnuk et al. 1984). This leads us to the conclusion that the requirement of more than one site occupied by Ca^{2+} on TnC does not play any role in the steep responses to Ca^{2+}.

With the simplest kinetic models, the increase in ATPase rate with [Ca^{2+}] for the crayfish system would be expected to be proportional to the fractional occupancy by Ca^{2+} of the regulatory site in its low affinity state (regulated actomyosin during steady state ATPase hydrolysis; see Sect. 7). However, this is not the case. When comparing the overall "weak" binding of Ca^{2+} with the ATPase activity as a function of [Ca^{2+}], it appears that the only common range of [Ca^{2+}] for both curves is the one where the Ca^{2+}-specific site begins to bind Ca^{2+} significantly and activation starts (Wnuk et al. 1984). At higher [Ca^{2+}], the regulated actomyosin ATPase rises more steeply with [Ca^{2+}] than does the bound Ca^{2+} at the regulatory site. This is reminiscent of observations indicating that the weak binding of S1 to the thin filament in the presence of ATP is Ca^{2+}-insensitive, whereas the thin filament-activated ATPase activity of S1 greatly increases upon addition of Ca^{2+} (Chalovich and Eisenberg 1982). The latter data imply that the binding of S1-ATP and S1-ADP.P_i to the thin filament does not affect the affinity of TnC for Ca^{2+}. Chalovich and Eisenberg (1982) postulated that a kinetic step, most likely P_i release, is the Ca^{2+}-sensitive part of the ATPase cycle. It is now clear from the subsequent experiments in vivo (Eisenberg and Hill 1985) that ATP binding to the myosin head does not necessarily cause it to dissociate from actin before the ATP is hydrolysed, as suggested by Lymn and Taylor (1971). Rather, when either ATP or ADP and P_i are on the head, it is weakly bound to actin, in rapid equilibrium between attachment and detachment. The attached myosin head swings round in a power stroke that moves the actin filament. The power stroke is associated with the Ca^{2+}-dependent release of P_i and drives the head into the strongly attached states (with bound ADP and without nucleotide). ATP binding returns the head to the weak-binding states rather than detaching it.

It has been shown for both skeletal (Godt 1974) and cardiac (Best et al. 1977) muscle that decreases in [ATP] in skinned fibers increase the sensitivity of Ca^{2+} activation. Similar observations were reported for rabbit skeletal myofibrils as to their ATPase activity (Murray and Weber 1980). Adding ADP to contracting solutions causes an increase in both maximal Ca^{2+}-activated tension and the calcium sensitivity of tension in skinned skeletal fibers (Kerrick and Hoar 1985). Also, the increasing ratio of myosin to regulated actin in the actomyosin preparations results in a shift in the midpoint of Ca^{2+}-activation curves to lower [Ca^{2+}] (Wnuk and Stein 1980; Murray and Weber 1980). All these experiments suggest that after P_i release in the actomyosin ATPase cycle, myosin strongly bound to the thin filament increases the affinity of TnC for Ca^{2+} (see Sect. 7). The interaction of actin and myosin may also contribute to the steep responses to Ca^{2+}. While the binding of Ca^{2+} with its low affinity may relieve the inhibition by Tn-Tm of the rate-limiting step in the ATPase cycle and then the strong attachment of myosin to the thin filament may augment the binding constant of Ca^{2+} to the Ca^{2+}-specific site(s), a rapid equilibrium between Ca^{2+} and the site(s) in the high affinity state may become (with increasing $[Ca^{2+}]$) more and more important in suppressing the inhibiting effect of Tn-Tm. This would explain the steep rise of the steady state ATPase activity with $[Ca^{2+}]$ as opposed to the equilibrium Ca^{2+} binding properties of regulated actomyosin which are essentially noncooperative (Wnuk et al. 1984).

In order to explain the cooperative steady-state ATPase activity of S1 on regulated actin and the sharp activation of isometric muscle contraction by Ca^{2+}, Hill and coworkers (Hill et al. 1981; Hill 1983) have developed models with two major ingredients in the regulation: (1) Ca^{2+} binds much more strongly to TnC if myosin is already strongly attached to the thin filament; (b) there is positive cooperativity in the system because of nearest-neighbor interactions between Tm-Tn units, each involving seven actin monomers. These interactions are responsable for the steep response to Ca^{2+}. The models postulate that Tn-Tm units can exist in two states characterized by weak and strong binding of myosin, respectively. State 1 is favored at low and state 2 at high myosin concentrations. The models permit the calculation of the fraction of actin sites in the two states as a function of Ca^{2+} and myosin heads. It appears that the favoring of actins in the strong binding state can be brought about by either Ca^{2+} or myosin or a combination of the two. Thus the system is not turned on solely by Ca^{2+}. The weak binding state is favored even in the presence of Ca^{2+} when myosin concentrations are low. In the absence of myosin, Ca^{2+} would be essentially bound to Tn-Tm units in state 1. One wonders, therefore, whether the models take sufficiently into account the Ca^{2+}-induced structural changes in the thin filament without myosin (see Sect. 7). It seems that extending the treatment to models with more states than two for the thin filament would be more realistic.

There are also some doubts as to whether a cooperative unit involves seven actins. A satisfactory agreement between the binding of S1 and its effect on ATPase activity was obtained without introducing interaction among Tn-Tm units, but only by taking values for the number of actin monomers within cooperative units that exceed seven (Nagashima and Asakura 1982). Brandt et al. (1984) have reported that extraction of as little as 5% of TnC from skinned skeletal fibers reduces the slope of the pCa/tension relation. They interpreted this to mean that Tn-Tm units along a thin filament are linked cooperatively so that a thin filament activates as a unit. In order to explain certain features of the thin filament, flexibility within a Tm strand was also taken into account (Murray et al. 1981). Wagner and Stone (1983) observed that single- and two-headed meromyosin bind to the thin filament (in the presence of ATP) with higher affinity in the presence of Ca^{2+} than in its absence. They also showed that the Ca^{+}-sensitive binding of myosin to regulated actin requires that both the head-tail junction and the light chain 2 be intact. These results suggest that two Ca^{2+}-dependent steps are involved in the ATPase cycle, the attachment of myosin to actin and a kinetic step which occurs after binding. It seems therefore that a precise relation between Ca^{2+} binding to TnC and the activation of muscle contraction cannot be established before more information becomes available on the actomyosin ATPase cycle and the interprotein interactions within the system.

Acknowledgment. This work was supported by the Swiss National Science Foundation Grant 3.639.087.

References

Adelstein RS, Eisenberg E (1980) Regulation and kinetics of the actin-myosin-ATP interaction. Ann Rev Biochem 49:921–956

Babu YS, Sack JS, Greehough TJ, Bugg CE, Means AR, Cook WJ (1985) Three-dimensional structure of calmodulin. Nature (Lond) 315:37–40

Benzonana G, Kohler L, Stein EA (1974) Regulatory proteins of crayfish tail muscle. Biochim Biophys Acta 638:247–258

Best PM, Donaldson SKB, Kerrick WGL (1977) Tension in mechanically disrupted mammalian cardiac cells: effects of magnesium adenosine triphosphate. J Physiol 265:1–17

Brandt PW, Cox RN, Kawai M (1980) Can the binding of Ca^{2+} to two regulatory sites on troponin C determine the steep pCa/tension relationship of skeletal muscle? Proc Natl Acad Sci USA 77: 4717–4720

Brandt PW, Diamond MS, Schachat FH (1984) The thin filament of vertebrate skeletal muscle cooperatively activates as a unit. J Mol Biol 180:379–384

Bremel RD, Weber A (1972) Cooperation within actin filament in vertebrate skeletal muscle. Nature New Biol 238:97–101

Brinley FJ, Scarpa A, Tiffert T (1977) The concentration of ionized magnesium in barnacle muscle fibers. J Physiol 266:545–565

Bullard B, Dabrowska R, Winkelman L (1973) The contractile and regulatory proteins of insect flight muscle. Biochem J 135:277–286

Chalovich JM, Eisenberg E (1982) Inhibition of actomyosin ATPase activity without blocking the binding of myosin to actin. J Biol Chem 257:2432–2437

Chantler P (1982) Retreats from the steric blocking of muscle contraction. Nature (Lond) 198: 120–121

Collins JH, Potter JD, Horn MJ, Wilshire G, Jackman N (1973) The amino acid sequence of rabbit skeletal muscle troponin C: gene replication and homology with calcium-binding proteins from carp and hake muscle. FEBS Lett 36:268–272

Collins JH, Greaser ML, Potter JD, Horn MJ (1977) Determination of the amino acid sequence of troponin C from rabbit skeletal muscle. J Biol Chem 252:6356–6362

Cox JA, Comte M, Stein EA (1981) Calmodulin-free skeletal-muscle troponin C prepared in the absence of urea. Biochem J 195:205–211

Dabrowska R, Sherry JMF, Aromatorio DK, Hartshorne DJ (1978) Modulator protein as a component of the myosin light chain kinase from chicken gizzard. Biochemistry 17:253–258

Donaldson SKB, Kerrick WGL (1975) Characterization of effects of Mg^{2+} on Ca^{2+}- and Sr^{2+}-activated tension generation of skinned skeletal muscle fibers. J Gen Physiol 66:427-444

Ebashi S, Endo M (1968) Calcium ions and muscle contraction. Progr Biophys Mol Biol 18:123–183

Ebashi S, Endo M, Ohtsuki I (1969) Control of muscle contraction. Quart Rev Biophys 2:351–384

Eisenberg E, Hill TL (1985) Muscle contraction and free energy transduction in biological systems. Science 227:999–1006

Eisenberg E, Kielley WW (1974) Troponin-tropomyosin complex. Column chromatography separation and activity of the three active troponin components with and without tropomyosin present. J Biol Chem 249:4742–4748

Endo T, Obinata T (1981) Troponin and its components from ascidian smooth muscle. J Biochem (Tokyo) 89:1599–1608

Evans JS, Levine BA, Leavis PC, Gergely J, Grabarek Z, Drabikowski W (1980) Proton magnetic resonance studies on proteolytic fragments of troponin C. Structural homology with the native protein. Biochim Biophys Acta 623:10–20

Fabiato A, Fabiato F (1975) Effects of magnesium on contractile activation of skinned cardiac cells. J Physiol 249:497–517

Fuchs F (1977) The binding of calcium to glycerinated muscle fibers in rigor. The effect of filament overlap. Biochim Biophys Acta 491:523–531

Fuchs F, Black B (1980) The effects of magnesium ions on the binding of calcium ions to glycerinated rabbit psoas muscle fibers. Biochim Biophys Acta 622:52–62

Godt RE (1974) Calcium-activated tension of skinned muscle fibers of the frog. Dependence on magnesium adenosine triphosphate concentration. J Gen Physiol 63:722–739

Goldberg A, Lehman W (1978) Troponin-like proteins from muscles of the scallop, *Aequipecten irradians*. Biochem J 171:413–418

Goodman M, Pechère JF, Haiech J, Demaille JG (1979) Evolutionary diversification of structure and function in the family of intracellular calcium-binding proteins. J Mol Evol 13:331–352

Grabarek Z, Drabikowski W, Vinokurov L, Lu RC (1981a) Digestion of troponin C with trypsin in the presence and absence of Ca^{2+}. Identification of cleavage points. Biochim Biophys Acta 671: 227–233

Grabarek Z, Drabikowski W, Leavis PC, Rosenfeld SS, Gergely J (1981b) Proteolytic fragments of troponin C. Interactions with the other troponin subunits and biological activity. J Biol Chem 256:13121–13127

Grabarek Z, Grabarek J, Leavis PC, Gergely J (1983) Cooperative binding to the Ca^{2+}-specific sites of troponin C in regulated actin and actomyosin. J Biol Chem 258:14098–14102

Grabarek Z, Leavis PC, Gergely J (1986) Calcium binding to the low affinity sites in troponin C induces conformational changes in the high affinity domain. A possible route of information transfer in activation of muscle contraction. J Biol Chem 261:608–613

Graeser ML, Gergely J (1971) Reconstitution of troponin activity from three protein components. J Biol Chem 246:4226–4233

Greene LE, Eisenberg E (1980) Cooperative binding of myosin subfragment-1 to the actin-troponin-tropomyosin complex. Proc Natl Acad Sci USA 77:2616–2620

Gupta RK, Moore RD (1980) ^{31}P-NMR studies of intracellular free Mg^{2+} in intact frog skeletal muscle. J Biol Chem 255:3987–3992

Hanson J (1968) Recent X-ray diffraction studies of muscle. Quart Rev Biophys 1:177–216

Hanson J, Lowy J (1962) The structure of F-actin and of actin filaments isolated from muscle. J Mol Biol 6:46–60

Hartshorne DJ, Siemankowski R (1981) Regulation of smooth muscle actomyosin. Ann Rev Physiol 43:519–530

Haselgrove JC (1972) X-ray evidence for a conformational change in the actin-containing filaments of vertebrate striated muscle. Cold Spring Harbor Symp Quant Biol 37:341–352

Herzberg O, James MNG (1985a) Structure of the calcium regulatory muscle protein troponin C at 2.8 Å resolution. Nature (Lond) 313:653–659

Herzberg O, James MNG (1985b) Common structural framework of the two Ca^{2+}/Mg^{2+} binding loops of troponin C and other Ca^{2+} binding proteins. Biochemistry 24:5298–5302

Herzberg O, Moult J, James MNG (1986) A model for the Ca^{2+}-induced conformational transition of troponin C. A trigger for muscle contraction. J Biol Chem 261:2638–2644

Hill TL (1983) Two elementary models for the regulation of skeletal muscle contraction by calcium. Biophys J 44:383–396

Hill TL, Eisenberg E, Chalovich JM (1981) Theoretical models for cooperative steady-state ATPase activity of myosin subfragment-1 on regulated actin. Biophys J 35:99–112

Hincke MT, Sykes BD, Kay CM (1981) Hydrogen-1 nuclear magnetic resonance investigation of bovine cardiac troponin C. Comparison of tyrosyl assignments and calcium-induced structural changes to those of two homologous proteins, rabbit skeletal troponin C and bovine brain calmodulin. Biochemistry 20:3286–3294

Hoar PE, Wnuk W, Kerrick WGL (1985) Crayfish troponin C can substitute for the endogenous troponin C of skinned rabbit skeletal muscle fibers. Biophys J 47:61a

Holroyde MJ, Robertson SP, Johnson JD, Solaro RJ, Potter JD (1980) The calcium and magnesium binding sites on cardiac troponin and their role in the regulation of myofibrillar adenosine triphosphatase. J Biol Chem 255:11688–11693

Huxley HE (1972) Structural changes in the actin- and myosin-containing filaments of vertebrate striated muscle. Cold Spring Harbor Symp Quant Biol 37:361–376

Irving M (1985) Weak and strong crossbridges. Nature (Lond) 316:292–293

Johnson JD, Charlton SC, Potter JD (1979) A fluorescence stopped-flow analysis of Ca^{2+} exchange with troponin C. J Biol Chem 254:3497–3502

Johnson JD, Robinson DE, Robertson SP, Schwartz A, Potter JD (1981) Ca^{2+} exchange with troponin and the regulation of muscle contraction. In: Grinnel A (ed) The regulation of muscle contraction: excitation-contraction coupling. Acadmic Press, New York, pp 241–259

Kawasaki Y, Van Eerd JP (1972) The effect of Mg^{2+} on the conformation of the Ca^{2+}-binding component of troponin. Biochim Biophys Res Commun 49:898–905

Kendrick-Jones J, Scholey JM (1981) Myosin-linked regulatory systems. J Muscle Res Cell Motil 2: 347–362

Kerrick WGL, Bolles LL (1981) Regulation of Ca^{2+}-activated tension in Limulus striated muscle. Pflügers Arch 392:121–124

Kerrick WGL, Hoar PE (1985) The effects of nucleotide diphosphate and inorganic phosphate on tension in skinned soleus and smooth muscle cells. Biophys J 47:296a

Klee CB, Vanaman TC (1982) Calmodulin. Adv Prot Chem 35:213–321

Konno K (1978) Two calcium regulation systems in squid muscle. Preparation of calcium-sensitive myosin and troponin-tropomyosin. J Biochem (Tokyo) 84:1431–1440

Leavis PC, Gergely J (1984) Thin filament proteins and thin filament-linked regulation of vertebrate muscle contraction. CRC Crit Rev Biochem 16:235–305

Leavis PC, Rosenfeld SS, Gergely J, Grabarek Z, Drabikowski W (1978) Proteolytic fragments of troponin C. J Biol Chem 253:5452–5459

Lehman W (1975) Hybrid troponin reconstituted from vertebrate and arthropod subunits. Nature (Lond) 255:424–426

Lehman W (1982) The location and periodicity of a troponin-T-like protein in the myofibril of the horseshoe crab. J Mol Biol 154:385–391

Lehman W, Ferrell M (1980) Phylogenetic diversity of troponin subunit-C amino acid composition. FEBS Lett 121:273–274

Lehman W, Szent-Györgyi AG (1975) Regulation of muscular contraction. Distribution of actin control and myosin control in the animal kingdom. J Gen Physiol 66:1–30

Lehman W, Head JF, Grant PW (1980) The stoichimetry and location of troponin I- and troponin C-like proteins in the myofibril of the bay scallop, *Aequipecten irradians.* Biochem J 187: 447–456

Levine BA, Thornton JM, Fernandes R, Kelly CM, Mercola D (1978) Comparison of the calcium- and magnesium-induced structural changes of troponin C. A proton magnetic resonance study. Biochim Biophys Acta 535:11–24

Lymn RW, Taylor EW (1971) Mechanism of adenosine triphosphate hydrolysis by actomyosin. Biochemistry 10:4617–4624

Miledi R, Parker I, Schalow G (1977) Measurements of calcium transients in frog muscle by the use of arsenazo (III). Proc R Soc Lond, Ser B 198:201–210

Moews PG, Kretsinger RH (1975) Refinement of the structure of carp muscle calcium binding parvalbumin by model building and difference Fourier analysis. J Mol Biol 91:201–228

Murray JM, Weber A (1980) Cooperativity of the calcium switch of regulated rabbit actomyosin system. Mol Cell Biochem 35:11–15

Murray JM, Weber A, Knox MK (1981) Myosin subfragments binding to relaxed actin filaments and steric model of relaxation. Biochemistry 20:641–649

Nagashima H, Asakura S (1982) Studies on a co-operative properties of tropomyosin-actin and tropomyosin-troponin-actin complexes by the use of N-ethylmaleimide-treated and untreated species of myosin subfragment 1. J Mol Biol 155:409–428

Nagy B, Gergely J (1979) Extent and localization of conformational changes in troponin C caused by calcium binding. Spectral studies in the presence and absence of 6 M urea. J Biol Chem 254: 12732–12737

Pearlstone JR, Carpenter MR, Johnson P, Smillie LB (1976) Amino-acid sequence of tropomyosin-binding component of rabbit skeletal muscle troponin. Proc Natl Acad Sci USA 73:1902–1906
Potter JD (1974) The content of troponin, tropomyosin, actin and myosin in rabbit skeletal muscle myofibrils. Arch Biochem Biophys 162:436–441
Potter JD, Gergely J (1975) The calcium and magnesium binding sites on troponin and their role in the regulation of myofibrillar adenosine triphosphatase. J Biol Chem 250:4628–4633
Potter JD, Johnson JD (1982) Troponin. In: Cheung WY (ed) Calcium and cell function Vol II. Academic Press, New York London, pp 145–173
Potter JD, Robertson SP, Johnson JD (1981) Magnesium and regulation of muscle contraction. Fed Proc 49:2653–2656
Regenstein JM, Szent-Györgyi AG (1975) Regulatory proteins of lobster striated muscle. Biochemistry 14:917–925
Reid RE, Hodges RS (1980) Cooperativity and calcium/magnesium binding to troponin C and muscle calcium binding parvalbumin: an hypothesis. J Theor Biol 84:401–444
Ridgway EB, Gordon AM, Martyn DA (1983) Histeresis in the force-caclium relation in muscle. Science 219:1075–1077
Robertson SP, Johnson JD, Potter JD (1981) The time-course of Ca^{2+} exchange with calmodulin, troponin, parvalbumin, and myosin in response to transient increases in Ca^{2+}. Biophys J 34: 559–569
Romero-Herrera AE, Cartillo O, Lehmann H (1976) Human skeletal muscle proteins. The primary structure of troponin C. J Mol Evol 8:251–270
Sin IL, Fernandes R, Mercola D (1978) Direct identification of the high and low affinity calcium binding sites of troponin C. Biochem Biophys Res Commun 82:1132–1139
Smillie LB (1979) Structure and functions of tropomyosins from muscle and non-muscle sources. Trends Biochem Sci 4:151–155
Solaro RJ, Shiner JS (1976) Modulation of Ca^{2+} control of dog and rabbit cardiac myofibrils by Mg^{2+}. Comparison with rabbit skeletal myofibrils. Circ Res 39:8–14
Sundaralingam M, Bergstrom R, Strasburg G, Rao ST, Roychowdhury P, Greaser M, Wang BC (1985a) Molecular structure of troponin C from chicken skeletal muscle at 3-angstrom resolution. Science 227:945–948
Sundaralingam M, Drendel W, Greaser M (1985b) Stabilization of the long central helix of troponin C by intrahelical salt bridges between charged amino acid side chains. Proc Natl Acad Sci USA 82:7944–7947
Szent-Györgyi AG, Szentkiralyi EM, Kendrick-Jones J (1973) The light chains of scallop myosin as regulatory subunits. J Mol Biol 74:179–203
Takagi T, Konishi K (1983) Amino acid sequence of troponin C obtained from ascidian (*Halocynthia roretzi*) body wall muscle. J Biochem (Tokyo) 94:1753–1760
Taylor EW (1979) Mechanisms of actomyosin ATPase and the problem of muscle contraction. CRC Crit Rev Biochem 6:103–164
Trybus KM, Taylor EW (1980) Kinetic studies of the cooperative binding of subfragment 1 to regulated actin. Proc Natl Acad Sci USA 77:7209–7213
Tsuchiya T, Head JF, Lehman (1982) The isolation and characterization of a troponin C-like protein from the mantle muscle of the squid *Loligo pealei*. Comp Biochem Physiol 71B:507–509
Van Eerd JP, Takahashi K (1976) Determination of the complete amino acid sequence of bovine cardiac troponin C. Biochemistry 15:1171–1180
Van Eerd JP, Capony JP, Ferraz C, Pechère JF (1978) The amino-acid sequence of troponin C from frog skeletal muscle. Eur J Biochem 91:231–242
Wagner P, Stone DB (1983) Calcium-sensitive binding of heavy meromyosin to regulated actin requires light chain 2 and the head-tail junction. Biochemistry 22:1334–1342
Wakabayashi T, Huxley HE, Amos LA, Klug A (1975) Three-dimensional image reconstruction of actin-tropomyosin complex and actin-tropomyosin-troponin T-troponin I complex. J Mol Biol 93:477–497
Wang CK, Cheung HC (1985) Energetics of the binding of calcium and troponin I to troponin C from rabbit skeletal muscle. Biophys J 48:727–739

Watanabe K, Kitaura T, Yamaguchi M (1982) Crayfish myosin has no Ca^{2+}-dependent regulation in actomyosin. J Biochem (Tokyo) 92:1635–1641

Watterson DM, Sharief F, Vanaman TC (1980) The complete amino acid sequence of the Ca^{2+}-dependent modulator protein (calmodulin) of bovine brain. J Biol Chem 255:962–975

Weber A, Murray JM (1973) Molecular control mechanisms in muscle contraction. Physiol Rev 53:612–673

Weeds AG, McLachlan AD (1974) Structural homology of myosin alkali light chains, troponin C and carp calcium binding protein. Nature (Lond) 252:646–649

Wegner Y, Walsh TP (1981) Interaction of tropomyosin-troponin with actin filaments. Biochemistry 20:5633–5642

Wilkinson JM (1976) The amino acid sequence of troponin C from chicken skeletal muscle. FEBS Lett 70:254–256

Wilkinson JM (1980) Troponin C from rabbit slow skeletal and cardiac muscle is the product of a single gene. Eur J Biochem 103:179–188

Wilkinson JM, Grand RJA (1975) The amino acid sequence of troponin I from rabbit skeletal muscle. Biochem J 149:493–496

Wilkinson JM, Grand RJA (1978) Comparison of amino acid sequence of troponin I from different striated muscles. Nature (Lond) 271:31–35

Wnuk W, Stein EA (1978) Evolution of the Ca-binding properties of troponin C. Experientia 34: 920

Wnuk W, Stein EA (1980) Does the cooperative response of myofibrils to Ca^{2+} result from multiple Ca^{2+}-sites on troponin C? In: Siegel FL, Carafoli E, Kretsinger RH, MacLennan DH, Wasserman RH (eds) Calcium binding proteins: structure and function. Elsevier North-Holland, New York, pp 343–344

Wnuk W, Cox JA, Stein EA (1982) Parvalbumins and other soluble high-affinity calcium-binding proteins from muscle. In: Cheung WY (ed) Calcium and cell function. Vol II. Academic Press, New York London, pp 243–278

Wnuk W, Schoechlin M, Stein EA (1984) Regulation of actomyosin ATPase by a single calcium-binding site on troponin C from crayfish. J Biol Chem 259:9017–9023

Wnuk W, Schoechlin M, Kobayashi T, Takagi T, Konishi K, Hoar PE, Kerrick WGL (1986) Two isoforms of troponin C from crayfish. Their characterization and a comparison of their primary structure with the tertiary structure of skeletal troponin C. J Muscle Res Cell Motil 7:67

Zot HG, Potter JD (1982) A structural role for the Ca^{2+}-Mg^{2+} sites on troponin C in the regulation of muscle contraction. Preparation and properties of troponin C depleted myofibrils. J Biol Chem 257:7678–7683

Zot HG, Potter JD (1984) The role of calcium in the regulation of the skeletal muscle contraction-relaxation cycle. In: Siegel H (ed) Metal Ions in biological systems. Vol XVII. Decker, New York Basel, pp 381–410

Zot HG, Iida S, Potter JD (1983) Thin filament interactions and Ca^{2+} binding to Tn. Chem Scr 21: 135–138

Caldesmon and Ca^{2+} Regulation in Smooth Muscles

S. MARSTON[1], W. LEHMAN[2], C. MOODY[3], K. PRITCHARD[1] and C. SMITH[4]

1 Introduction

It has been known since the pioneering study of Filo et al. (1965) that the contractile proteins of smooth muscles, actin and myosin, are controlled by Ca^{2+} levels in the micromolar range as they are in striated muscles. Ca^{2+} control of myosin activity via the calmodulin-dependent myosin light chain kinase was the first to be discovered and characterised. Subsequently Ca^{2+} control of the actin-based thin filaments was also shown, first in crude actomyosin preparations using the myosin competition test and then by the isolation of native smooth muscle thin filaments which were Ca^{2+}-regulated (Marston et al. 1979; Marston and Smith 1984). The major components of these thin filament preparations are actin, tropomyosin and caldesmon. Caldesmon is a protein of Mr 120,000, first isolated by Kakiuchi as a calmodulin-binding protein, but subsequently shown to also bind to actin and to have an inhibitory effect on actin-activation of ATPase activity (Kakiuchi and Sobue 1983). We have studied caldesmon in some detail, since it is an obvious candidate for the regulatory protein function in smooth muscle thin filaments. In this article we shall describe the mechanism by which caldesmon, together with calmodulin controls actin filament activity in response to Ca^{2+} and then we shall consider the structural basis of caldesmon regulation and whether such regulation plays a role in the control of contractility in intact smooth muscles.

2 A Mechanism for Ca^{2+}-Regulation of Smooth Muscle of Thin Filaments by Caldesmon and Calmodulin

Native thin filament preparations (Fig. 1) contain actin, tropomyosin and caldesmon in molar ratios 28:4:1 (Marston and Smith 1984; Marston and Lehman 1985). There is also probably a calcium binding protein which has not yet been properly characterised

1 Cardiothoracic Institute, 2 Beaumont Street, London W1N 2DX, U.K.
2 Boston University Medical Center 80 E. Concord Street, Boston, Ma 02118, USA
3 Present address: University of Massachusetts Medical School, Department of Anatomy, Worcester, Ma 01605, USA
4 Present address: Children's Hospital, Department of Cardiology, 300 Longwood Avenue, Boston, Ma 02115, USA

Ch. Gerday, R. Gilles, L. Bolis (Eds.)
Calcium and Calcium Binding Proteins

A

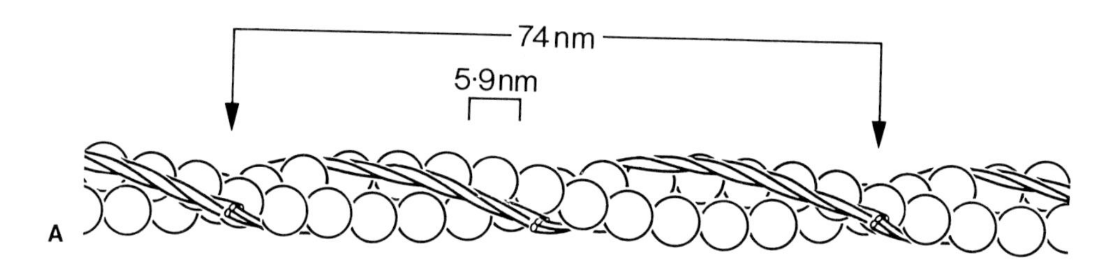

B

Fig. 1A, B. Structure of smooth muscle thin filaments. **A** Scale diagram of the basic structure of smooth muscle thin filaments made up of actin and tropomyosin. **B** Electron micrograph of native thin filaments from sheep aorta

or quantitated. We have developed a procedure by which the components of the thin filament can be separated from each other and purified (Smith and Marston 1985). Using the separated pure components it is possible to reconstitute in part or in full the thin filament and thus determine how the proteins interact.

2.1 Actin and Tropomyosin

Smooth muscle actin is obtained directly in the polymeric form (F-actin) (Marston and Smith 1984). This actin interacts with myosin in the usual way: it forms strong "rigor" links in the absence of ATP, and in the presence of MgATP the actin activates myosin MgATPase activity. Activation has been observed with myosin, HMM or S1 from both smooth and striated muscles and its characteristics differ only a little from skeletal muscle F-actin (Marston and Smith 1985). Tropomyosin is always found bound to actin filaments in vivo. Isolated smooth muscle F-actin binds tropomyosin from smooth or striated muscles: binding is cooperative with a stoichiometry of one per seven actins and it is believed that the tropomyosin is located in the grooves of the actin helix (Fig. 1). Binding is optimal in the presence of >1 mM Mg^{2+} and at ionic strengths approaching physiological (Sanders and Smillie 1984). Tropomyosin may be dissociated from actin at very high or very low ionic strengths. Tropomyosin modifies actin's interaction with myosin: it may enhance actin-activation of ATPase (known as potentiation) or it may inhibit activation. This property differs between tropomyosins from different sources: smooth muscle tropomyosin usually potentiates actin activation under the experimental conditions we commonly use (Marston and Smith 1985). Tropomyosin, however, should not be considered a regulatory protein per se since none of these effects are Ca^{2+} dependent and it does not interact with calmodulin or troponin-C.

2.2 Actin, Caldesmon and Tropomyosin

Caldesmon is a basic protein of molecular weight around 120,000 (Marston and Lehman 1985; Bretscher 1984) found in association with actin filaments in all smooth muscles and also many non-muscle motile cells such as platelets [non-muscle caldesmon often appears as a low molecular weight – 80,000, isoform (reviewed by Marston and Smith 1985)]. Pure caldesmon is an F-actin-binding protein which typically has an affinity of greater than 10^6 M^{-1}. This binding affinity is enhanced about fivefold when tropomyosin is also bound to the F-actin (Smith and Marston 1985). The binding has been measured under a wide range of conditions and protein concentrations. Analysis of many binding curves has shown us that binding is not simple: there are two classes of binding sites which we term "tight" and "weak". It is the "tight" binding that is most interesting because it saturates at a stoichiometry of 0.035 ± 0.007 CD/A (mean ± SD of 29 measurements) whether or not tropomyosin is present (Smith et al. 1987). This stoichiometry corresponds to the amount of caldesmon present in native thin filaments.

The effect of caldesmon on F-actin alone is variable. Most of our measurements have been made with excess Mg^{2+}, KCl 60–120 mM and at 25–37°C. Under these conditions no consistent effect of caldesmon on an ATPase has been observed even at caldesmon concentrations that produced full occupancy of the "tight" sites (Fig. 2). However, we (Moody et al. 1985) and others (Chalovitch et al. 1987) have noted some inhibitory effects of caldesmon on actin activation, at lower temperatures or low ionic strength: in some cases inhibition seemed to be almost as good as when tropomyosin was present.

When caldesmon binds to F-actin-tropomyosin at low stoichiometries it always inhibits. The greatest percentage inhibition is obtained when the initial actomyosin ATPase is well potentiated by tropomyosin (up to 95% see Fig. 2), but potentiation is not necessary for inhibition to be observed since caldesmon is still an inhibitor under conditions where actin-tropomyosin activates myosin MgATPase less than actin does (e.g. Fig. 2). Inhibition is never complete but seems to reach a basal level which we find to be consistently about 1/4 the ATPase activation due to F-actin alone (Smith et al. 1987).

The quantity of caldesmon required to inhibit actin-tropomyosin was found to vary with temperature and [KCl], but this is merely a reflection of the variation of binding affinity of caldesmon for actin tropomyosin. We determined the relationship between inhibition by caldesmon and caldesmon binding under many different conditions and found that inhibition to the basal level consistently correlated with occupation of the

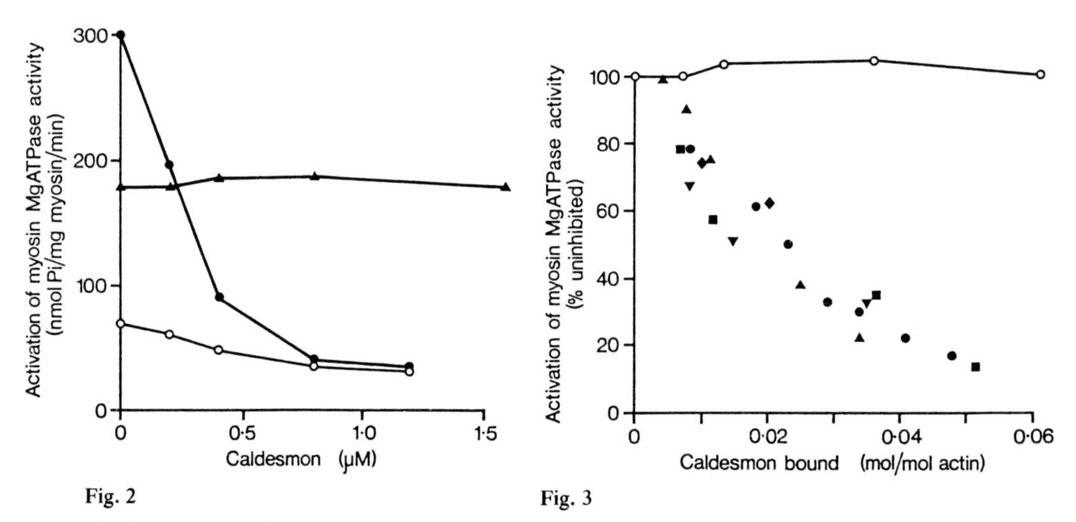

Fig. 2 **Fig. 3**

Fig. 2. Inhibition of actin activation of myosin MgATPase activity by caldesmon. MgATPase activity was measured at 25°C, 60 mM KCl with 0.125 mg/ml skeletal muscle myosin, 0.5 mg/ml (12 μM) aorta actin or 0.5 mg/ml actin + 0.125 mg/ml (1.8 μM) tropomyosin, and 0–0.2 mg/ml (1.7 μM) aorta caldesmon. ▲ actin; ● actin and aorta tropomyosin; ○ actin and skeletal muscle tropomyosin. Myosin MgATPase activity (12 nmol/mg/min) was subtracted from total MgATPase activity

Fig. 3. The relationship of caldesmon binding to actin activation of myosin MgATPase. Parallel measurements were made of caldesmon binding to actin and actin activation of skeletal muscle myosin MgATPase (described in Fig. 2) with the same protein preparation under identical conditions (except for the additional presence of myosin and MgATP in the ATPase measurements). *Conditions:* ○ aorta actin, 25°C, 60 mM KCl; ● aorta actin-tropomyosin, 4°C, 50 mM KCl; ■ aorta actin-tropomyosin, 25°C, 60 mM KCl; ▲ aorta actin-tropomyosin, 37°C, 70 mM KCl; ◆ aorta actin-tropomyosin, 37°C, 120 mM KCl; ▼ aorta actin and skeletal muscle tropomyosin, 25°C, 60 mM KCl.

Data is plotted as relative ATPase: $= \frac{(\text{ATPase} - \text{caldesmon})}{(\text{ATPase} + \text{caldesmon})}$ as a function of caldesmon bound per actin

"tight" sites (Fig. 3). This correlation is independent of whether any of the three proteins (actin, tropomyosin, myosin) involved were from skeletal or smooth muscle (Smith et al. 1987).

The inhibitory action of caldesmon is consistent with a regulatory role, but caldesmon alone does not constitute a regulatory system since tropomyosin is also required for full expression of the inhibitory action and inhibition is totally independent of Ca^{2+} concentrations. Caldesmon does bind to tropomyosin (Smith et al. 1987) and it may be that conformational changes transmitted via tropomyosin are important for inhibition to occur at very low caldesmon: actin ratios. Caldesmon also binds to calmodulin and troponin C and it is through this form of interaction that the system becomes calcium-regulated.

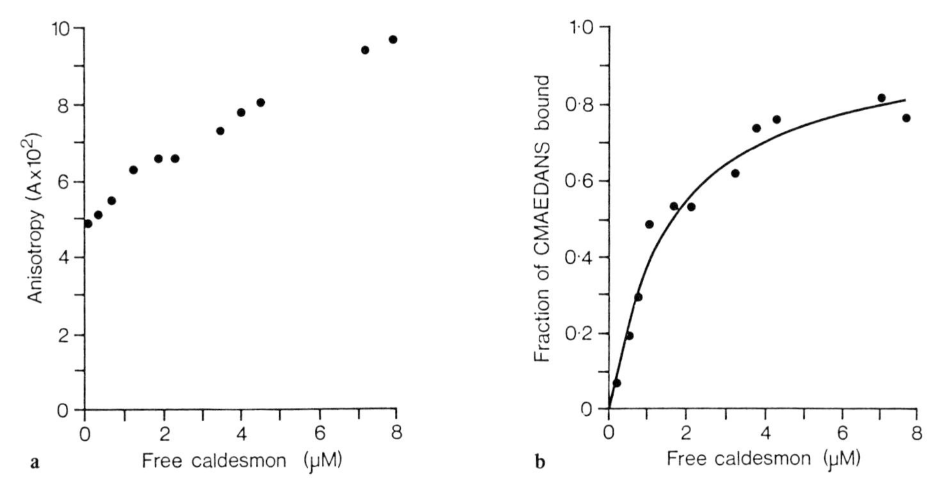

Fig. 4. **a** Fluorescence anisotropy changes on binding of calmodulin AEDANS to caldesmon. Anisotropy changes were measured using a Perkin-Elmer polarization accessory.

Anisotropy $A = \frac{I_{\parallel} - I_{\perp}}{I_{\parallel} + 2 \cdot I_{\perp}}$, where $I_{\parallel} \& \perp$ are the intensities recorded when the planes of polarization of the excitation polarizer and emission analyzer are parallel, and at 90°, respectively. In practice, $I_{\parallel}$ was measured using vertically orientated polarizers, and $I_{\perp}$ was corrected for "grating anomalies" by substituting (G.I) where the ratio $G = \frac{I_{90;0}}{I_{90;90}}$.

Conditions: 0.7 μM calmodulin-AEDANS (CMAEDANS), 10^{-4} M Ca^{2+}, 60 mM KCl; pH 7.0 at 25°C. **b** Binding curve derived from data of **a**. The fraction of CMAEDANS bound was calculated as: $f_B = \frac{A - A_F}{A(1-q) + q \cdot A_B - A_F}$

where A = anisotropy of sample; A_F = initial anisotropy (no CD added); A_B = final anisotropy; q = ratio of intensities of bound/free species.

Background anisotropies due to caldesmon (CD) were subtracted to yield anisotropy due to calmodulin thus: CMAEDANS A = (total $A - A_{CD} \cdot I_{CD})/(1 - I_{CD})$. Where A_{CD} was obtained from a CD-only "background" cuvette, and I_{CD} is the fraction of the total intensity due to CD fluorescence. In this experiment, $K = 0.54 \pm 0.11 \times 10^6\ M^{-1}$

2.3 Caldesmon and Calmodulin

Caldesmon binds to Ca^{2+}.calmodulin; indeed this was the property by which caldesmon was originally isolated and characterised (Sobue et al. 1981). We have developed fluorescent techniques to measure this interaction. Calmodulin is covalently labelled with the fluorophore 1,5,IAEDANS (Olwin and Storm 1984). The fluorescence is sensitive to Ca^{2+}-binding to calmodulin and there is a further increase of fluorescence when Ca^{2+}.calmodulin binds to caldesmon. Furthermore, when the small, compact Ca^{2+}.calmodulin binds to the large and extended caldesmon molecule, it becomes less mobile and therefore the fluorescence is more polarised. We have used both these properties to determine caldesmon-Ca^{2+}.calmodulin binding constants (Fig. 4). Binding constants were in the region of $10^6\ M^{-1}$. Binding affinity was not much affected by temperature or ionic strength. Relative to the binding constant at 25°C, 60 mM KCl, the constant

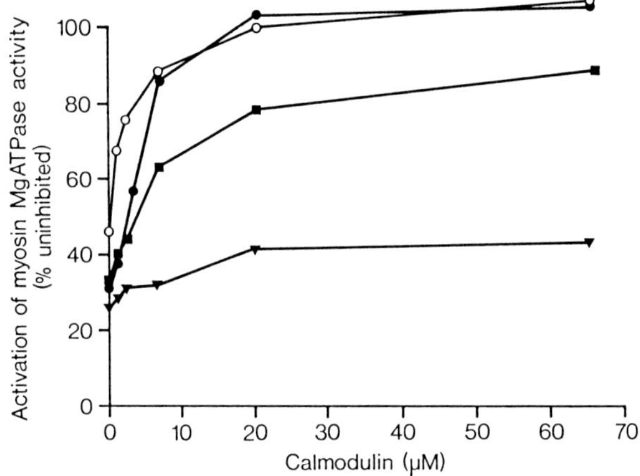

Fig. 5. Release of caldesmon inhibition by Ca^{2+}.calmodulin. The ability of Ca^{2+}.calmodulin to release caldesmon inhibition of actin-tropomyosin activated myosin MgATPase was measured under a range of conditions. ATPase activities were determined at 10^{-4} M $CaCl_2$. Protein concentrations were: actin, 0.5 mg/ml (12 μM): skeletal muscle myosin 0.125 mg/ml (0.27 μM); aorta tropomyosin 0.125 mg/ml (1.8 μM); calmodulin 0–10 mg/ml (60 μM); caldesmon was added to 0.1–0.4 mg/ml (0.8–3.2 μM) to give similar levels of inhibition according to conditions. ■ 25°C 60 mM KCl, skeletal actin, caldesmon, 0.2 mg/ml; ▼ 25°C 60 mM KCl, aorta actin, caldesmon, 0.1 mg/ml; ● 37°C 70 mM KCl, aorta actin, caldesmon, 0.14 mg/ml; ○ 37°C 120 mM KCl, aorta actin, caldesmon, 0.4 mg/ml. In 1 mM EGTA calmodulin never had any effect (data not shown)

at 37°C was decreased by 22 ± 39%, whilst at 37°C, 120 mM KCl the constant was decreased by 52 ± 7% (n = 3 in each case).

This interaction is not specific, since AEDANS-labelled skeletal muscle troponin C also binds to caldesmon with an affinity similar to calmodulin in the presence of Ca^{2+} and there is a little evidence that Ca^{2+}-free calmodulin can also bind to caldesmon weakly ($K < 10^5$ M^{-1}). By comparison with other Ca^{2+}.calmodulin binding proteins such as myosin light chain kinase ($K = 10^9$ M^{-1}, Adelstein and Klee 1981) binding is surprisingly weak.

2.4 Actin, Tropomyosin, Caldesmon and Calmodulin

Under suitable conditions, Ca^{2+}.calmodulin releases the inhibitory effect of caldesmon on actin-tropomyosin activation of myosin MgATPase activity, but Ca^{2+}-free calmodulin does not (Fig. 5). It is presumed but not directly proven that the release of caldesmon inhibition occurs when Ca^{2+}.calmodulin binds to caldesmon. This binding is associated with changes in the caldesmon-actin-tropomyosin interaction, currently of an unknown nature, which result in release of inhibition. Kakiuchi proposed that Ca^{2+}.calmodulin binding to caldesmon was directly coupled to dissociation of caldesmon from actin-tropomyosin (Kakiuchi and Sobue 1983). In fact this is only observed under certain circumstances, generally low ionic strength and temperature. As reaction conditions approach the physiological, we find that Ca^{2+}.calmodulin can readily bind without dissociating caldesmon from actin tropomyosin with consequent formation of a quaternary

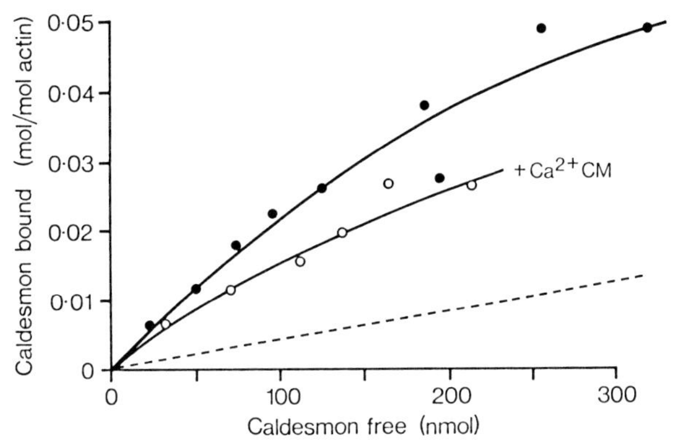

Fig. 6. Effect of Ca^{2+}.calmodulin on caldesmon binding to actin tropomyosin. Caldesmon binding was measured in ATPase buffer, 37°C, 70 mM KCl, 10^{-4} M $CaCl_2$. Protein concentrations were: v.s.m. actin, 0.2 mg/ml (4.5 μM), v.s.m. tropomyosin, 0.1 mg/ml (1.4 μM), caldesmon 0–0.125 mg/ml (0–1.1 μM). ● no calmodulin; ○ 0.8 mg/ml (50 μM) calmodulin. *Solid lines* are fitted curves with the strong binding parameters $K = 6.6 \times 10^6\ M^{-1}$ in the absence and $K = 4.0 \times 10^6\ M^{-1}$ in the presence of Ca^{2+}.calmodulin, max. binding 0.038 caldesmon/actin. Weak binding was $K = 3 \times 10^5\ M^{-1}$, max.binding 0.22 caldesmon/actin in both conditions. *Dotted line* shows the contribution to binding due to the weak sites alone

complex Ca^{2+}.CM-CD-A-Tm (see Fig. 6). Caldesmon must be in a non-inhibitory state in this quaternary complex since Ca^{2+}.calmodulin causes full release of caldesmon inhibition under these conditions (Fig. 5). One important consequence of the absence of a competition between Ca^{2+}.calmodulin and actin-tropomyosin for caldesmon under near physiological conditions is that release of inhibition can be obtained with much lower Ca^{2+} calmodulin concentrations. This may be demonstrated by the experimental results shown in Fig. 5. At 60 mM KCl, 25°C no quaternary complex is formed and even 60 μM Ca^{2+}.calmodulin does not greatly release inhibition; at 70 mM KCl, 37°C (conditions of Fig. 6.) 50% release of inhibition requires only 8 μM calmodulin, whilst at 120 mM KCl 37°C, 50% release of inhibition requires only 1 μM Ca^{2+}.calmodulin in excess of stoichiometric.

2.5 Ca^{2+} Regulation of Actin-Tropomyosin by Caldesmon and Calmodulin

It is now appropriate to gather together the experimental results discussed in the previous sections into a concise statement of how we believe caldesmon and calmodulin can regulate thin filament activity.

1. Caldesmon binds to actin with a high affinity at a stoichiometry of 1 per 28 actin monomers.
2. In the presence of tropomyosin the bound caldesmon inhibits actin activation of myosin MgATPase by up to 95%. Neither the binding nor the inhibition are Ca^{2+} sensitive.
3. In the presence of Ca^{2+}, calmodulin binds to the caldesmon and weakens its interaction with actin tropomyosin as a result of which there may be a partial dissociation of Ca^{2+}.calmodulin-caldesmon from the actin tropomyosin.

Low Ca^{2+} *inhibited* $>10^{-6}M\ Ca^{2+}$ *activated*

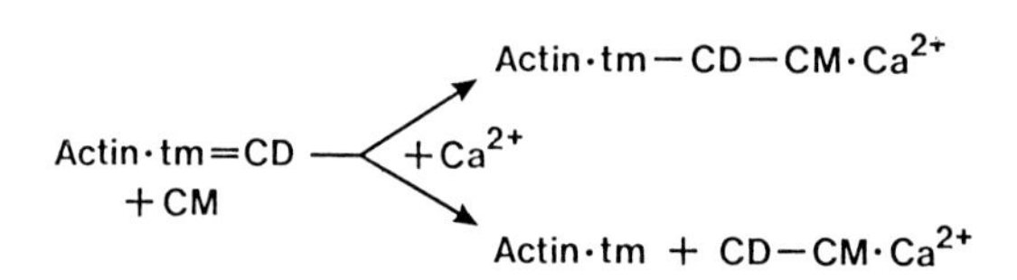

Fig. 7. A model for the Ca^{2+}-dependent release of caldesmon inhibition by calmodulin

4. This Ca^{2+}.calmodulin binding releases caldesmon's inhibition of actomyosin MgATPase activity. Ca^{2+} regulation of the thin filaments is therefore achieved by the conversion of the inhibited caldesmon-actin-tropomyosin complex to the active complexes actin-tropomyosin and Ca^{2+}.calmodulin-caldesmon-actin-tropomyosin, as illustrated in Fig. 7.

Since we have measured binding constants for the various protein-protein interactions over a range of conditions it is possible to assess this scheme quantitatively using the observation that inhibition is linearly related to the fraction of actin in the actin-tropomyosin-caldesmon complex (Fig. 3). Model calculations were made to predict the inhibition by caldesmon and its release by Ca^{2+}.calmodulin under the various conditions used in the experiments in Fig. 5. The results in Fig. 8 show a reasonable match to the experimental data and therefore are strongly in support of our model.

3 Structure of the Actin-Tropomyosin-Caldesmon System

Biochemical studies on caldesmon action in native thin filaments and reconstituted systems have produced a mechanism which we presume reflects structural changes in the thin filament and which emphasises the functional importance of the 28 actin:4 tropomyosin:1 caldesmon stoichiometry. We are therefore very interested in determining the structure of the regulated filament. The arrangement of actin monomers in a double helix and of tropomyosin stretched along the grooves of the helix is reasonably well established (Fig. 1) but what about caldesmon? So far negatively stained E.M. pictures of thin filaments or actin-tropomyosin-caldesmon are indistinguishable from actin-tropomyosin (e.g. Fig. 1). If it were a compact molecule of Mr 120,000, caldesmon ought to be large enough to be visible. Evidence is now building up that caldesmon is, in fact a thin extended molecule which is spread along the actin filament.

1. Rotary shadowed images of pure caldesmon have shown it to be a very long, randomly structured molecule (Furst et al. 1986) and this has been supported by gel filtration studies (Bretscher 1984).
2. Bivalent anti-caldesmon polyclonal antibody crosslinks thin filaments and causes them to form parallel aggregates in which several antibody molecules are bound to every caldesmon. Negatively stained images of these bundles show no periodicity,

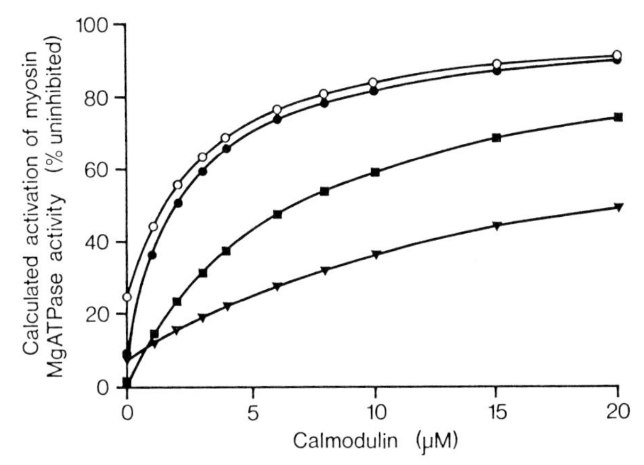

Fig. 8. Calculated curves for the release of inhibition by Ca^{2+}.calmodulin. Calculations are based on the model (Fig. 7) and the observation that inhibition is linearly related to the fraction of ATm which is in the ATm-CD complex (Fig. 3). The calculations are intended to describe the experiment in Fig. 5 and the binding constants are those measured under the appropriate conditions. Constants are given in units 10^{-6} K, M^{-1}.

	$K_{Atm\text{-}CD}$	$K_{CD\text{-}CM}$	$K_{Atm\text{-}CDCM}$
○ 120 mM KCl, 37°C	1.1	0.4	0.9
● 70 mM KCl, 37°C	7	0.66	4
▲ 60 mM KCl, 25°C	28	0.83	0
■ 60 mM KCl, 25°C skeletal actin.	28	0.83	4

Basal activation is not allowed for in this calculation

suggesting that caldesmon is not located at discrete intervals on the filament like troponin is in striated muscles (Lehman and Marston 1986; Lehman 1983) (Fig. 9). The position of the antibodies may be determined by streptavidin-gold labelling of biotinylated antibody. Such experiments confirm that the antibody, and hence the antigenic sites on caldesmon, are distributed all along the filament.

3. A single or a few defined points in the caldesmon molecule may be labelled by attaching biotin covalently using N-hydroxysuccinimide-biotin. When this is done, that point may be visualised on E.M. images of reconstituted filaments by labelling with streptavidin gold. The number and distribution of gold particles on filaments was rather variable, however filaments containing biotin-caldesmon showed at least double the labelling observed in controls omitting biotin. Some filaments showed a regular pattern of gold labelling but in general there was insufficient labelling for periodicity to be confirmed. In the example shown (Fig. 10) analysis of 80 gold particles indicated a high frequency (3 x background) of spacing at about 80, 140, and 240 nm (cf. Fig. 1).
4. Although the data is mainly circumstantial, when it is taken together the most likely structure of caldesmon in the smooth muscle thin filament is that of a long thin, flexible molecule which is located in the groove of the actin helix where it may interact with both actin and tropomyosin. In this way caldesmon can control

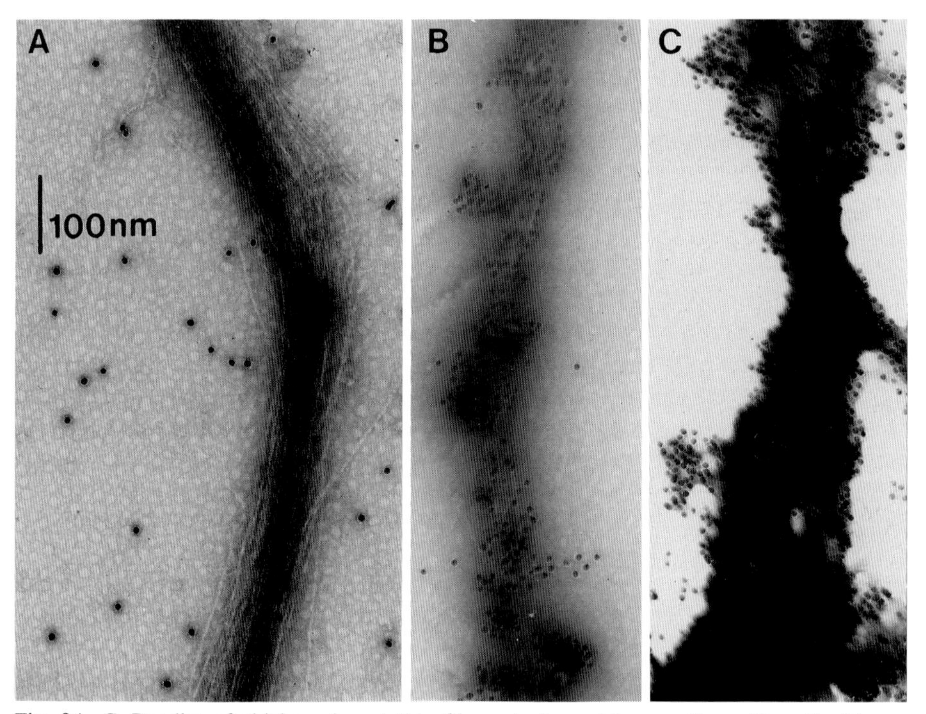

Fig. 9A–C. Bundles of chicken gizzard thin filaments formed by reaction with anti-caldesmon antibody. **A** 1 mg/ml thin filaments was mixed with 10 mg/ml anti-caldesmon IgG. After 15′ bundles of filaments were isolated by sedimentation 5′ at 5000 *g*. Pellets were redispersed and placed on E.M. grids, then negatively stained. Filaments were aggregated into long bundles with no apparent periodicities. **B, C** The location of antibody molecules in the bundles was revealed using biotinylated anticaldesmon IgG. Redispersed bundles were incubated with an equal volume of streptavidin-gold particles (15 nm, BRL) before being examined by E.M. **A** shows that there is no gold in bundles when the antibody is not biotinylated. **C** shows a bundle of size comparable to **A** – the gold particles are densely bound so that the bundle appears black. **B** shows a smaller and less ordered bundle where individual gold particles can be distinguished. In neither case is there any periodic distribution of the gold-antibody conjugate

activation of up to 28 actins and by occupying sites on 56 adjacent actins per 2 caldesmons (one in each groove) the observed binding stoichiometry is obtained (Fig. 1). We are currently very interested in obtaining positive evidence for such a structure. Questions as to where calmodulin binds to caldesmon and why small fragments (down to 20,000 Mr) of caldesmon retain its biochemical activities (Szpacenko and Dabrowska 1986) have yet to be tackled.

Caldesmon is capable of promoting filament-filament interactions of at least two types: a readily dispersed network observed at low caldesmon:actin ratios and low ionic strengths (Furst et al. 1986; Lehman 1986; Chalovitch et al. 1987) and large compact bundles observed at high CD:actin ratios (Bretscher 1984; Moody et al. 1985). Neither of these interactions correlate with caldesmon's regulatory properties (Moody et al. 1985; Chalovitch et al. 1987).

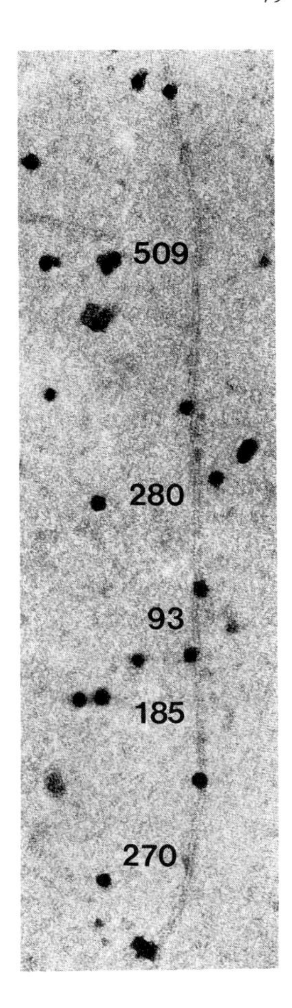

Fig. 10. Localization of caldesmon in reconstituted smooth muscle thin filaments. Sheep aorta thin filaments were reconstituted from pure F-actin, tropomyosin and caldesmon, covalently labelled with [^{3}H]-NHS-biotin. The quantity of NHS-biotin per caldesmon was 0.25 and 0.04 caldesmons were bound per actin monomer. On the E.M. grid, caldesmon was labelled by streptavidin-gold and the filaments were negatively stained. The figure shows a single filament with six bound gold-streptavidin particles. Distance between gold particles is given in nm

4 A Physiological Role for Caldesmon-Based Regulation

In smooth muscle cells caldesmon has a concentration of around 10 μM and is the most abundant calmodulin-binding protein (Marston and Smith 1985). Relative to actin its molar ratio in the cell is 28:1 as it is in isolated thin filaments, thus caldesmon is likely to be all bound to the thin filaments (Marston and Lehman 1985). Estimates of muscle cell calmodulin are not very reliable, but suggest a concentration of around 30 μM (Grand et al. 1979). If the thin filaments in the smooth muscle cell behave in the same way as the reconstituted model filaments (e.g. Fig. 5) then under physiological conditions of temperature and ionic strength the caldesmon system ought to be functioning: caldesmon would inhibit activation at low Ca^{2+} and inhibition would be released by Ca^{2+}.calmodulin at micromolar Ca^{2+} concentrations. At the moment it is not at all easy to devise experimental approaches that could unequivocally demonstrate that this system did operate to control contractility. The caldesmon system would, presumably, operate in tandem or synergystically with the regulation of contraction by Ca^{2+}.calmodulin-dependent myosin phosphorylation and ways must be devised to distinguish the two systems.

At the moment we are concerning ourselves with considering the most basic questions which must be answered before moving on to determining caldesmon's role in regulating contraction. These are:

1. Does the native thin filament of smooth muscle behave in the same way as the reconstituted systems we have investigated? If not why not? and does this exclude a regulatory role for caldesmon or merely modify its supposed mechanism.
2. Can caldesmon and calmodulin control contractility in smooth (or indeed any) muscles as well as actin activation of myosin MgATPase activity.

A number of approaches which we are pursuing to answer these questions promise conclusions in the near future.

Acknowledgments. Our work has been supported by: the British Heart Foundation, the Medical Research Council, the Royal Society, and the National Institute of Health (U.S.A.).

References

Adelstein RS, Klee CB (1981) Purification and characterisation of smooth muscle myosin light chain kinase. J Biol Chem 256:7501–7509

Bretscher A (1984) Smooth muscle caldesmon. Rapid purification and F-actin crosslinking properties. J Biol Chem 259:12873–12880

Chalovitch JM, Cornelius P, Benson CE (1987) Caldesmon inhibits skeletal actomyosin subfragment-l ATPase activity and the binding of myosin subfragment-l to action. J Biol Chem 262: 5711–5716

Filo RS, Bohr DF, Ruegg JC (1965) Glycerinated skeletal and smooth muscle: calcium and magnesium dependence. Science 147:1581–1583

Furst D, Cross R, DeMey J, Small JV (1986) Caldesmon is an elongated, flexible molecule localised in the actomyosin domains of smooth muscle. EMBO J 5:251–257

Grand RJA, Perry SV, Weeks RA (1979) TnC like proteins from mammalian smooth muscle and other tissues. Biochem J 177:521–529

Kakiuchi S, Sobue K (1983) Control of the cytoskeleton by calmodulin and calmodulin-binding proteins. Trends Biochem Sci 8:59–62

Lehman W (1983) The distribution of troponin like proteins on thin filaments of the bay scallop Aequipecten irradiens. J Muscle Res 4:379–389

Lehman W (1986) The effect of calcium on the aggregation of chicken gizzard thin filaments. J Muscle Res 7:537–549

Lehman W, Marston S (1986) Caldesmon associated with smooth muscle thin filaments. Biophys J 49:67a

Marston SB, Lehman W (1985) Caldesmon is a Ca^{2+}-regulatory component of native smooth muscle thin filaments. Biochem J 231:517–522

Marston SB, Smith SCJ (1984) Purification and properties of Ca^{2+}-regulated thin filaments and f-actin from sheep aorta smooth muscle. J Muscle Res 5:582–597

Marston SB, Smith CWJ (1985) The thin filaments of smooth-muscles. J Muscle Res 6:669–708

Marston SB, Trevett RM, Walters M (1979) Calcium-ion regulated thin filaments from vascular smooth muscle. Biochem J 185:355–365

Moody C, Marston SB, Smith CSJ (1985) Bundling of actin filaments by aorta caldesmon is not related to its regulatory function. FEBS Lett 191:107–112

Olwin BB, Storm DR (1984) Preparation of fluorescent labeled calmodulins. Meth Enzymol 102: 148–153

Sanders C, Smillie LB (1984) Chicken gizzard tropomyosin: head-to-tail assembly and interaction with f-actin and troponin. Can J Biochem Cell Biol 62:443–448

Smith CWJ, Marston SB (1985) Disassembly and reconstitution of the Ca^{2+} sensitive thin filaments of vascular smooth muscle. FEBS Lett 184:115–119

Smith CWJ, Pritchard K, Marston SB (1987) The mechanism of Ca^{2+} regulation of vascular smooth muscle thin filaments by caldesmon and calmodulin. J Biol Chem 262:116–122

Sobue K, Muramoto Y, Fujita M, Kakiuchi S (1981) Purification of a calmodulin binding protein from chicken gizzard that interacts with F-actin. Proc Natl Acad Sci USA 78:5652–5655

Szpaceko A, Dabrowska R (1986) Functional domains of caldesmon. FEBS Lett 202:182–186

Thin Filament-Linked Regulation in Vertebrate Smooth Muscle

M. P. WALSH, G. C. SCOTT-WOO, M. S. LIM, C. SUTHERLAND and PH. K. NGAI[1]

1 Introduction

It is widely accepted that myosin-linked regulation, specifically the reversible phosphorylation of myosin, represents the primary Ca^{2+}-dependent mechanism for the regulation of smooth muscle contraction. This mechanism is illustrated schematically in Fig. 1. Stimulation (nervous or hormonal) of the smooth muscle cell leads to an elevation of the cytosolic Ca^{2+} concentration (Somlyo 1985) whereupon Ca^{2+} binds to calmodulin (CaM) to form $Ca_4^{2+} \cdot$CaM. This binding of Ca^{2+} to calmodulin induces a conformational change in the calmodulin molecule (Klee 1977); only in this altered conformation can calmodulin bind to the target enzyme, myosin light chain kinase (MLCK), to form the ternary complex $Ca_4^{2+} \cdot$CaM $\cdot$ MLCK. The apoenzyme of MLCK is inactive, whereas the ternary complex is fully active (Walsh and Hartshorne 1982). In this activated state, MLCK catalyzes the phosphorylation of myosin, specifically at serine-19 of each of the two 20,000-Da light chains (Pearson et al. 1984). Only in the phosphorylated state is myosin capable of interaction with actin filaments,

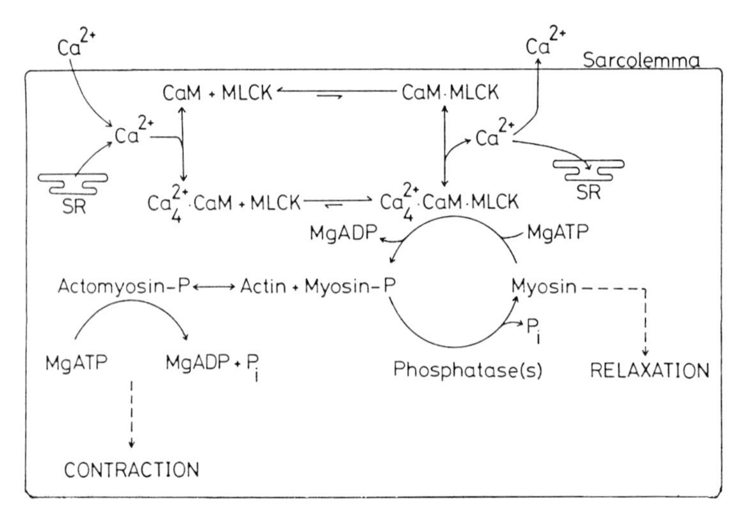

Fig. 1. The regulation of smooth muscle contraction by reversible phosphorylation of myosin. *CaM* calmodulin; *MLCK* myosin light chain kinase; *SR* sarcoplasmic reticulum

1 Department of Medical Biochemistry, University of Calgary, Calgary, Alberta, Canada T2N 4N1

Ch. Gerday, R. Gilles, L. Bolis (Eds.)
Calcium and Calcium Binding Proteins

whereupon the Mg^{2+}-ATPase activity of myosin is greatly enhanced (Sobieszek 1977) providing the driving force for contraction, which apparently occurs by a crossbridge cycling-sliding filament mechanism similar to that described for striated muscle (H.E. Huxley and Hanson 1954; A.F. Huxley and Niedergerke 1954). Relaxation is thought to occur when the cytosolic Ca^{2+} concentration returns to resting levels. Myosin phosphorylation then ceases and myosin which was phosphorylated during the activation phase of the cycle becomes dephosphorylated by the action of one or more phosphatases (Pato 1985). Myosin crossbridges dissociate from actin filaments and the muscle relaxes.

Two major contributions have led recently to a revival of interest in the possibility of secondary Ca^{2+}-dependent regulation in vertebrate smooth muscle. Firstly, the demonstration that tension in smooth muscle fibers can be maintained in a Ca^{2+}-dependent manner while myosin is dephosphorylated, the so-called latch state (Kamm and Stull 1985); and, secondly, the isolation of Ca^{2+}-regulated thin filaments from smooth muscle (Marston and Lehman 1985). This latter development corresponded with the discovery and characterization of caldesmon, a major actin- and calmodulin-binding protein of smooth muscle (Sobue et al. 1981; Ngai and Walsh 1985a). Marston and Lehman (1985) showed that caldesmon is a major component of smooth muscle thin filaments.

We have reinvestigated the possibility of actin-linked regulation in smooth muscle by, firstly, using the myosin competition test devised by Lehman and Szent-Györgyi (1975) which has given contradictory results in the past when applied to vertebrate smooth muscle (Bremel 1974; Marston et al. 1980); and, secondly, isolating native thin filaments from chicken gizzard smooth muscle and comparing the Ca^{2+}-sensitivity of their activation of phosphorylated gizzard myosin Mg^{2+}-ATPase with that achieved by purified gizzard actin and tropomyosin.

2 The Myosin Competition Test

Table 1 shows the results of the myosin competition test. Purified, and therefore unregulated, rabbit skeletal muscle myosin was hybridized with either gizzard thin filaments (present in a crude actomyosin preparation) or purified actin plus tropomyosin, and the Ca^{2+}-sensitivities of the Mg^{2+}-ATPases were compared. As expected, the Mg^{2+}-ATPases of gizzard actomyosin and skeletal myosin alone were very low. Substantial activation was obtained when skeletal myosin was hybridized with gizzard actomyosin or gizzard actin plus tropomyosin. However, only in the case of the gizzard actomyosin-skeletal myosin hybrid was the Mg^{2+}-ATPase activity Ca^{2+}-sensitive. Similar results were obtained in six independent experiments.

Table 1. Myosin competition test of thin filament-linked regulation in vertebrate smooth muscle[a]

	nmol P_i/mg myosin (or actomyosin)/min				Ca^{2+}-sensitivity (%)
	$+Ca^{2+}$	$-Ca^{2+}$	$+Ca^{2+}$	$-Ca^{2+}$	
Gizzard actomyosin	5.7	0.0	–	–	–
Skeletal myosin	5.0	5.2	–	–	–
Gizzard actomyosin + skeletal myosin	257.2	119.8	246.5[b]	114.6[b]	53.5
Gizzard actin + tropomyosin + skeletal myosin	302.0	299.3	297.0[b]	294.1[b]	1.0

[a] Hybrids between rabbit skeletal muscle myosin and either gizzard actomyosin or purified gizzard actin plus tropomyosin were formed as described by Marston et al. (1980). The Mg^{2+}-ATPase activities of these hybrids and the individual protein components were measured in the presence and absence of Ca^{2+}.
[b] Values corrected for ATPase rates of individual components.

3 Properties of Isolated Thin Filaments

We then isolated native thin filaments from chicken gizzard as described recently (Marston and Lehman 1985). Figure 2A shows the protein composition of a typical preparation. The thin filaments contain actin, tropomyosin, caldesmon and an unidentified 32,000-Da protein; caldesmon was identified by immunoblotting (Fig. 2B) using anti-chicken gizzard caldesmon (Ngai and Walsh 1985b). A trace of contaminating myosin is usually evident.

When reconstituted with phosphorylated gizzard myosin, these thin filaments conferred Ca^{2+}-sensitivity on the myosin Mg^{2+}-ATPase (Fig. 3). In this experiment, gizzard myosin was prephosphorylated by Ca^{2+}/calmodulin-dependent myosin light chain kinase for 5 min prior to addition of EGTA and thin filaments. After measurement of

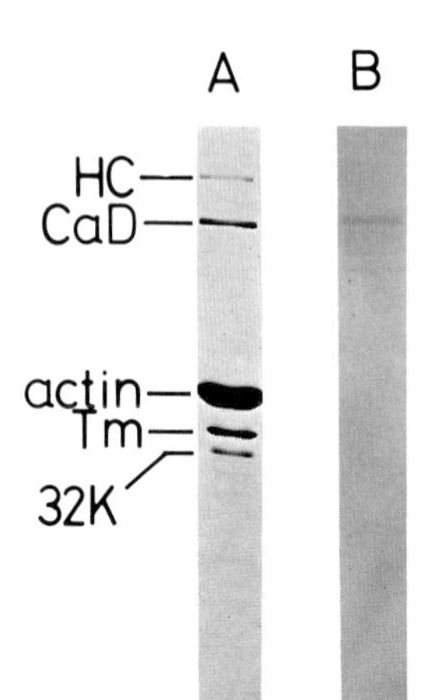

Fig. 2. The protein composition of chicken gizzard thin filaments. Thin filaments (30 μg protein) were electrophoresed in duplicate on 0.1% SDS, 7.5-20% polyacrylamide gradient gels. One gel (**A**) was stained with Coomassie Brilliant Blue R-250 and the other (**B**) subjected to immunoblotting with anti-chicken gizzard caldesmon. Proteins were identified by molecular weight, determined with the use of M_r marker proteins electrophoresed simultaneously, and by comigration with isolated proteins (myosin, caldesmon, actin and tropomyosin). *HC* myosin heavy chain; *CaD* caldesmon; *Tm* β subunit of tropomyosin (the γ subunit comigrates with actin in this electrophoretic system); *32K* an unidentified 32 kDa protein

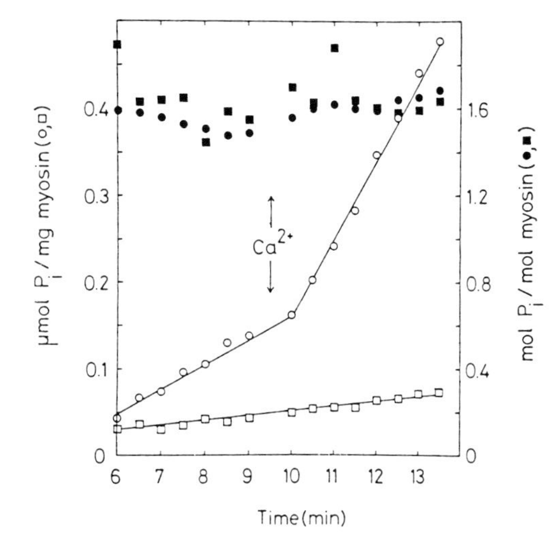

Fig. 3. Ca^{2+}-dependence of the Mg^{2+}-ATPase of phosphorylated gizzard myosin reconstituted with gizzard thin filaments. Gizzard myosin (1.0 mg/ml) was phosphorylated by incubation at 30°C in 60 mM KCl, 25 mM Tris-HCl (pH 7.5), 6 mM $MgCl_2$, 0.1 mM $CaCl_2$, 15 μg/ml calmodulin, 15 μg/ml myosin light chain kinase and 2 mM [γ-^{32}P]ATP (∿12,600 cpm/nmol). Reactions were initiated by the addition of [γ-^{32}P]ATP at t = 0. Reaction volume = 4.0 ml. At t = 5 min, EGTA (1 mM excess over Ca^{2+}) was added to both reaction mixtures and, at t = 5.5 min, thin filaments (1.0 mg protein/ml final concentration) were added to one reaction mixture (○, ●) but not the other (□, ■). Aliquots (0.25 ml) of reaction mixtures were withdrawn at the indicated times for quantification of $^{32}P_i$ release (○, □) and myosin phosphorylation (●, ■). At t = 9.5 min, $CaCl_2$ (0.1 mM excess over EGTA) was added to each reaction mixture. The ATPase rates of the reconstituted phosphorylated myosin-thin filament system determined from these data were 25.7 nmol P_i/mg myosin/min in the absence of Ca^{2+} and 87.5 nmol P_i/mg myosin/min in the presence of Ca^{2+} (values corrected for the ATPase rate of myosin alone of 5.7 nmol P_i/mg myosin/min)

the ATPase rate under these conditions, excess Ca^{2+} was added and the ATPase rate redetermined. The Ca^{2+}-sensitivity (70.6%) is clearly apparent. Furthermore, myosin remained stably phosphorylated throughout the course of the experiment, indicating that the observed Ca^{2+}-sensitivity was not due to dephosphorylation of myosin in the absence of Ca^{2+} followed by rephosphorylation upon subsequent readdition of Ca^{2+}. The open squares indicate the low ATPase rate of phosphorylated myosin alone, i.e., in the absence of thin filaments.

Figure 4 illustrates that, unlike the phosphorylated myosin-thin filament complex, a system reconstituted from phosphorylated gizzard myosin and either gizzard actin or gizzard actin plus tropomyosin does not exhibit a Ca^{2+}-sensitive Mg^{2+}-ATPase. Again, gizzard myosin was prephosphorylated and then incubated in the absence of Ca^{2+} either alone or with gizzard actin, gizzard actin plus tropomyosin or thin filaments. Excess Ca^{2+} was added at the indicated time. Only in the presence of thin filaments was there any effect of the addition of Ca^{2+} on the ATPase rate.

We conclude from these observations that gizzard thin filaments contain a factor(s) which confers Ca^{2+}-sensitivity on the Mg^{2+}-ATPase of phosphorylated gizzard myosin. Caldesmon in conjunction with a Ca^{2+}-binding protein (possibly calmodulin) may

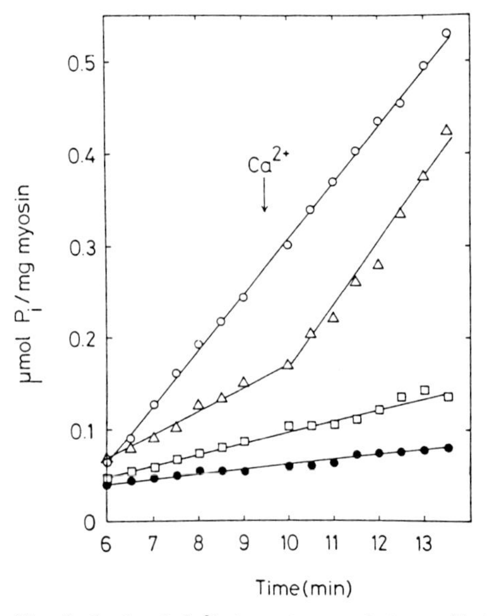

Fig. 4. Lack of Ca^{2+}-dependence of the Mg^{2+}-ATPase of phosphorylated gizzard myosin reconstituted with gizzard actin or gizzard actin and tropomyosin. Gizzard myosin was prephosphorylated as described in the legend to Fig. 3 prior to addition of EGTA (1 mM excess over Ca^{2+}) at t = 5 min and either thin filaments (1.0 mg/ml) (△), actin (0.65 mg/ml) (□), actin (0.65 mg/ml) + tropomyosin (0.175 mg/ml) (○) or no addition (●) at t = 5.5 min. Aliquots (0.25 ml) of reaction mixtures were withdrawn at the indicated times for quantification of $^{32}P_i$ release. At t = 9.5 min, $CaCl_2$ (0.1 mM excess over EGTA) was added to each reaction mixture. The following ATPase rates were calculated from these data: phosphorylated myosin-thin filament complex = 22.1 nmol P_i/mg myosin/min in the absence of Ca^{2+} and 65.9 nmol P_i/mg myosin/min in the presence of Ca^{2+} (i.e., 66.5% Ca^{2+}-sensitivity); phosphorylated myosin-actin complex = 8.0 nmol P_i/mg myosin/min; phosphorylated myosin-actin-tropomyosin complex = 61.9 nmol P_i/mg myosin/min (values corrected for the ATPase rate of myosin alone of 5.7 nmol P_i/mg myosin/min). Simultaneous measurements of myosin phosphorylation (cf. Fig. 3) verified that myosin remained stably phosphorylated throughout the course of the reactions and specific phosphorylation of myosin LC_{20} was indicated by quenching the reactions at t = 14 min followed by SDS-polyacrylamide gel electrophoresis and autoradiography

constitute this thin filament-associated Ca^{2+} regulatory mechanism (Marston and Smith 1985). To test this possibility directly, we devised a method for the selective removal of caldesmon from thin filaments which involved washing of the thin filaments with 25 mM Mg^{2+} (Fig. 5). Untreated thin filaments are shown in lanes 1 and 6. Lanes 2–4 show the supernatants following successive extractions with 25 mM Mg^{2+} and lane 5 the final pellet of caldesmon-free thin filaments. Scanning densitometry using a laser densitometer revealed removal of >92% of the caldesmon and the 32,000-Da protein component.

With this preparation we were able to examine directly the possible involvement of caldesmon in the observed Ca^{2+}-regulation of phosphorylated gizzard myosin Mg^{2+}-ATPase by thin filaments. Figure 6 shows that the same Ca^{2+}-sensitivity of Mg^{2+}-ATPase activity was observed whether phosphorylated gizzard myosin was recon-

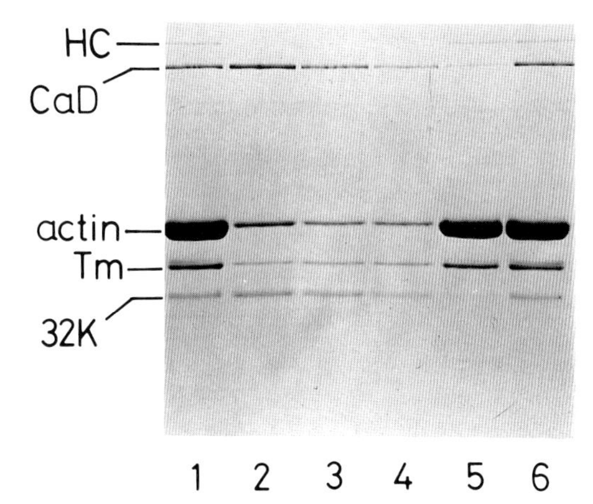

Fig. 5. Removal of caldesmon from gizzard thin filaments by treatment with 25 mM $MgCl_2$. Thin filaments (5.0 mg protein/ml) in 10 mM sodium phosphate (pH 7.0), 1 mM EGTA, 0.1 M NaCl, 1 mM NaN_3, 1 mM dithiothreitol, 25 mM $MgCl_2$ (Buffer A) were centrifuged at 200,000 *g* for 60 min. The supernatant was recovered and the pellet was resuspended in Buffer A and centrifuged as before. The cycle was repeated a third time. The final pellet was resuspended in Buffer A. Supernatants and the final pellet were compared with untreated thin filaments by SDS-polyacrylamide gel electrophoresis. Key to gel lanes: *1*, *6* untreated thin filaments; *2* first supernatant; *3* second supernatant; *4* third supernatant; *5* final pellet (caldesmon-free thin filaments)

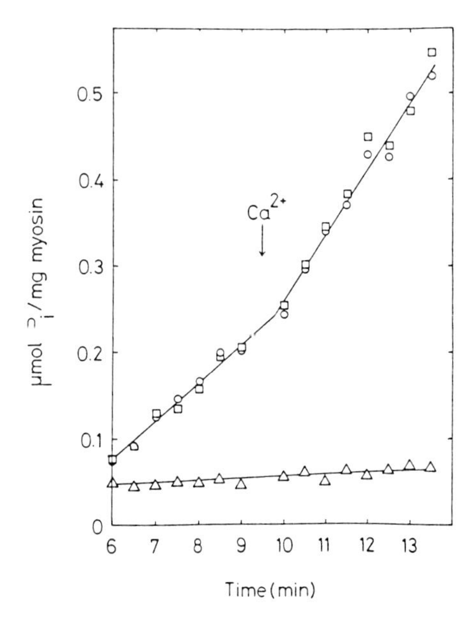

Fig. 6. Activation of the Mg^{2+}-ATPase activity of phosphorylated gizzard myosin by thin filaments and caldesmon-depleted thin filaments. Gizzard myosin (1.0 mg/ml) was prephosphorylated as described in the legend to Fig. 3 prior to addition of EGTA (1 mM excess over Ca^{2+}) at t = 5 min and either thin filaments (1.0 mg/ml; ○), caldesmon-free thin filaments (1.0 mg/ml; □) or no further addition (△) at t = 5.5 min. Aliquots (0.25 ml) of reaction mixtures were withdrawn at the indicated times for quantification of $^{32}P_i$ release. At t = 9.5 min, $CaCl_2$ (0.1 mM excess over EGTA) was added to each reaction mixture

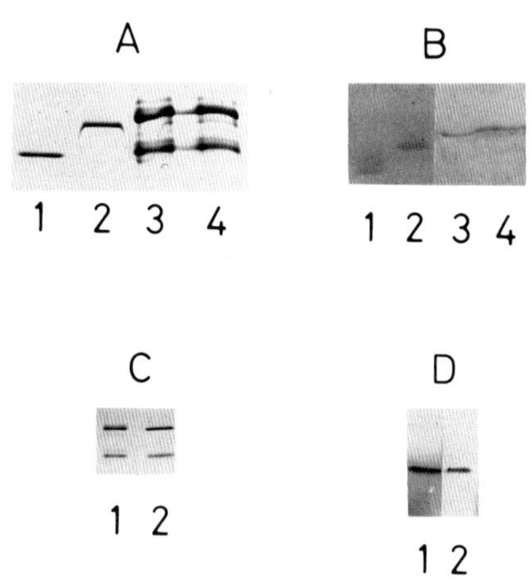

Fig. 7A–D. Detection of heat-stable Ca^{2+}-binding proteins in thin filaments. Native thin filaments were heat-treated in the presence of Ca^{2+} and the heat-stable proteins concentrated by lyophilization. Samples (100 μl) of heat-treated thin filaments were electrophoresed on 0.1% SDS, 7.5–20% polyacrylamide gradient slab gels. **A** Coomassie Blue-stained gel. Lane *1* calmodulin (3 μg) + 10 mM $CaCl_2$; lane *2* calmodulin (3 μg) + 10 mM EGTA; lane *3* heat-treated thin filaments + 10 mM $CaCl_2$; lane *4* heat-treated thin filaments + 10 mM EGTA. **B** ^{45}Ca autoradiogram. Samples as in **A**. **C** Coomassie Blue-stained gel. Lane *1* heat-treated thin filaments + Ca^{2+}; lane *2* gizzard myosin (5 μg) + Ca^{2+}. **D** immunoblot (anti-myosin LC_{20}). Samples as in **C**. Only the regions of the gels containing Ca^{2+}-binding proteins are shown for clarity

stituted with native thin filaments or caldesmon-free thin filaments: 42.7% and 45.1% Ca^{2+}-sensitivity, respectively. We have obtained similar results in six independent experiments using three different preparations of native and caldesmon-free thin filaments.

An important component of the Ca^{2+}-sensitive effects observed in all these experiments must be a Ca^{2+}-binding protein. We have been unable to detect significant binding of $^{45}Ca^{2+}$ to native thin filaments by equilibrium dialysis. However, we have detected two Ca^{2+}-binding proteins present in thin filament preparations in trace amounts by heat-treating native thin filaments in the presence of Ca^{2+} followed by centrifugation to remove denatured proteins, concentration of the resultant supernatant, electrophoresis, transblotting, $^{45}Ca^{2+}$ overlay and autoradiography (Fig. 7). Two low molecular weight Ca^{2+}-binding proteins were detected by ^{45}Ca autoradiography. Panel A shows the Coomassie blue-stained gel and panel B the autoradiogram. Lanes 1 and 2 contained isolated gizzard calmodulin in the presence and absence of Ca^{2+}, respectively. Lanes 3 and 4 contained the heat-treated thin filament supernatant in the presence and absence of Ca^{2+}, respectively. We have identified the two Ca^{2+}-binding proteins present in trace amounts in the thin filaments as calmodulin and the 20,000-Da light chain of myosin. Calmodulin was identified by comigration with authentic calmodulin in the presence and absence of Ca^{2+} (Fig. 7A,B) and by its ability to activate cyclic nucleotide phosphodiesterase in a Ca^{2+}-dependent manner. The 20,000-Da light chain of myosin was identified by comigration with the authentic polypeptide in a purified gizzard myosin preparation in the presence and absence of

Ca^{2+} (Fig. 7C) and by immunological crossreactivity with a specific antibody to the 20,000-Da light chain of gizzard myosin (Fig. 7D). Any other Ca^{2+}-binding protein present in thin filament preparations must be labile and present in very small amounts.

4 Conclusions

The following conclusions can be made from these studies. Firstly, vertebrate smooth muscle (chicken gizzard) contains a Ca^{2+}-dependent thin filament-linked regulatory mechanism. Secondly, this Ca^{2+} regulation does not involve caldesmon. Thirdly, thin filaments contain very small amounts of two Ca^{2+}-binding proteins: calmodulin and the 20,000-Da light chain of myosin. Fourthly, the Ca^{2+}-sensitivity observed in an in vitro system reconstituted from gizzard thin filaments and either skeletal myosin or phosphorylated gizzard myosin is due to calmodulin and/or an unidentified minor protein component of the thin filaments, possibly an actin-binding protein which regulates actin filament structure in a Ca^{2+}-dependent manner (Ebisawa and Nonomura 1985; Ebisawa et al. 1985; Hinssen et al. 1984; Kanno et al. 1985; Strzelecka-Golaszewska et al. 1984).

Acknowledgements. This work was supported by grants (to M.P.W.) from the Medical Research Council of Canada and the Alberta Heritage Foundation for Medical Research. M.P.W. is recipient of an MRC Development Grant and AHFMR Scholarship. G.C.S-W., M.S.L. and P.K.N. are recipients of Studentships from the AHFMR.

References

Bremel RD (1974) Myosin-linked calcium regulation in vertebrate smooth muscle. Nature (Lond) 252:405–407

Ebisawa K, Nonomura Y (1985) Enhancement of actin-activated myosin ATPase by an 84K M_r actin-binding protein in vertebrate smooth muscle. J Biochem (Tokyo) 98:1127–1130

Ebisawa K, Maruyama K, Nonomura Y (1985) Ca^{2+}-regulation of vertebrate smooth muscle thin filaments mediated by an 84K M_r actin-binding protein: purification and characterization of the protein. Biomed Res 6:161–173

Hinssen H, Small JV, Sobieszek A (1984) A Ca^{2+}-dependent actin modulator from vertebrate smooth muscle. FEBS Lett 166:90–95

Huxley AF, Niedergerke R (1954) Structural changes in muscle during contraction. Nature (Lond) 173:971–973

Huxley HE, Hansen J (1954) Changes in the cross-striations of muscle during contraction and stretch and their structural interpretation. Nature (Lond) 173:973–976

Kamm KE, Stull JT (1985) The function of myosin and myosin light chain kinase phosphorylation in smooth muscle. Ann Rev Pharmacol Toxicol 25:593–620

Kanno K, Sasaki Y, Hidaka H (1985) A Ca^{2+}-sensitive actin regulatory protein from smooth muscle. FEBS Lett 184:202–206

Klee CB (1977) Conformational transition accompanying the binding of Ca^{2+} to the protein activator of 3′, 5′-cyclic adenosine monophosphate phosphodiesterase. Biochemistry 16:1017–1024

Lehman W, Szent-Györgyi AG (1975) Regulation of muscular contraction. Distribution of actin control and myosin control in the animal kingdom. J Gen Physiol 66:1–30

Marston SB, Lehman W (1985) Caldesmon is a Ca^{2+}-regulatory component of native smooth-muscle thin filaments. Biochem J 231:517–522

Marston SB, Smith CWJ (1985) The thin filaments of smooth muscles. J Muscle Res Cell Motil 6:669–708

Marston SB, Trevett RM, Walters M (1980) Calcium ion-regulated thin filaments from vascular smooth muscle. Biochem J 185:355–365

Ngai PK, Walsh MP (1985a) Properties of caldesmon isolated from chicken gizzard. Biochem J 230:695–707

Ngai PK, Walsh MP (1985b) Detection of caldesmon in muscle and non-muscle tissues of the chicken using polyclonal antibodies. Biochem Biophys Res Commun 127:533–539

Pato MD (1985) Properties of the smooth muscle phosphatases from turkey gizzards. Adv Prot Phosphatases 1:367–382

Pearson RB, Jakes R, John M, Kendrick-Jones J, Kemp BE (1984) Phosphorylation site sequence of smooth muscle myosin light chain (M_r = 20,000). FEBS Lett 168:108–112

Sobieszek A (1977) Vertebrate smooth muscle myosin. Enzymatic and structural properties. In: Stephens NL (ed) The biochemistry of smooth muscle. University Park Press, Baltimore, pp 413–443

Sobue K, Muramoto Y, Fujita M, Kakiuchi S (1981) Purification of a calmodulin-binding protein from chicken gizzard that interacts with F-actin. Proc Natl Acad Sci USA 78:5652–5655

Somlyo AP (1985) Excitation-contraction coupling and the ultrastructure of smooth muscle. Circ Res 57:497–507

Strzelecka-Golaszewska H, Hinssen H, Sobieszek A (1984) Influence of an actin-modulating protein from smooth muscle on actin-myosin interaction. FEBS Lett 177:209–216

Walsh MP, Hartshorne DJ (1982) Actomyosin of smooth muscle. In: Cheung WY (ed) Calcium and cell function. Vol III. Academic Press, New York, pp 223–269

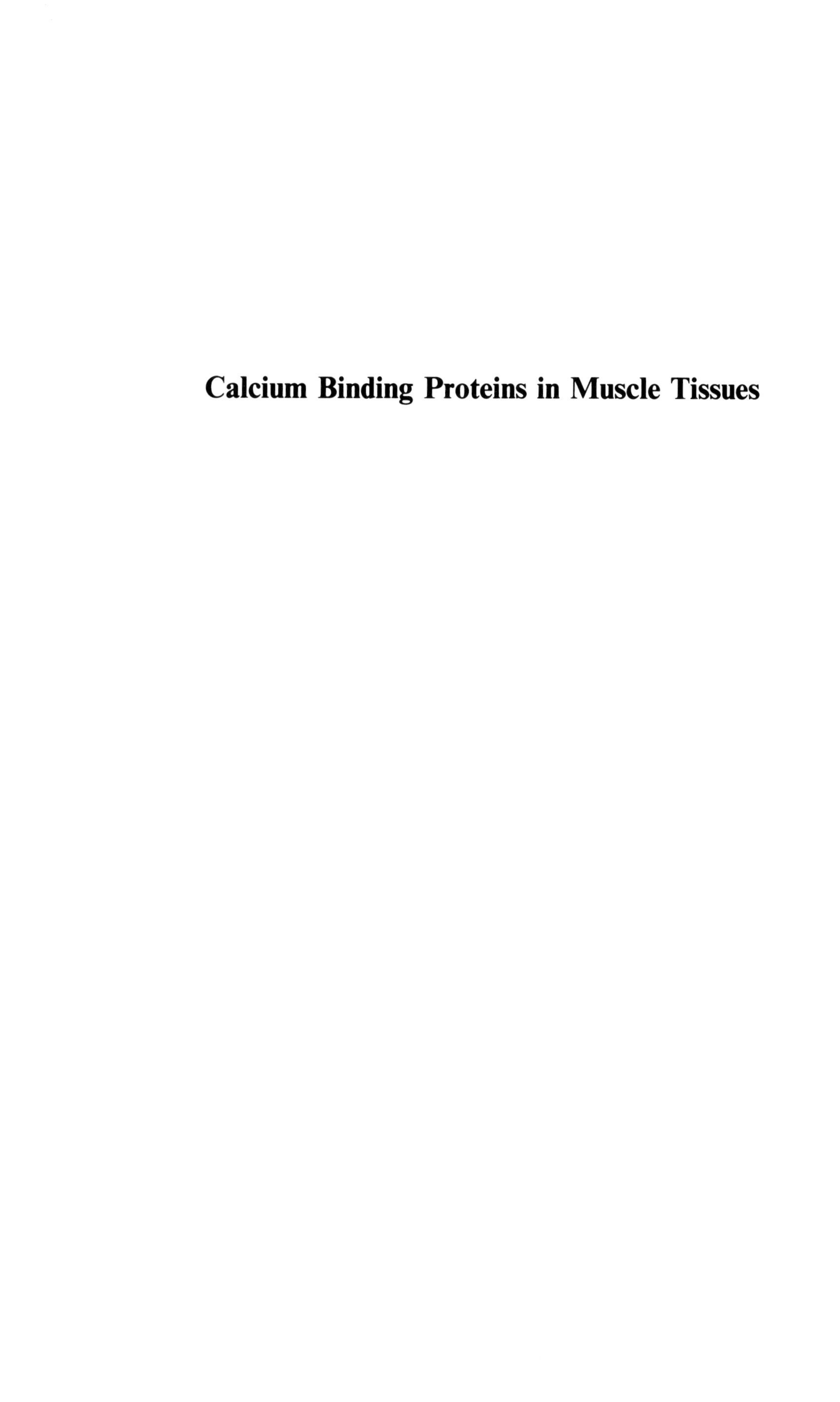

Calcium Binding Proteins in Muscle Tissues

Parvalbumin in Non-Muscle Cells

C. W. HEIZMANN[1]

1 Introduction

Cells have developed an elaborate system of proteins which specifically interact with calcium ions regulating transmission and reception of the calcium signal. The major task of these proteins which bind calcium with high affinity is to mediate the calcium signal intracellularly and to fine-tune the calcium levels. In some of these proteins an α-helix-loop-α-helix arrangement of the peptide chain has been found and the calcium binding residues are located in the loop connecting the two α-helices. This arrangement is referred to as the "EF-hand" structure, and is derived from the first three-dimensional structure analysis done on carp parvalbumin 4.25 by Kretsinger and Nockolds (1973). This carp parvalbumin structure is the basis of our understanding of how proteins bind calcium with high affinity. The essence of Kretsinger's hypothesis is that when calcium is acting as a second messenger, its targets are proteins which contain EF-hand structures. In proteins which belong to this family the residues involved in calcium binding are especially well conserved. This has been recently confirmed by X-ray analysis of calmodulin (Babu et al. 1985), troponin C (Herzberg and James 1985; Sundaralingam et al. 1985) and intestinal calcium-binding protein (calbindin-D-10K; Szebenyi et al. 1981; Szebenyi and Moffat 1986).

Parvalbumins were first isolated from muscles of lower vertebrates and their primary structures and biochemical properties are now well documented (for reviews see Hamoir 1968; Pechere et al. 1971; Kretsinger 1980; Wnuk et al. 1982; Gerday 1982).

More recently, parvalbumins were isolated from skeletal muscles of higher vertebrates, and the rabbit and rat parvalbumins have been sequenced (for review see Heizmann 1984). Parvalbumins from lower and higher vertebrates share common properties; they are all acidic, monomeric proteins with a molecular weight of approximately 12,000 and bind two calcium ions with high affinity. Fish fast muscles, however, contain up to five different iso-parvalbumins, whereas in mammalian muscles only one parvalbumin has been detected so far.

Most recently the structural organization of the rat parvalbumin gene has been elucidated (Berchtold et al. 1987; Heizmann and Berchtold 1987). The rat parvalbumin transciption unit is 15.5 kilobases in length and contains five exons interrupted by four introns. The last three exons encode most of the protein and are similar in size

1 Institute for Pharmacology and Biochemistry, University of Zürich-Irchel, Winterthurerstr. 190, CH–8057 Zürich, Switzerland

Ch. Gerday, R. Gilles, L. Bolis (Eds.)
Calcium and Calcium Binding Proteins

and organization to the coding region of genes encoding four-domain calcium-binding proteins (such as calmodulin and myosin light chains). From these data it was concluded that the parvalbumin gene would maintain the structural pattern of the four-domain calcium-binding protein gene family even though it would exhibit deletion of one domain. The availability of parvalbumin cDNA clones (Berchtold and Means 1985; Epstein et al. 1986) and knowledge about the parvalbumin gene structure can now be used to explore the cell-specific expression and development-dependent regulation of the parvalbumin gene in muscle and non-muscle tissues.

The single copy parvalbumin gene has been assigned to human chromosome 22, from pter to q11.2. This segment is deleted in some patients with DiGeorge syndrome. The parvalbumin gene may therefore serve as an useful marker in this immunodeficiency disease. Recently also a loss of genes on chromosome 22 has been reported in tumorigenesis of human acoustic neuroma (Seizinger et al. 1986).

The possible role(s) of parvalbumin in fast-twitch skeletal muscles has been discussed in previous reviews (Gerday 1982; Heizmann 1984; Gillis 1985; Heizmann and Berchtold 1987).

In this review I would like to summarize studies investigating the physiological role(s) of parvalbumins in non-muscle tissues. Some years ago, parvalbumin-immunoreactivity was found in extracts of carp brain, various non-muscle tissues of the pike (Gosselin et al. 1978) and in extracts of non-muscle tissues from the rabbit (Baron et al. 1975). However, when tissues under investigation are not completely freed from any underlying muscle or adipose tissue (containing sizeable amounts of parvalbumin) the results may be misleading. Parvalbumin must therefore be isolated from each tissue separately and its precise localization studied in tissue sections using monospecific antibodies which do not cross-react with other structurally related calcium-binding proteins. This information can then be used to design experiments exploring the physiological role(s) of parvalbumin and related proteins in non-muscle tissues.

So far, parvalbumins have been isolated from brain, kidney, adipose tissue and testis. As the biochemical and immunological properties of these parvalbumins were found to be indistinguishable from those of their muscle counterparts, monospecific antibodies against muscle parvalbumins can therefore be used to localize parvalbumin in non-muscle tissues.

2 Parvalbumin in the Central Nervous System

One way to understand the biological role of parvalbumin in the brain was to determine its precise localization in brain sections and follow its expression during the maturation of the tissue in vivo and in vitro.

Parvalbumin was found to be a marker for a distinct subpopulation of neurons in the central nervous systems of the rat, cat, zebra finch, monkey and man (for review see Heizmann and Celio 1987).

The intensity of labeling varied among individual neurons (as it does in individual muscle fibres) and staining of some (but not all) cell nuclei was observed. The Purkinje cells in the cerebellum of the rat and zebra finch stain stronger than basket or stellate cells, and basket cells of the cerebral cortex of the rat are more reactive than bipolar cells. Thus some neurons seem to synthesize more parvalbumin than others. Most parvalbumin-positive cells in the central nervous system are probably interneurons.

Parvalbumin and the inhibitory neurotransmitter-aminobutyric acid (GABA) are similarly distributed in the cortical areas of the rat brain. However, there are some GABA-containing cells which lack parvalbumin. The percentage of neurons which contain both parvalbumin and GABA varies considerably in different areas of the brain (Kosaka et al. 1987). For example, 20–30% of the GABAergic neurons in the stratum oriens, 50% in the stratum pyramidale but less than 5% in the stratum radiatum also contain parvalbumin. It is possible, therefore, that parvalbumin may regulate calcium-dependent metabolic and electrical activities in a subpopulation of GABAergic neurons. For example, it has been reported recently that calcium-ions decrease the affinity of the GABA receptor (Inoue et al. 1986). It was demonstrated that the apparent affinity of the GABA receptor is controlled rapidly (within 100 ms) by intracellular calcium. Furthermore, suppression of GABA sensitivity by calcium is an important factor in epileptogenesis. It might therefore be interesting to investigate the effect of micro-injected parvalbumin (or corresponding antibodies) on the activities of neurons. All nuclei of the auditory, vocal motor and visual systems of adult zebra finches (Braun et al. 1985a,b) contain high levels of parvalbumin and cytochrome oxidase activity. Electron microscopy of immunostained neurons of the song system of the zebra finch (Zuschratter et al. 1985) and neurons of the visual cortex of cats (Stichel et al. 1987) showed parvalbumin-reaction product present in the amorphous material of perikarya, dendrites and axons, in most nuclei, and in association with microtubules, postsynaptic densities and intracellular membranes.

The presence of parvalbumin in neurons that are metabolically and electrically very active suggest that cell-type specific calcium-regulated mechanisms are present in these cells.

Another way to explore the physiological role of parvalbumin was to study its distribution in the mammalian retina (Röhrenbeck et al. 1987). By virtue of its well-defined neuroanatomy, synaptic organization and electrophysiology, the retina offers an unique opportunity to investigate the distribution of parvalbumin in well-characterised cell types. In addition, many neurotransmitters present in the brain are also found in the retina. Six different cells can be distinguished in the retina: receptor cells, horizontal cells, bipolar cells, amacrine cells, interplexiform cells and ganglion cells. Preliminary results show a staining of the horizontal, amacrine and ganglion cell layers. In the horizontal cell layer the type A and B cells were parvalbumin-immunoreactive. In future experiments it will be possible to perform intracellular recording on individual retinal neurons with and without microinjected parvalbumin to test if parvalbumin can modulate electrical activity in these neurons.

There is some recent evidence (collaboration with the group of Dr. A. Hermann, University of Salzburg, Austria) that parvalbumin may be involved in the regulation of electrical activity in individual neurons isolated from the abdominal ganglion of the Aplysia Californica. In the abdominal ganglion there are several types of cells which are identified electrophysiologically as silent-, beating-, and bursting cells. Silent cells are cells which are silent in the absence of an external input, beating cells discharge action potentials continuously at low frequences, and bursting cells are neurons which discharge action potentials in regular and irregular oscillatory cycles. Bursting activity is probably generated by oscillation of the intracellular free calcium and this in turn may be modified by specific calcium-binding proteins. We therefore isolated individual neurons, ^{14}C-labeled their proteins by reductive methylation and separated them on two-dimensional polyacrylamide gelelectrophoresis. Bursting cells contained an acidic low molecular weight protein (Mr = 10–12K, pI = 4.5) at a high concentration. This protein, which co-migrates with carp muscle parvalbumin, was present in very low amounts in beating cells and absent from silent cells. Thus this parvalbumin-like protein may be involved in the regulation of bursting cell activity.

3 Parvalbumin in the Peripheral Nervous System

In the peripheral nervous system parvalbumin-immunoreactivity has been found in large cells of the spinal ganglia and their peripheral processes. As in the central nervous system, only a certain percentage of the peripheral nerve fibres are positive. Very small immunoreactive neurons and positive punctate structures could be seen in lamina IIb (Rexed) of the dorsal horn of the spinal cord. Labeled terminals surrounded motor neurons in the ventral horn, and heavily immunoreactive myelinated axons coursed through the dorsal funiculi. Immunolabeling was restricted to large cells in spinal ganglia and to large diameter axons in the peripheral nerves (Berchtold et al. 1984).

Recently it was found (Endo and Onaya 1986) that the levels of parvalbumin in the sciatic nerve of streptozotocin-induced diabetic rats were significantly decreased when compared with those of normal rats. It was suggested that the decreased levels of parvalbumin in the peripheral nerve might contribute to the genesis of diabetic neuropathy.

In the future it will also be interesting to measure parvalbumin levels in the cerebrospinal fluid of patients with various disorders to see if the appearance of parvalbumin can give any useful information about the extent and/or location of neuronal damage elsewhere in the brain.

4 Parvalbumin in Endocrine Glands

Calcium is involved in the regulation of the secretion of endocrine substances. Calmodulin is one of the calcium receptors thought to be involved in stimulus-secretion coupling in endocrine cells (for review see Brown et al. 1985). In the testis, other calcium-binding poteins are also present, but little is known about them or their physiological roles. However, several key processes in the testis are regulated by calcium. Fore example, it has been shown that calcium and cAMP are required for maximum LH-stimulated production of testosterone in rat Leydig cells (Janszen et al. 1976; Sullivan and Cook 1986). The evidence obtained suggests, that at submaximal rates of testosterone synthesis calcium, rather than cAMP, is the second messenger, whereas for maximum steroidogenesis both calcium and cAMP-dependent pathway may be functional. The detailed mechanism of these processes and the identity of the calcium-binding proteins involved in their regulation is currently unknown.

We have isolated parvalbumin from rat testis and shown that it is indistinguishable from muscle parvalbumin by several biochemical and immunological methods (Kägi et al. 1986, 1987).

A strong parvalbumin-immunoreaction was also observed in the Leydig cells of the rat. In order to examine a possible involvement of parvalbumin in the production of testosterone in Leydig cells we then studied the expression of parvalbumin during testis development.

The rat testis is strongly active in steroid hormone production at two developmental stages, the first occuring from prenatal day –3 to birth, when sexual differentiation takes place, the second lasting from postnatal day 25 to 55, during maturation of the adult testis. The rate of testicular growth in the rat is greatest between days 25 and 60, and the serum level of luteinizing hormone (LH) steadily rises in parallel. At the same time, Leydig cell number, LH receptor content and testosterone secretion also increase.

Parvalbumin concentrations in the rat testis were measured using a specific RIA, and the time course of changes in parvalbumin content was followed through the developmental period from postnatal days 14 to 60. Parvalbumin mRNA levels were measured by dot blot and Northern blot analysis using the parvalbumin-clone 9f (Berchtold and Means 1985).

Parvalbumin concentrations were low around day 30, increased steadily thereafter until day 45, and finally showed a sharp maximum at day 50 parallel to the increase of LH and testosterone in serum. Parvalbumin mRNA levels also reached a maximum around postnatal day 50 and declined later on.

Earlier stages of development were examined by immunohistochemistry only because of the very small size of the testis in fetal and neonatal rats. High parvalbumin-immunoreactivity was observed in all fetal Leydig cells, and this prominent staining persisted until birth. Thereafter, as the testicular steroidogenic activity decreased before puberty, the parvalbumin-immunoreactivity also decreased (postnatal day 7) and finally dropped to background staining (postnatal day 14). During the second phase of testicular steroidogenesis, starting with puberty, when the formation of the adult Leydig cell population occurs, parvalbumin-immuno-reactivity reappeared in Leydig cells. Maximum staining was obtained around day 50, in agreement with the highest parvalbumin and parvalbumin mRNA levels measured during the same time period. These results demonstrate that parvalbumin is synthesized in hormonally active Leydig cells of the rat testis. Our developmental study indicates a strong correlation between the appearance of parvalbumin and Leydig cell activity with maximum parvalbumin concentrations and mRNA levels around day 50. Experiments have been initiated to measure parvalbumin and mRNA levels in the testis of hypophysectomized rats before and after treatment with FSH and LH. Preliminary results have already demonstrated a decrease of parvalbumin-immunoreactivity in the Leydig cells of hypophysectomized rats.

Parvalbumin co-exists with calbindin-D-28K and S-100 proteins in Leydig cells (Kägi et al. 1988). However, the developmental appearances of these proteins is different from parvalbumin, suggesting that they are involved in calcium-dependent processes other than steroid production.

Parvalbumin-immunoreactivity was obtained not only in the Leydig cells but also in maturing spermatids, whereas staining of the Sertoli cells was never observed (Kägi et al. 1987, 1988). Four germinal cell types can generally be distinguished in the testis of mature animals: spermatogonia, spermatocytes, spermatids, and spermatozoa. In maturing spermatids (from postnatal day 26 onwards) a strong parvalbumin-immunoreaction was observed. The immunostaining was first seen at the beginning of spermatogenesis after postnatal day 25, at sites where the acrosome is formed (spermatids stage 1; in middle and late spermatids) and was maintained in the maturing and adult testis. Staining of spermatogonia and spermatocytes was never observed.

Parvalbumin-immunoreactivity in differentiating spermatides indicates a possible involvement of parvalbumin in calcium-regulated events during spermatogenesis. Parvalbumin-immunohistochemistry thus seems a suitable tool for the study of Leydig cell development and spermatogenesis, and may help in the diagnosis of testicular disturbances.

Parvalbumin concentrations in several other endocrine glands have also been measured by RIA (Endo et al. 1985). A detailed study of the immunohistochemical localization of parvalbumin in ovary, pituitary, thyroid and adrenal glands is currently under way.

5 Parvalbumin in the Kidney

Parvalbumin has been purified from rat kidney by immunaffinity chromatography (Schneeberger and Heizmann 1986). Its biochemical and immunological properties were found to be indistinguishable from those of muscle and brain parvalbumins. Parvalbumin was localized in the distal tubules and proximal collecting duct with a distribution similar to that of calbindin-D-28K. In contrast to calbindin, parvalbumin concentrations were found to be independent of the vitamin-D status in the rat kidney.

6 Parvalbumin in Rat Skin

There is a controversy about a skin calcium-binding protein (SCaBP) reported to be present in the proliferating cells of the epidermis. On the one hand, SCaBP was reported to be immunologically and biochemically distinguishable from parvalbumin (Rinaldi et al. 1982), but on the other hand the complete amino acid sequence of rat SCaBP showed identity with parvalbumin (MacManus et al. 1985). Furthermore, antibodies against SCaBP cross-reacted with parvalbumin. We therefore prepared epidermis (free of dermis and the underlying muscle layer which contains parvalbumin) from adult and newborn rats (the latter have no parvalbumin in the muscle layer). From these extracts two proteins were purified and characterized. One was found to be indistinguishable from parvalbumin and the second (Mr = 12,000) was different in its amino acid composition, peptide maps and immunological properties (Schelling et al. 1987). This protein incorporated 45calcium to a lesser extent than parvalbumin, when assayed with the 45calcium transblot electrophoresis, thus indicating a lower affinity for calcium. By immunohistochemical methods, parvalbumin was found to be located excusively in the striated cutaneous muscle of the skin. This type of muscle (panniculus carnosus) is found in all vertebrates and persists in man only in the neck and face as the platysma muscle. By analogy with its function in skeletal muscle, parvalbumin is probably involved in its fast relaxation. No parvalbumin staining, however, was found in the epidermis. Antisera raised against the protein isolated from newborn skin showed a staining of the epidermis but not of the muscle layers of newborn or adult rat skins. The identity of this protein is currently unknown, although it shares biochemical properties with the epidermal thiol proteinase inhibitor (Takio et al. 1984).

Generally, parvalbumin levels are high in muscle cells, neurons and endocrine cells of high activities and even increase during periods when the cells display maximal physiological and/or functional activities. The latter correlation suggests that special calcium-dependent mechanisms are important determinants in the specific functions of these cells.

Acknowledgements. I should like to thank Dr. A. Rowlerson for valuable discussions. This work was supported by the Swiss National Science Foundation (grant 3.147.-085), Wilhem Sander Stiftung (FRG), Krebsliga des Kantons Zürich, Hartmann-Müller Stiftung, Bundesamt für Veterinärwesen, Jubiläumsspende für die Universität Zürich and Schweizerische Mobiliar Versicherungsgesellschaft.

References

Babu YY, Sack JS, Greenhough TJ, Bugg CE, Means AR, Cook WJ (1985) Three-dimensional structure of calmodulin. Nature (Lond) 316:37–40

Baron G, Demaille J, Dutruge E (1975) The distribution of parvalbumin in muscle and other tissues. FEBS Lett 56:156–160

Berchtold MW, Celio MR, Heizmann CW (1984) Parvalbumin in non-muscle tissues of the rat. Quantitation and immunohistochemical localization. J Biol Chem 259:5189–5196

Berchtold MW, Means AR (1985) The calcium-binding protein parvalbumin: molecular cloning and developmental regulation of mRNA abundance. Proc Natl Acad Sci 82:1414–1418

Berchtold MW, Epstein P, Beaudet AL, Payne EM, Heizmann CW, Means AR (1987) The structural organization of the rat parvalbumin gene. J Biol Chem 262:8696–8701

Braun K, Scheich H, Schachner M, Heizmann CW (1985a) Distribution of parvalbumin, cytochrome oxidase activity and ^{14}C-2-deoxyglucose uptake in the brain of the zebra finch. I Auditory and vocal motor system. Cell and Tissue Res 240:101–115

Braun K, Scheich H, Schachner M, Heizmann CW (1985b) Distribution of parvalbumin, cytochrome oxidase activity and ^{14}C-2-deoxyglucose uptake in the brain of the zebra finch. II Visual system. Cell and Tissue Res 240:117–127

Brown BL, Walker SW, Tomlinson S (1985) Calcium, calmodulin and hormone-secretion. Clin Endocrinol 23:201–218

Endo T, Onaya T (1986) Parvalbumin is reduced in the peripheral nerves of diabetic rats. J Clin Invest 78:1161–1164

Endo T, Takazawa K, Onaya T (1985) Parvalbumin exists in rat endocrine glands. Endocrinology 117:527–531

Epstein P, Means AR, Berchtold MW (1986) Isolation of a rat parvalbumin gene and full length cDNA. J Biol Chem 261:5886–5891

Gerday C (1982) Soluble calcium-binding proteins from fish and invertebrate muscle. Molec Physiol 2:63–67

Gillis JM (1985) Relaxation of vertebrate skeletal muscle. A synthesis of the biochemical and physiological approaches. Biochem Biophys Acta 811:97–145

Gosselin-Rey C, Piront A, Gerday C (1978) Polymorphism of parvalbumins and tissue distribution. Characterization of component I, isolation from red muscle of Cyprinus carpio L. Biochem Biophys Acta 532:284–304

Hamoir G (1968) The comparative biochemistry of fish sarcoplasmic proteins. Acta Zool Pathol Antwerpen 46:69–76

Heizmann CW (1984) Parvalbumin, an intracellular calcium-binding protein; distribution, properties and possible roles in mammalian cells. Experientia (Basel) 40:910–921

Heizmann CW, Berchtold MW (1987) Expression of parvalbumin and other calcium-binding proteins in normal and tumor cells: a topical review. Cell Calcium 8:1–41

Heizmann CW, Celio MR (1987) Immunolocalization of parvalbumin. Methods Enzymol 139: 552–570

Herzberg O, James MNG (1985) Structure of the calcium regulatory muscle protein troponin-C at 2.8 A resolution. Nature (Lond) 313:653–659

Inoue M, Oomura Y, Yakushiji T, Akaike N (1986) Intracellular calcium ions decrease the affinity of the GABA receptor. Nature (Lond) 324:156–158

Janszen FHA, Cook BA, Van Driel MJA, van der Molen HJ (1976) The effect of calcium ions on testosterone production in Leydig cells from rat testis. Biochem J 160:433–437

Kägi U, Berchtold MW, Heizmann CW (1986) Expression of the calcium-binding parvalbumin during development of rat testis. In: Stefanini et al. (eds) Molecular and cellular endocrinology of the testis. Elsevier, Amsterdam, pp 165–171

Kägi U, Berchtold MW, Heizmann CW (1987) Calcium-binding parvalbumin in rat testis. Characterization, localization and expression during development. J Biol Chem 262:7314–7320

Kägi U, Chafouleas JG, Norman AW, Heizmann CW (1988) Developmental appearance of the Ca^{2+}-binding proteins parvalbumin, calbindin D-28K, S-100 proteins and calmodulin during development of the testis in the rat. Cell and Tissue Res (in press)

Kosaka T, Katsumaru H, Hama K, Wu JY, Heizmann CW (1987) GABAergic neurons containing the calcium-binding protein parvalbumin in the rat hippocampus and dendate gyrus. Brain Res 419:119–130

Kretsinger RH (1980) Structure and evolution of calcium-modulated proteins. CRC Crit Rev Biochem 8:119–174

Kretsinger RH, Nockolds CE (1973) Carp muscle calcium-binding protein. II. Structure determination and general description. J Biol Chem 248:3313–3326

MacManus JP, Watson DC, Yaguchi M (1985) Rat skin calcium-binding protein is parvalbumin. Biochem J 229:39–45

Pechère JF, Demaille J, Capony JF (1971) Muscular parvalbumins: preparative and analytical methods of general applicability. Biochim Biophys Acta 236:391–408

Rinaldi ML, Haiech J, Pavlovitch J, Rizk M, Ferraz C, Derancourt J, Demaille JG (1982) Isolation and characterization of a rat skin parvalbumin-like calcium-binding protein. Biochemistry 21: 4805–4810

Röhrenbeck J, Wässle H, Heizmann CW (1987) Immunocytochemical labelling of horizontal cells in mammalian retina using antibodies against calcium-binding proteins. Neurosci Lett 77:255–260

Schelling CP, Didierjean L, Rizk M, Pavlovitch JH, Heizmann CW (1987) Calcium-binding proteins in rat skin. FEBS Lett 214:21–27

Schneeberger PR, Heizmann CW (1986) Parvalbumin in rat kidney. Purification, characterization and localization. FEBS Lett 201:51–56

Seizinger BR, Martuza RL, Gusella JF (1986) Loss of genes on chromosome 22 in tumorigenesis of human acoustic neuroma. Nature (Lond) 322:644–647

Stichel CC, Singer W, Heizmann CW, Norman AW (1987) Immunohistochemical localization of calcium-binding proteins, parvalbumin and calbindin D-28K, in the adult and developing visual cortex of cats: A light and electron microscopic study. J Comp Neurol 262:563–577

Sullivan MH, Cook BA (1986) The role of calcium in steroidogenesis in Leydig cells. Stimulation of intracellular free calcium by lutropin (LH), luberin (LHRH) agonist and cyclic AMP. Biochem J 236:45–51
Sundaralingam M, Bergstrom R, Strasburg G, Rao SP, Roychowdihury P (1985) Molecular structure of troponin C from chicken skeletal muscle at 3-angstrom resolution. Science 227:945–948
Szebenyi DME, Moffat K (1986) The refined structure of vitamin D-dependent calcium-binding protein from bovine intestine. Molecular details, ion binding and implications for the structure of other calcium-binding proteins. J Biol Chem 261:8761–8777
Szebenyi DME, Obendorf SK, Moffat K (1981) Structure of vitamin D-dependent calcium binding protein from bovine intestine. Nature (Lond) 294:327–332
Takio K, Kominami E, Baudo Y, Katanuma N, Titani K (1984) Amino acid sequence of the rat epidermal thiol proteinase inhibitor. Biochem Biophys Res Commun 121:149–154
Wnuk W, Cox JA, Stein EA (1982) Parvalbumins and other soluble high-affinity calcium-binding proteins from muscle. In: Cheung WY (ed) Calcium and cell function. Vol II. Academic Press, New York, pp 243–278
Zuschratter W, Scheich H, Heizmann CW (1985) Ultrastructural localization of the calcium-binding protein parvalbumin in neurons of the song system of the zebra finch Peophila guttata. Cell Tissue Res 241:77–83

S100 Proteins: Structure and Calcium Binding Properties

J. Baudier[1]

1 Introduction

The highly acidic water-soluble protein, S100 (Moore 1965) is a group of closely related proteins which, when purified, appear as non-covalent dimers with subunit composition $\alpha\alpha$ (S100$\alpha\alpha$), $\alpha\beta$ (S100a) and $\beta\beta$ (S100b). The amino acid sequences of the α subunit (93 residues) and the β subunit (91 residues) are highly homologous, with 54 identical residues and 23 replacements compatible with single nucleotide base substitution (Isobe and Okuyama 1978, 1981). Immuno-histochemical as well as biochemical studies demonstrated the ubiquitous nature of S100 protein (Suzuki et al. 1982; Hidaka et al. 1983; Molin et al. 1984, 1985; Kato et al. 1983) and differences in the distribution of the α and β subunits in the brain and peripheral tissues are now clearly established (Isobe et al. 1984; Hidaka et al. 1983; Molin et al. 1984, 1985; Kato et al. 1985). However, the reason for the heterogeneity of S100 proteins is still unclear. The intracellular location of S100 protein is generally admitted, although some studies also showed that cells secrete S100 proteins (Ishikawa et al. 1983; Kato et al. 1987, 1983; Shashowa et al. 1984). In brain tissue, a membrane-bound form of S100 protein was characterized (Haglid et al. 1974; Donato et al. 1975).

It was suggested that S100 proteins might be involved in the maturation and differentiation of glial cells (Labourdette and Mandel 1980) or regulation of protein phosphorylation (Patel et al. 1983; Qi and Kuo 1984). Moreover, S100 protein was reported to interact with microtubule proteins (Baudier et al. 1982; Donato 1983; Deinum et al. 1983). However, the exact role of S100 proteins in all those different biological systems remains to be elucidated.

The amino acid sequence of the α and β subunits revealed the structural relationship of S100 with the calcium-binding proteins of the "E.F. Hand" type (Isobe and Okuyama 1978, 1981). Both subunits have one 30-residue putative "E.F. Hand" calcium binding domain (site I) in the N-terminal part and one typical 28-residue domain (site II) in the C-terminal part (Szebenyi et al. 1981) (see Table 1). By analogy with other Ca^{2+}-binding proteins, such as calmodulin, one might suppose that the biological functions of S100 protein are related to some Ca^{2+}-dependent process within the cells, perhaps a regulatory one. Therefore, a first approach to elucidation of the functions of S100 protein might be a detailed study of their Ca^{2+}-binding properties

1 Université Louis PASTEUR, Faculté de Pharmacie, Laboratoire de Physique, U.A. CNRS 491, B.P.10, 67048 Strasbourg, France

Ch. Gerday, R. Gilles, L. Bolis (eds.)
Calcium and Calcium Binding Proteins

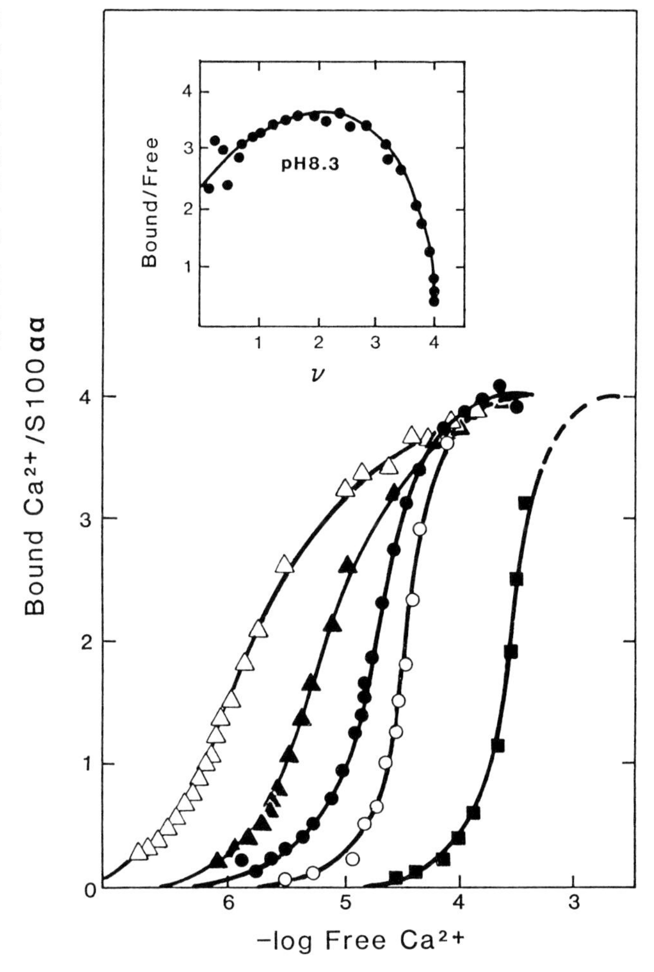

Fig. 1. The effects of pH and KCl on the affinity of native S100αα and Bimane-S100αα protein for calcium. S100αα concentration was 30 μM in 20 mM Tris-HCl, pH 7.5 (○); pH 8.3 (●); plus 120 mM KCl, pH 8.3 (■). Bimane-S100αα concentration was 35 μM in 20 mM Tris-HCl, pH 7.5 (△) plus 120 mM KCl (▲). *Inset* shows the scatchard plot of the binding data of native S100αα at pH 8.3

and Ca^{2+}-induced conformational changes, in order to correlate them with behavior in vivo. It emerged from these studies that the calcium binding properties of purified S100 proteins depend greatly on the conformational states of the proteins. It could be postulated that S100 proteins in vivo interact with other cellular components (ions, proteins, membranes) which affect their conformation and modulate their calcium binding properties. The aim of this review is to summarize our recent findings on the conformational and ion-binding properties of the S100αα and S100b (ββ) which support such a postulate.

2 Calcium Binding to S100 Protein Dimers Studied by the Flow Dialysis Technique

The calcium binding properties of bovine brain S100αα and S100b protein dimer were studied with the flow dialysis technique. The experiments were performed at ambient temperature and observations were completed within 1–2 h.

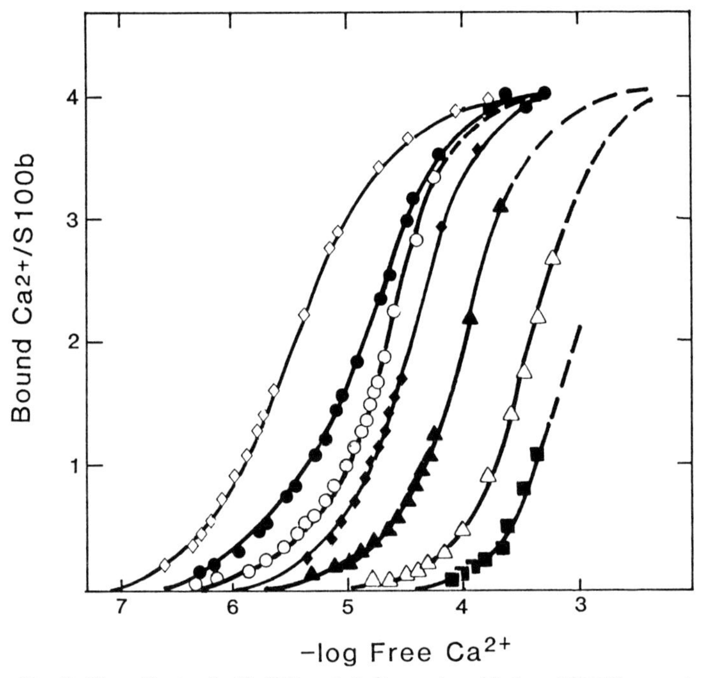

Fig. 2. The effects of pH, KCl and Zn^{2+} on the affinity of S100b protein for calcium. S100b concentration was 50–60 μM in 20 mM Tris-HCl; pH 7.5 (○), plus 10 mM KCl (▲), 50 mM KCl (△), 120 mM KCl (■); pH 8.3 (●); plus 4 Zn^{2+}/S100b molar ratio (◇) plus 4 Zn^{2+}/S100b molar ratio, 120 mM KCl (◆)

The calcium binding isotherms for S100$\alpha\alpha$ in 20 mM Tris buffer at pH 7.5 and 8.3 are presented in Fig. 1 and the binding constants are reported in Table 2. At both pH's, the S100$\alpha\alpha$ protein dimer bound four calcium ions. There was a slight increase in the protein affinity for calcium at basic pH. In 120 mM KCl, a large decrease in calcium affinity occurred and in these conditions, the affinity of the protein for calcium ranged around 10^{-4} to 10^{-3} M Ca^{2+}. As with S100$\alpha\alpha$, the S100b protein bound four Ca^{2+} ions per β dimer with a slight increase of calcium affinity at basic pH (Table 3). Moreover, a strong antagonistic effect of KCl on calcium binding was observed. In Fig. 2, the effects of increasing KCl concentration on calcium binding to S100b are reported. As KCl concentration increased there was a general decrease in calcium affinity and in 120 mM KCl the S100b affinity for calcium was too low to be determined accurately with the flow dialysis technique.

From the above results, two conclusions emerged: i) As expected from the amino acid sequences (Table 1), both α and β subunits bind two Ca^{2+} ions; ii) the marked antagonistic effect of K^+ on Ca^{2+} binding to S100 protein dimers called into question the ability of S100 proteins to bind Ca^{2+} in physiological conditions.

Table 1. Comparison between the test sequence proposed by Tufty and Kretsinger (1975) for the EF hand structure and the typical and rearranged Ca^{2+} binding domains of S100 proteins. In the test sequence, the asterisk indicates an oxygen-containing residue (D, E, N, Q, S, and T) corresponding to the Ca^{2+} coordinating ligands named X, Y, Z, –Y, –X, and –Z in the calcium binding sites (underlined sequences). L indicates a hydrophobic residue (L, V, I, F, and M), and G is a Gly residue

		Helix		Loop		Helix	
Test sequence		L- -L- -L		* – * – * G	* – * – – *	L- -LL-L	
(EF Hand)							
S100 protein (bovine)							
α subunit	GSELETAMET	LINVFHAH	S	G K E G D K	Y K L S K K E	LKELLQTE	LSGFLDAQKDAD$_{52}$
	53^{A}	VDKVMKEL		D E D G D G	E R D F Q E	YVVLVAAL	TVACNNFFWENS$_{93}$
β subunit	SELEKAVVA	LIDVFHQY	S	G R E G D K	H K L K K S E	LKELINNE	LSHFLEEIKEQE$_{51}$
	52^{V}	VDKVMETL		D S D G D G	E C D F Q E	FMAFVAMI	TTACHEFF EHE$_{91}$
				X Y Z	–Y –X –Z		

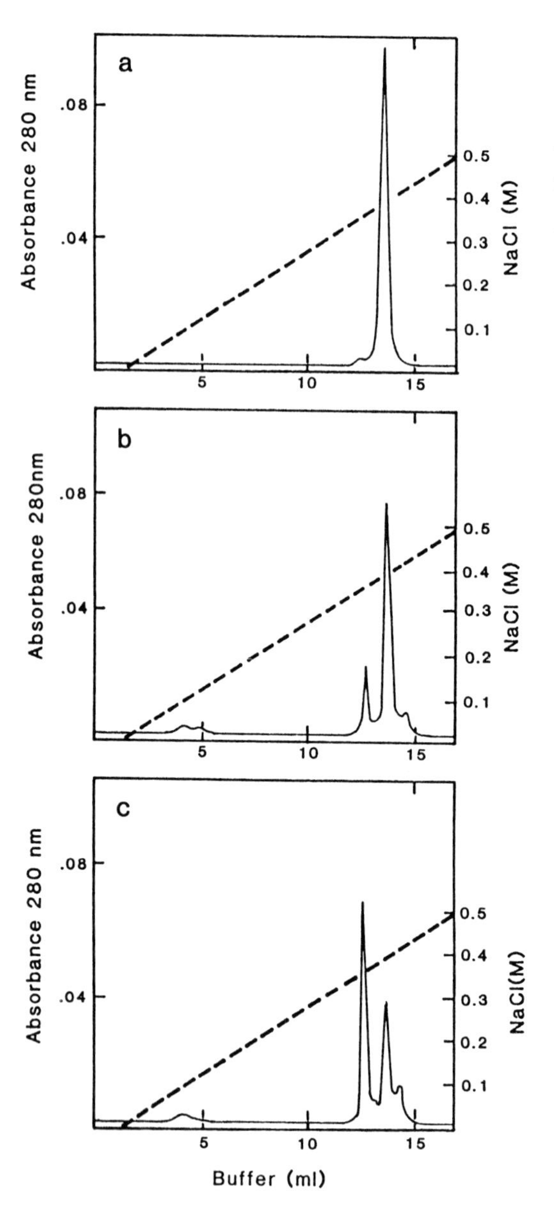

Fig. 3a–c. Effects of Ca^{2+} incubation on the FPLC elution of S100a ($\alpha\beta$) on a Mono Q column. The Mono Q column was equilibrated with 40 mM Tris-HCl buffer, pH 7.5, 1 mM EGTA. The elution profiles correspond to the following protein samples: S100a incubated in the presence of 1 mM Ca^{2+} at 40°C for 0 min (**a**); 5 min (**b**); and 30 min (**c**)

3 The Effect of Ca^{2+} Incubation on S100a ($\alpha\beta$) Conformation

In the presence of 120 mM KCl, the Ca^{2+}-affinity of S100 proteins was drastically reduced. However, in a more precise study on the effect of KCl on Ca^{2+} binding to S100a ($\alpha\beta$) dimer, we observed that the antagonistic effect of KCl on Ca^{2+} binding was reversible. Indeed, if KCl is able to reverse most of the Ca^{2+}-induced conformational changes in S100a when added subsequently to Ca^{2+}, in agreement with its strong antagonistic effect on Ca^{2+} binding, KCl failed to reverse the conformational changes when the protein was incubated with Ca^{2+} prior to KCl addition (Baudier and Gerard 1986). We concluded that some time-dependent conformational changes in the Ca^{2+}-bound S100a protein occurred that inhibited the antagonistic effect of KCl on Ca^{2+}-binding. These time-dependent conformational changes were confirmed by studying the FPLC chromatography on a Mono Q column of S100a protein after incubation with calcium. In Fig. 3, the elution profiles of S100a protein incubated with calcium for 0, 5, and 30 min at 40°C are presented. At time 0, the protein eluted as a single peak corresponding to the $\alpha\beta$ dimer. As the incubation time with calcium increased, we observed a gradual decrease in the S100a protein peak associated with the emergence of two new protein peaks that eluted before and after the S100a peak. Analysis in urea-SDS gel electrophoresis as well as spectroscopic characterization proved that the two new peaks corresponded to $\alpha\alpha$ and $\beta\beta$ protein dimers, respectively. The different absorbances for the $\alpha\alpha$ and $\beta\beta$ dimers, shown on the chromatogram, were due to the difference in molar extinction coefficients of S100$\alpha\alpha$ ($\epsilon_{280\ nm}$ = 18,600/M/cm) and S100b ($\beta\beta$) ($\epsilon_{280\ nm}$ = 3400/M/cm). From this experiment, we concluded that in the presence of calcium, time-dependent conformational changes probably resulted in destabilization of the quaternary $\alpha\beta$ dimer structure which allowed subunits to exchange and to form $\alpha\alpha$ and $\beta\beta$ dimers. This destabilization of the quaternary structure accounts for the inhibition of the antagonistic effect of KCl on calcium binding. This is the first indication of a dependence of the Ca^{2+}-binding properties of S100 proteins on their conformational states.

4 Calcium Binding to S100 Protein Chemically Alkylated on Cys 85α and Cys 84β

Since the calcium binding properties of the S100 protein dimer apparently depend on the interaction between subunits, we tried to find some experimental conditions where the S100 subunits would dissociate without possible denaturation. Then we could study the calcium binding properties of the individual α or β subunits. This was made possible by the observation that alkylation of Cys 85α and Cys 84β in S100$\alpha\alpha$ and S100b induced conformational changes in the protein dimers that destabilized the quaternary structure and allowed a slow monomer-dimer equilibrium at high protein concentration and total subunit dissociation at μM protein concentration (Baudier et al. 1986b). Cys 85α and Cys 84β in S100$\alpha\alpha$ and S100b were alkylated with the thiol-specific fluorescence probe, Bimane (Kosower et al. 1979) and the calcium binding properties of the derivatized protein studied by flow dialysis.

The calcium binding isotherm for Bimane-S100$\alpha\alpha$ in the absence and presence of 120 mM KCl are shown in Fig. 1. We observed a large increase in the calcium affinity for the alkylated protein as compared to native S100$\alpha\alpha$ and the antagonistic effect of KCl on calcium binding was considerably reduced. Bimane-S100$\alpha\alpha$ in 120 mM KCl had even higher calcium affinity than did the native protein in the absence of KCl. In Table 2, the calcium binding constants for Bimane-S100$\alpha\alpha$ protein are reported. In the presence of 120 mM KCl, calcium binding constants remain largely compatible with intracellular Ca^{2+} concentrations. Similar results were obtained with Bimane S100b

Table 2. Macroscopic constants describing Ca^{2+}-binding to S100αα protein dimer in 20 mM Tris-HCl buffer[a,b]

Conditions	K_1 (10^{-5}) M^{-1}	K_2 (10^{-5}) M^{-1}	K_3 (10^{-5}) M^{-1}	K_4 (10^{-5}) M^{-1}
pH 7.5	0.13	0.29	0.29	0.6
pH 8.3	1.09	0.23	0.48	0.9
Bimane S100αα pH 7.5	12.2	9.2	4.3	0.4
Bimane S100αα pH 7.5 plus 120 mM KCl	1.6	2.13	1.83	0.2

[a] Baudier et al. (1986a) J Biol Chem 261:8192–8203.
[b] Baudier et al. (1986b) Biochemistry 25:6934–6941.

Table 3. Macroscopic constants describing Ca^{2+}-binding to S100b (ββ) protein dimer in 20 mM Tris-HCl buffer[a,b]

Conditions	K_1 (10^{-5}) M^{-1}	K_2 (10^{-5}) M^{-1}	K_3 (10^{-5}) M^{-1}	K_4 (10^{-5}) M^{-1}
pH 7.5	1.5	0.34	0.25	0.5
pH 8.3	2.8	1.57	0.18	0.58
4 Zn^{2+}/S100b pH 8.3	9.93	3.63	3.17	0.56
Bimane S100b pH 7.5	5.69	1.17	0.19	0.10

[a] Baudier et al. (1986a) J Biol Chem 261:8192–8203.
[b] Baudier et al. (1986b) Biochemistry 25:6934–6941.

protein, although the increase in Bimane-S100 affinity for calcium was slightly less than for Bimane-S100αα protein (Table 3).

5 Calcium Binding on Zn^{2+}-Bound S100b Protein

Zinc binding to S100b protein is, so far, the major physico-chemical property that differentiates β subunit from α subunit. Bovine, human, and rat S100b proteins are able to bind zinc ions with high affinity, whereas the α subunit has only weak zinc binding sites (Baudier et al. 1983, 1984, 1985, 1986a). Bovine brain S100b protein has four high affinity zinc binding sites with affinity ranging around 10^{-8} to 10^{-7} M and other lower affinity sites. Zinc binding to the high affinity sites induces conformational changes in the S100b protein structure that are similar to those induced upon calcium binding (Baudier et al. 1986a). However, calcium binding studies on the Zn^{2+}-bound S100b proved the Ca^{2+} and Zn^{2+}-binding sites to be different. Figure 2 shows the Ca^{2+}-binding isotherm for the Zn^{2+}-bound S100b protein in the absence and presence of 120 mM KCl and they are compared to the Zn^{2+}-free protein. In the presence of 4 Zn^{2+} equivalents per mole S100b dimer, four calcium binding sites remain titratable, but with much higher affinity than for the protein in the absence of zinc (Table 3). Furthermore, zinc binding to S100b decreases the antagonistic effect of KCl on calcium binding. Zinc binding to S100b apparently has the same effect that alkylation of cysteine 84β does on the calcium binding properties of S100b protein. Recent data also indicated that zinc binding to the high affinity sites might have destabilized the quaternary dimer structure and allowed subunits to dissociate at low protein concentrations as does alkylation of cysteine 84β (unpublished).

6 Conclusion

The calcium binding properties of S100 protein are largely dependent on the protein conformation and more particularly, on their quaternary structure. A resume of our work on the conformation and calcium binding properties of S100 protein is presented in Fig. 4 as a schematic model for a slow monomer-dimer equilibrium between S100-subunits that might explain the variation in calcium affinity observed for the different conformational states of the proteins.

In non-denaturing solvent, S100 proteins exist as dimers. The α and β subunits in S100$\alpha\alpha$ and S100b are symmetrically positioned as revealed on the H-NMR spectra of the proteins (Baudier, Lefévre, Angstrom, in preparation). The subunits are held together by hydrogen bonds, which probably involve chemical groups that ionize

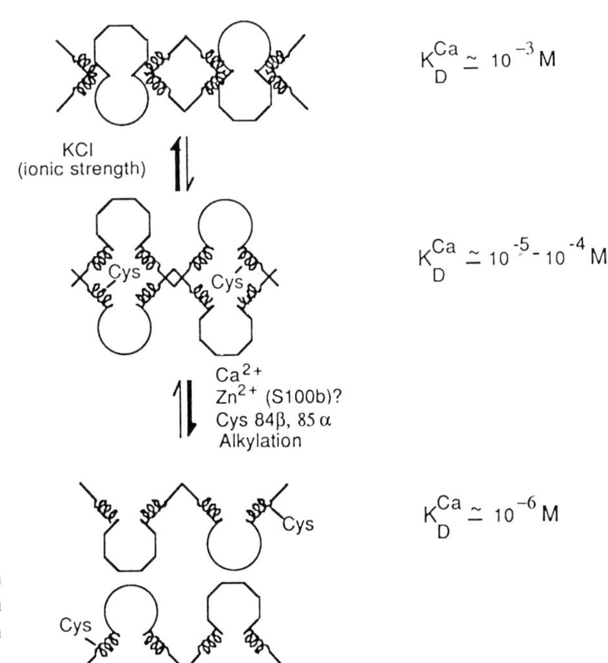

Fig. 4. Schematic representation of the monomer-dimer equilibrium in S100 proteins and its effect on the protein affinity for calcium

below pH 3 (Moreno and Weber 1982; Baudier and Gerard 1986) and hydrophobic forces resulting from the interactions between aromatic amino acid side-chains that are buried inside the protein matrix (Baudier and Gerard 1986). In this conformational state, the S100 protein dimers have a calcium affinity ranging between 10^{-5} to 10^{-4} M Ca^{2+}.

When KCl concentration (or ionic strength in general) increases, the hydrophobic interactions between subunits increase, rendering the S100 proteins more compact (Mani and Kay 1984). Under these conditions, the calcium binding sites become less accessible to solvent and the apparent protein affinity for calcium markedly decreases. However, upon extenced incubation with Ca^{2+}, or with alkylation of Cys 85α or 84β, or upon zinc binding in the special case of S100b, the aromatic residues become fully exposed to solvent (Baudier and Gerard 1983, 1986; Baudier et al. 1986a,b) and the quaternary protein structure becomes less stable. This allows a slow monomer-dimer equilibrium between subunits and total subunit dissociation at low protein concentrations. In these conformational states, the Ca^{2+}-binding sites are fully exposed to solvent and easily accessible to Ca^{2+} ions. The affinity of the S100 proteins for calcium apparently increases. Preliminary results have shown that the affinity affected by protein conformation is mainly on the typical Ca^{2+} binding site (site II) of the α and β subunits (Baudier and Gerard 1986; Baudier et al. 1986b; Baudier, Lefévre, Angstrom, in preparation).

Such a model, indeed, needs more experimental data for confirmation and completion. However, it offers new perspectives in searching for biological functions of S100 proteins. It suggests that S100 proteins might exist, in certain conditions, as monomers with high calcium affinities and that protein dimers might interact with other cellular components that in turn regulate their calcium binding properties. Several recent reports argue for such a hypothesis. A membrane-bound form of S100 protein was identified (Haglid et al. 1974; Donato et al. 1975), but in aqueous media S100 protein dimers are highly hydrophilic and in this physical state their interaction with lipid membranes seems unlikely. However, upon alkylation of Cys 85α, the dissociated a subunit which has its calcium affinity dramatically increased exposes highly hydrophobic domain to solvent (Baudier et al. 1986b). Thus, it is plausible that in a membrane-bound form, α-subunit will take on a conformation that resembles that observed upon alkylation of Cys 85α. A Ca^{2+}-dependent function for S100 proteins bound to membrane is now only hypothetical and obviously needs further investigation. The possibility that S100 proteins might have their Ca^{2+}-binding affinity regulated upon interaction with target proteins should also be taken into consideration. Direct evidence for such regulation in vitro was reported for S100b when it interacted with the bee venom polypeptide melittin (Baudier et al. 1987). Recently, S100's subunits were shown to be related in sequence to the P10 protein, the regulatory chain in the P38 kDa substrate complex of viral tyrosine-specific protein kinase (Gerke and Weber 1985a; Glenny and Tack 1985). P10 protein exists as a monomer in the P38 kDa complex, but associates as dimers when it is purified (Gerke and Weber 1985b). The physico-chemical relationship between S100 and P10 could also argue for a similarity in their interaction with target proteins as monomer. We recently found that S100b is able to regulate phosphorylation of the microtubule associated tau proteins by protein kinase C and Ca^{2+}/calmodulin-dependent kinase, but it does so by interacting with the substrate tau protein rather than with the kinases. The effect of S100b on tau phosphorylation was potentiated in 10 μM–100 μM Ca^{2+} (Baudier and Cole, in preparation).

Finally, it was reported that S100b protein stimulated the differentiation of chicken embryo cerebral cortical neurons, but a disulfide form of the protein was required for biological activity; the activity was lost after reduction of cysteine residues in the protein (Kligman and Marshak 1985). A cautionary note needs to be included. In simple solutions in vitro, the disulfide form of S100b was shown to have conformational properties (investigated by UV spectroscopy) nearly identical to those of the protein alkylated on Cys 84β (Baudier et al. 1986a,b). This suggests that for in vivo activity on neuron differentiation, S100 protein probably adopts conformations that are different from those of purified native protein in aqueous solvent.

Acknowledgments. This work was supported by grants from INSERM, CNRS and Université Louis Pasteur. The author wishes to thank Professor R.D. Cole and Dr. M. Almagor, University of California, Berkeley, for critical reading of the manuscript, and Karen Ronan for editorial assistance.

References

Baudier J, Gerard D (1983) Calcium and zinc ion binding to S100 proteins. Structural changes induced by calcium and zinc on S100a and S100b. Biochemistry 22:3360–3369

Baudier J, Gerard D (1986) Conformational studies and calcium-induced conformational changes in S100αα protein: The effect of acidic pH and calcium incubation on subunit exchange in S100a (αβ) protein. J Biol Chem 261:8204–8212

Baudier J, Briving C, Deinum J, Haglid K, Sorshog L, Wallin M (1982) Effects of S100 proteins and calmodulin on Ca^{2+}-induced disassembly of brain microtubule proteins in vitro. FEBS Lett 147:165–167

Baudier J, Haglid K, Haiech J, Gerard D (1983) Zinc ion binding to human brain calcium binding proteins, calmodulin and S100b protein. Biochem Biophys Res Commun 114:1138–1146

Baudier J, Glasser N, Haglid K, Gerard D (1984) Purification, characterization and ion binding properties of human brain S100b protein. Biochim Biophys Acta 790:164–173

Baudier J, Labourdette G, Gerard D (1985) Rat brain S100b proteins: purification, characterization and ion-binding properties. A comparison with bovine S100b protein. J Neurochem 44: 76–84

Baudier J, Glasser N, Gerard D (1986a) Calcium and zinc-binding properties of bovine brain S100αα, S100a (αβ) and S100b (ββ) protein: Zn^{2+} regulates Ca^{2+} binding on S100b protein. J Biol Chem 261:8192–8203

Baudier J, Glasser N, Duportail G (1986b) Bimane and acrylodan-labelled S100 proteins. The role of cysteine 85α and 84β in the conformation and calcium binding properties of S100αα and S100b (ββ) protein. Biochemistry 25:6934–6941

Baudier J, Mochly-Rosen D, Newton A, Lee S-H, Koshland DE Jr, Cole RD (1987) A comparison of S100b protein with calmodulin: Interactions with melittin and microtubule associated proteins; inhibition of phosphorylation of tau proteins by protein kinase C. Biochemistry 26: 2886–2893

Deinum J, Baudier J, Briving C, Rosengreen L, Wallin M, Gerard D, Haglid K (1983) The effect of S100a and S100b proteins and Zn^{2+} on the assembly of brain microtubule proteins in vitro. FEBS Lett 163:287–291

Donato R (1983) Effect of S100 protein on assembly of brain microtubule proteins in vitro. FEBS Lett 162:310–313

Donato R, Michetti F, Miani N (1975) Soluble and membrane-bound S100 protein in cerebral cortex synaptosomes. J Neurochem 36:1698–1705

Gerke V, Weber K (1985a) The regulatory chain in the p36 kD substrate complex of viral tyrosine-specific protein kinases is related in sequence to the S100 protein of glial cells. EMBO J 4: 2917–2920

Gerke V, Weber K (1985b) Calcium-dependent conformational changes in the 36 kDa subunit of intestinal protein I related to the cellular 36 kDa target of Rous sarcoma virus tyrosine kinase. J Biol Chem 260:1688–1695

Glenney JR, Tack BF (1985) Amino-terminal sequence of p36 and associated pI10: Identification of the site of tyrosine phosphorylation and homology with S100. Proc Natl Acad Sci USA 82: 7884–7888

Haglid K, Hamberger A, Hansson HA, Hyden H, Persson L, Rönnbäck L (1974) S100 protein in synapses of the central nervous system. Nature (Lond) 251:532–534

Hidaka H, Endo T, Kawamoto S, Yamada E, Unekawa H, Tanabe K, Hara K (1983) Purification and characterization of adipose tissue S100b protein. J Biol Chem 258:2705–2709

Ishikawa H, Nagami H, Shirasawa N (1983) Novel clonal strains from adult rat anterior pituitary producing S100 protein. Nature (Lond) 303:711–713

Isobe T, Okuyama T (1978) The amino-acid sequence of s100 protein (PAP Ib protein) and its relation to the calcium-binding proteins. Eur J Biochem 89:379–389

Isobe T, Okuyama T (1981) The amino-acid sequence of the α subunit in bovine brain S100a protein. Eur J Biochem 116:79–86

Isobe T, Takahashi K, Okuyama T (1984) S100a$_0$ ($\alpha\alpha$) protein is present in neurons of the central and peripheral nervous system. J Neurochem 43:1494–1496

Kato K, Kimura S (1985) S100a$_0$ ($\alpha\alpha$) protein is mainly located in the heart and striated muscles. Biochim Biophys Acta 842:146–150

Kato K, Suzuki F, Nakajima T (1982) Decrease in adipose S100 protein levels by epinephrine in rat. J Biochem 93:311–313

Kato K, Kimura S, Sermba R, Suzuki F, Nakajima T (1983) Increase in S100 protein levels in blood plasma by epinephrine. J Biochem 94:1009–1011

Kilgman D, Marshak DR (1985) Purification and characterization of a neurite extension factor from bovine brain. Proc Natl Acad Sci USA 82:7136–7139

Kosower NS, Kosower EM, Newton GL, Ranney HM (1979) Bimane fluorescence labels: Labeling of normal human red cells under physiological conditions. Proc Natl Acad Sci USA 76:3382–3386

Labourdette G, Mandel P (1978) Effect of norepinephrine and dibutyryl cyclic AMP on S100 protein levels in C_6 glioma cells. Biochem Biophys Res Commun 96:1702–1709

Mani RS, Kay CM (1984) Hydrodynamic properties of bovine brain S100 proteins. FEBS Lett 166:258–262

Molin SO, Rosengreen L, Haglid K, Baudier J, Hamberger A (1984) Differential localization of "brain specific" S100 and its subunits in rat salivary glands. J Histochem Cytochem 32:805–814

Molin SO, Rosengreen L, Baudier J, Hamberger A. Haglid K (1985) S100α-like immunoreactivity in tubules of rat kidney. A clue to the function of a "brain specific" protein. J Histochem Cytochem 33:367–374

Moore BW (1965) A soluble protein characteristic of the nervous system. Biochem Biophys Res Commun 6:739–744

Moreno R, Weber G (1982) Properties of S100 protein studied by fluorescence methods. Biochim Biophys Acta 703:231–240

Patel J, Marangos PJ, Heydorn WE, Chang G, Verna A, Jacobowitz D (1983) S100-mediated inhibition of brain protein phosphorylation. J Neurochem 41:1040–1045

Qi DF, Kuo JF (1984) S100 modulates Ca^{2+}-independent phosphorylation of an endogenous protein (Mr = 19 K) in brain. J Neurochem 43:256–260

Shashoua VE, Hesse GW, Moore BW (1984) Proteins of brain extracellular fluid: evidence for release of S100 protein. J Neurochem 42:1536–1541
Suzuki F, Nakajima T, Kato K (1982) Peripheral distribution of nervous system-specific S100 protein in rat. J Biochem 92:835–838
Szebenyi DM, Obendorf SK, Moffat K (1981) Structure of vitamin D-dependent calcium-binding protein from bovine intestine. Nature (Lond) 294:327–332

Mechanisms of Action of the S100 Family of Calcium Modulated Proteins

L. J. VAN ELDIK and D. B. ZIMMER [1]

1 Introduction

Eukaryotic cells require calcium ions for optimal growth and functioning. The intracellular actions of calcium as a biological second messenger appear to be a result of its interaction with a set of calcium binding proteins referred to as calcium-modulated proteins (Van Eldik et al. 1982). Calcium modulated proteins are characterized by the ability to bind calcium reversibly with dissociation constants in the nanomolar to micromolar range under physiological conditions. Examples of calcium modulated proteins include calmodulin, parvalbumin, troponin C, oncomodulin, vitamin D-dependent calcium binding protein, and S100 proteins. The calcium modulated proteins characterized to date possess a calcium binding structure called an EF-hand (Kretsinger 1980). The EF-hand structure is based on the calcium binding sites found in parvalbumin and is characterized by an α-helix, loop, α-helix arrangement of the peptide chain. The calcium-ligating residues are found within the peptide loop.

In this chapter, we will concentrate on the S100 family of proteins, and emphasize studies addressing the mechanisms of action of the S100 proteins. S100 was originally isolated from bovine brain as a brain-specific protein fraction (Moore 1965) and called S100 to denote its partial solubility in 100% saturated ammonium sulfate. The bovine brain S100 fraction is composed of two small, acidic proteins called S100α and S100β (Isobe and Okuyama 1981). These proteins share approximately 50% identity in amino acid sequence (Fig. 1) and can form homologous and heterologous dimers referred to as S100α_0 ($\alpha\alpha$), S100a ($\alpha\beta$), and S100b ($\beta\beta$) (Isobe et al. 1981, 1983; Masure et al. 1984). S100α and S100β have two regions of amino acid sequence that could form calcium binding structures: a putative EF-hand calcium binding structure and a calcium binding structure similar to that found in the vitamin D-dependent calcium binding protein (Szebenyi et al. 1981; Van Eldik et al. 1982). More recent studies (Baudier et al. 1986) have demonstrated that S100 also binds zinc and the presence of zinc may influence the calcium binding properties of S100.

Bovine brain is an excellent tissue for isolation of S100 because large amounts of both S100α and S100β can be obtained (total S100: 200–250 mg/kg, Moore 1982; Masure et al. 1984). All three S100 dimers ($\alpha\alpha$, $\alpha\beta$, $\beta\beta$) can be found in bovine brain (Isobe et al. 1981; Masure et al. 1984) and the S100β subunit makes up approximately

1 Departments of Pharmacology and Cell Biology, and Howard Hughes Medical Institute, Vanderbilt University, Nashville, TN 37232, USA

Ch. Gerday, R. Gilles, L. Bolis (Eds.)
Calcium and Calcium Binding Proteins

50–80% of the total S100 (Isobe et al. 1983; Jensen et al. 1985). In contrast, in rat brain 90–95% of the total purified S100 is S100β (Baudier et al. 1985) and in human brain, 80–96% of the total S100 is S100β (Baudier et al. 1984; Jensen et al. 1985). Rat brain S100β differs in amino acid sequence from bovine brain S100β in only four positions (Kuwano et al. 1984). Similarly, human S100β has only three amino acid changes when compared to bovine S100β (Jensen et al. 1985) and these changes are compatible with minimum single base changes in codon structures. Finally, bovine adipose S100β is identical in amino acid sequence to bovine brain S100β (Marshak et al. 1985). Thus, while there appear to be species differences in the relative amount of S100β in brain, the S100β polypeptide is structurally conserved in mammals. Less information is available regarding the structural conservation of the S100α polypeptide. A recent report (Kuwano et al. 1986) of the isolation of a S100α cDNA from bovine brain shows that the predicted amino acid sequence derived from the nucleotide sequence is identical to the amino acid sequence elucidated for the bovine brain S100α polypeptide, except for one position (residue 64 in Fig. 1 is an aspartic acid by amino acid sequence and an asparagine by nucleotide sequence). S100α has been purified from human pectoral muscle and has been shown to have gel mobility, calcium binding properties, and immunochemical properties similar to those of bovine brain S100α (Kato et al. 1986). However, elucidation of the primary amino acid sequence of S100α from human pectoral muscle and from other tissues will be required to determine the extent of structural conservation of the S100α polypeptide among species and tissues.

While brain is one of the richest sources of S100, it is known now that S100 proteins are not brain-specific. S100 proteins have been isolated from or detected in various tissues, and there is an extensive literature on the immunocytochemical distribution of S100 proteins, especially reports about the use of S100 localization in diagnostic surgical pathology. For example, a computer search of the literature for the period of January, 1984 – September 1986 revealed approximately 500 publications on S100, with greater than 85% of these papers dealing with immunohistochemical localization studies of S100. In spite of the plethora of literature on S100, little is known about the molecular mechanisms of action of S100 proteins. In this article, we will emphasize studies addressing potential functions of S100 proteins. We will discuss three areas: (1) the distribution of S100 proteins in relation to possible functions, (2) reported activities for S100, and (3) recent evidence of multiple, S100-like proteins and their potential significance. Several comprehensive reviews on the chemistry and localization of S100 proteins are available (Moore 1982; Zomzely-Neurath and Walker 1980; Donato 1986), and the reader is referred to these reviews for additional information that is not discussed in this article.

2 S100 Distribution

Studies on the tissue distribution of the individual S100 polypeptides, S100α and S100β, and their cellular and subcellular localization have provided insight into potential function(s). In the central nervous system, S100 is located primarily in glial cells and appears to be diffusely located throughout the cytoplasm including the cell body and extensions, and associated with glial filaments. Some investigators have reported localization of S100 in the nucleus of glial cells (Hyden and McEwen 1966; Cocchia 1981; Sviridov et al. 1972; Michetti et al. 1974); however, others have not confirmed this finding (Tabuchi and Kirsch 1975; Ghandour et al. 1981). These early immuno-localization studies used antibodies prepared against an S100 fraction from bovine brain, which contains both S100α and S100β. Therefore, it is not known whether the observed localization patterns were due to S100α and/or S100β immunoreactivity. More recent studies using antibodies which are specific for S100α and S100β (Isobe et al. 1984) have shown that S100β is located primarily in astrocytes and ependymal cells while S100α is located in neurons and astrocytes. In addition to their differing cellular localizations, S100α and S100β have different intracellular localizations; S100β is located diffusely throughout the cytoplasm while S100α has a granular distribution.

S100 has also been localized in the peripheral nervous system. Early studies found S100 in Schwann cells of myelinated and unmyelinated peripheral nerves, in satellite cells of dorsal root ganglia and in the adrenal medulla (Stefansson et al. 1982). Subsequent studies utilizing S100α- and S100β-specific antibodies have demonstrated that S100β is localized in Schwann cells and satellite cells of the automatic ganglia while S100α is localized in a granular pattern in neurons but not in Schwann cells or satellite cells (Isobe et al. 1984). As in the central nervous system, the S100β staining pattern in the periphal nervous system was diffuse while the S100α staining was granular.

Other studies using immunocytochemistry have demonstrated that S100 is not brain-specific and can be detected in the pineal gland (Moller et al. 1978), adrenal gland (Cocchia and Michetti 1981), hypophysis (Nakajima et al. 1980), melanocytes and Langerhans cells of skin (Cocchia et al. 1981), chondrocytes (Takahashi et al. 1984), lymph node macrophages (Takahashi et al. 1984), kidney (Molin et al. 1985), salivary gland (Molin et al. 1984), testis (Michetti et al. 1985), and adipocytes (Michetti et al. 1983). Studies which have used S100α- and S100β-specific antibodies (Takahashi et al. 1984; Molin et al. 1984, 1985) have found differential cellular distributions of these proteins in salivary gland and kidney.

While the localization studies have provided significant information regarding the cellular distribution of S100α and S100β in central and peripheral nervous tissue and non-nervous tissues, they do not provide quantitative estimates regarding the levels of S100 in tissues. Using an immunoassay and antibody which detects predominantly S100β, Suzuki et al. (1982) examined the levels of S100 in various rat tissues. These investigators found that central nervous tissue contained significantly more S100/mg protein than did peripheral tissues. However, they did find high levels of S100 in adipose tissue and trachea. When rat tissues were examined by radioimmunoassay using S100α- and S100β-specific antibodies (Zimmer and Van Eldik 1987), nervous tissue (brain) was found to contain the highest levels of both S100α and S100β. All tissues examined contained S100, but the relative distribution of S100α and S100β differed among tissues. The ratio of S100β and S100α varied from 18- to 40-fold more S100β than S100α in brain and testes, about equivalent ratios in skin and liver, and 8- to 75-fold more S100α than S100β in kidney, spleen, and heart. In addition, kidney contained high levels of S100α, almost as much as detected in brain. The results of S100β distribution in rat tissue as determined by radioimmunoassay are in good agreement with other studies on S100β (Suzuki et al. 1982; Hidaka et al. 1983; Kato et al. 1986) and S100α (Kato et al. 1986) distribution as determined by enzyme immunoassay.

The results on S100 distribution in rat tissue are similar to the S100α and S100β distribution seen in human tissues (Kato and Kimura 1985). However, lower levels of S100α were observed in rat heart (Zimmer and Van Eldik 1987) than those found in human heart (Kato and Kimura 1985) suggesting that there may be species differences in S100 levels. Subsequent studies (Kato et al. 1986) have demonstrated that S100α and S100β levels in brain, heart, and muscle vary from species to species. Since all these studies measured the immunoreactive levels of S100α and S100β by using antibodies made against bovine brain S100α and S100β, it is possible that there may be a population of S100 proteins in certain tissues that do not cross-react with these antibodies. Therefore, the purification and characterization of S100 proteins from individual tissues and species will be necessary for unequivocal determination of the total levels of S100α and S100β.

S100 is considered to be a soluble cytoplasmic protein; however, it can be found associated with the particulate fraction of cells. The particulate-associated S100 resists extraction with EDTA and high salt (1.0 M NaCl) and can be solubilized by n-pentanol (Haglid and Stavrou 1973) or non-ionic detergents such as Triton X-100 (Rusca et al. 1972; Donato et al. 1975) suggesting that some of the S100 interacts strongly with membrane components. It has been shown (Donato 1977) that synaptosomal membrane preparations have ^{125}I-S100 binding activity which can be solubilized by Triton X-100. This interaction is specific and saturable with respect to S100 concentration, suggesting that the membranes contain a "receptor" for S100. In addition to synaptosomal S100-binding sites, binding sites for S100 have been demonstrated in isolated brain nuclei (Donato and Michetti 1981), and I^{125}-S100 has been shown to concentrate in the nuclear membrane and nucleolus (Michetti and Donato 1981). S100 purified from isolated brain nuclei represents 0.55% of the soluble S100 in the cytosol and occurs as free S100, labile-bound S100 found in the deoxyribonucleoprotein fraction, and stable-bound S100 found in the nucleolar fraction (Michetti et al. 1974).

Immunocytochemical studies and indirect immunofluorescence microscopy (Hyden and Ronnback 1975, 1978; Cocchia 1981; Haglid et al. 1977) have demonstrated that in adult brain S100 is located in two fractions: a cytoplasmic fraction represented by diffuse staining throughout the cytoplasm and a membranous fraction with staining at the plasma membrane, outer mitochondrial membrane, endoplasmic reticulum, golgi apparatus and nuclear membranes. Subsequent studies have demonstrated that the cytoplasmic and membrane-bound S100 proteins from bovine and rat brains are identical (Donato et al. 1986b). Additional studies on the properties of membrane-associated S100 and its "receptor(s)" will be required before the function of membrane-associated S100 can be elucidated. Several studies have shown a shift from membrane-bound to soluble S100 in response to malignancy (Haglid and Stavrou 1973) and norepinephrine and dibutyryl cAMP treatment of cervical sympathetic ganglia (Nagata et al. 1984). Further investigation of the factors which regulate S100 subcellular distribution may provide insights into the function(s) of particulate and soluble S100.

3 S100 Function

Very little is known about the function of S100 even though it was isolated over 20 years ago. Two approaches have been utilized in studying the function of S100: (1) examination of the S100 interaction with isolated cellular components in vitro, and (2) the isolation, characterization and identification of unknown S100 target proteins. The first approach has provided information regarding the interaction of S100 with cytoskeletal elements, in particular microtubules, and the nucleus. The localization of S100 in the nucleus and its presence in isolated nuclei suggest that S100 may function in gene regulation via interaction with proteins and enzymes which regulate gene expression. The addition of S100 to isolated brain nuclei resulted in an increase in the RNA-polymerase I activity and differences in the S100 activation occurred at different stages of brain maturation (Miani et al. 1973). Further studies have revealed that S100 stimulates brain RNA-polymerase I activity (Michetti et al. 1976).

The effect of S100 on protein kinase activities in the nucleus has also been examined. An increase in total protein phosphorylation by brain nuclear kinases in the presence of S100 has been reported (Perumal 1976). Subsequent studies have demonstrated that this effect was brain-specific, i.e. isolated liver nuclear kinases were not affected by S100 (Perumal and Rapport 1978). In addition, a brain protein kinase (protein kinase X) has been detected whose activity is stimulated in the presence of S100 and other calcium modulated proteins in a calcium independent manner (Qi et al. 1984). An endogenous substrate protein (Mr = 19K) for protein kinase X has recently been isolated which exhibits S100-stimulated phosphorylation in a calcium independent manner (Qi and Kuo 1984). In addition to activation of this brain kinase, S100 has also been shown to affect the activity of the catalytic subunit of cAMP dependent protein kinase (Kuo et al. 1986a). Other investigators have found that S100 inhibits the phosphorylation of a 73,000 molecular weight protein in rat cell extracts in a calcium dependent manner (Patel et al. 1983). The phosphorylation of three additional proteins (56K, 50K, and 47K) was also inhibited at higher S100 concentrations. It is interesting to note that the phosphorylation of these three proteins was increased in the presence of calmodulin, suggesting that S100 and calmodulin may function antagonistically in brain. More recent studies (Kuo et al. 1986b) have shown that the activity of phosphoprotein phosphatases can be stimulated by S100. These data suggest that S100 may regulate neural cell function via inhibition or stimulation of phosphorylation in the cytoplasm and nucleus. The identification of the substrates which appear to be phosphorylated or dephosphorylated in response to S100 will provide significant information regarding the neural cell functions which may be regulated by S100. In addition, the recent isolation of protein kinases and phosphoprotein phosphatases from muscle whose activities are altered in the presence of S100 (Kuo et al. 1986b) suggest that S100 function via protein phosphorylation/dephosphorylation may occur in tissues other than neural tissue.

The subcellular localization of S100 along microtubules has suggested that S100 may interact with microtubules. Baudier and coworkers (Baudier et al. 1982) reported that S100 induced microtubule disassembly in vitro at mM Ca^{2+} concentrations more effectively than calmodulin. Subsequent studies have suggested a direct interaction between S100 proteins and tubulin (Endo and Hidaka 1983; Donato 1984) or between S100 and a microtubule associated protein, tau (Fujii et al. 1986). Biochemical (Donato 1984) and ultrastructural studies (Donato et al. 1986a) suggest that S100 inhibits the nucleation of new microtubules and elongation of existing microtubules. The different S100 isoforms have similar effects on microtubule assembly in the presence of Ca^{2+}, but different effects in the presence of Zn^{2+} (Donato et al. 1985; Deinum et al. 1983). Recent studies (Fujii et al. 1986; Donato 1985) have examined in greater detail the effects of Ca^{2+} on S100 inhibition of microtubule assembly and have found that S100 increases the calcium sensitivity of microtubules.

A second approach which has proven successful in studying S100 function is the detection and identification of potential S100 target proteins, i.e. S100 binding proteins. Using affinity chromatography techniques, Gopalakrishna et al. (1985) detected calmodulin- and S100-binding proteins from bovine brain. They found that while some binding proteins interact with both S100 and calmodulin, others interact with calmodulin or S100 exclusively. In addition, they found that some proteins bound to S100 in a calcium-dependent manner and others in a calcium-independent manner. None of the S100-binding proteins were identified in that study. More recent experiments have compared the S100α-, S100β, and calmodulin-binding proteins in various rat tissues, including brain, with a gel overlay technique (Zimmer and Van Eldik 1987). These studies indicated that each tissue possessed its own complement of S100-binding proteins, suggesting that S100 may have tissue specific functions as a result of differential distribution of S100-binding proteins. In addition, the calmodulin-binding protein profile did not resemble the S100-binding protein profile, suggesting that S100 and calmodulin interact with different molecular targets.

While the comparison of S100- and calmodulin-binding profiles from different tissues and species is informative, it does not provide definitive information regarding S100 function. Such information would come from identification and characterization of the binding proteins detected by the gel overlay techniques. Using classical protein purification techniques in combination with microchemical methods, S100-binding proteins have been isolated from rat brain (Zimmer and Van Eldik 1986). Based on partial amino acid sequence data and examination of protein sequence databases, one S100-binding protein in rat brain was tentatively identified as fructose-1,6-bisphosphate aldolase. Amino acid analysis, peptide maps, and electrophoretic analysis confirmed that this binding protein was rat brain aldolase (Zimmer and Van Eldik 1986). Aldolase enzyme activity is stimulated by S100 in vitro and the stimulation of the brain-specific form of aldolase is calcium dependent. Additional studies will be required to ascertain the physiological significance of the S100-aldolase interaction, but these studies demonstrate the feasibility of using a direct biochemical approach to study S100 function.

In addition to the intracellular activities of S100, it appears that S100 may also have extracellular functions. It has recently been reported (Kligman and Marshak 1985) that bovine brain extracts or conditioned media from rat C6 glioma cells stimulate neurite outgrowth from chicken embryo cerebral cortical neurons. This neurite extension factor (NEF) activity has been purified and appears to be identical to S100β dimer. NEF activity seems to require disulfide bonds, but how the disulfide bonds are arranged in S100β to produce biologically active NEF is not known. S100β contains two cysteines at residues 68 and 84 (Fig. 1), so there is the possibility for the formation of parallel dimers of S100β, antiparallel dimers, or intrachain disulfide bonds. The exact structural features of S100β that result in NEF activity must be determined in future studies.

Potential NEF activity of S100β is consistent with previous studies that have suggested a role for S100 in nervous system development. For example, studies on the levels of S100 mRNA and protein during brain development have shown that S100 levels increase during the time that neuronal processes are elongating (Zomzely-Neurath and Walker 1980). It is attractive to speculate that during development of the central nervous system, S100β is secreted from glial cells and then acts in a paracrine fashion to stimulate neurite outgrowth. What structural modifications of S100β or what cellular signals trigger secretion, and the mechanisms by which the neuron responds to extracellular S100 will be important areas for future research.

It is enlightening, however, to examine some of the earlier studies of S100 in light of potential paracrine functions for S100. For example, it has been reported that an S100 receptor can be detected on neuronal cells (Donato et al. 1975), that a percentage of S100 can be found tightly associated with particulate fractions (Donato et al. 1986b), and that there are S100-binding sites on neuronal membranes (Hyden and Ronnback 1975). Observations that S100 can be released into the extracellular space of the brain (Shashoua et al. 1984) and detected in the conditioned media of C6 glioma cells (Van Eldik and Zimmer 1987) suggest that S100 can be secreted from brain cells. Although these observations are circumstantial and must be interpreted with caution, they are consistent with a paracrine function of S100 in neural cells.

Interestingly, there have been several studies that suggest that S100 can be secreted from various cell types. A clonal cell strain derived from rat anterior pituitary, JH-S3, produces S100 and releases the protein into the conditioned medium (Ishikawa et al. 1983). This released S100 appears to stimulate prolactin secretion from another clonal cell in a dose-dependent manner, and antibody to S100 inhibits secretion (Ishikawa et al. 1983). Although the active form of S100 was not defined, nor whether the S100 in the media was a result of secretion or cell lysis, these results are consistent with another potential paracrine activity for S100; i.e., stimulation of prolactin secretion.

The second system examined for S100 release is rat epididymal adipose tissue. Treatment of adipose tissue with catecholamines resulted in a decrease in the levels of soluble, intracellular S100 with no effect on membrane-bound S100 (Suzuki et al. 1984). This decrease in intracellular S100 was reflected in an increase in S100 released into the media (Suzuki et al. 1983), and the S100 release has been postulated to be linked to increases in cAMP content of the cells (Suzuki and Kato 1985). The mechanism by which S100 release occurs in adipocytes and the function of the released S100 are not known. Nevertheless, these studies suggest that S100 may be secreted under certain conditions.

Little is known about the signals that regulate S100 secretion. The observation that S100 is synthesized by both free and membrane-bound polysomes (Cosgrove et al. 1983) is consistent with both intracellular and secreted forms of S100. Studies of cell-free translation of S100 (Mahony and Brown 1980) and examination of S100 mRNA with cDNA probes for S100α (Kuwano et al. 1986) or S100β (Kuwano et al. 1984; Dunn et al. 1987) have found no evidence for a precursor or signal peptide for

S100. There is a region of basic amino acids in S100 from residues 20 to 33 that had been suggested in an early paper (Cosgrove et al. 1983) as a sequence that could potentially interact with negatively charged phospholipids in the membrane to target S100 for secretion. However, unequivocal determination of the structural features of S100 that allow secretion will require further study.

4 S100-Like Proteins

One of the most interesting developments in the field of S100 research are the recent reports of proteins with homology to S100. For example, Baserga and co-workers (Calabretta et al. 1986) have identified a cDNA clone, called 2A9, that is preferentially expressed when quiescent cells are stimulated by serum or growth factors to enter the G1 phase of the cell cycle. A comparison of the amino acid sequence of S100β with the amino acid sequence deduced from the nucleotide sequence of 2A9 is shown in Fig. 1. It is evident that there is a high degree of homology between the two amino acid sequences (37 identical residues), and the homologies in the putative calcium binding sites (residues 20–32 and residues 62–73 in Fig. 1) are especially striking. While the 2A9 protein has not been isolated yet, the homologies at the nucleotide level suggest that the product of 2A9 (which Baserga and colleagues call calcyclin) may be a member of the S100 family of calcium modulated proteins. It will be interesting to characterize calcyclin and determine whether it is indeed a calcium binding protein with functional similarities to S100.

Another protein that is homologous to S100 is a polypeptide of 95 residues that has been referred to as p10 (Glenney and Tack 1985), p11 (Gerke and Weber 1985), 10,000-M_r polypeptide (Hexham et al. 1986), and calpactin light chain (Glenney et al. 1986). This protein is found in some tissues as a subunit of the p36 tyrosine kinase substrate. The amino acid sequence of p10 (Fig. 1) shows a high degree of homology with the S100 proteins, being most similar to S100α (43 identical residues). However, p10 does not appear to bind calcium. The inability of p10 to bind calcium is consistent with the observed amino acid sequence differences in the regions corresponding to the putative calcium binding sites for S100 (Fig. 1). For example, the sequence alignment in the first calcium binding site (residues 20–32 in Fig. 1) shows that p10 lacks three residues in the putative calcium binding loop region, and p10 shows six amino acid sequence differences in the second calcium binding loop (residues 62–73 in Fig. 1). Thus, although p10 is structurally an S100-like protein, it is probably not a calcium modulated protein.

The function of p10 is not known. The p10 protein binds to the amino terminal region of p36 producing a tetrameric species containing two p36 heavy chains and two p10 light chains, which has been called the calpactin I complex (Glenney et al. 1986). The p36 protein binds calcium and phospholipids, is a substrate for src kinase, and has phospholipase A2 inhibitory activity (Glenney 1986; Saris et al. 1986). Whether p10 regulates any of the known activities of p36 is not clear. Another calcium/phospholipid binding protein, the p35 substrate of the epidermal growth factor receptor kinase (Fava and Cohen 1984), does not have an associated p10 light chain. However, it is not

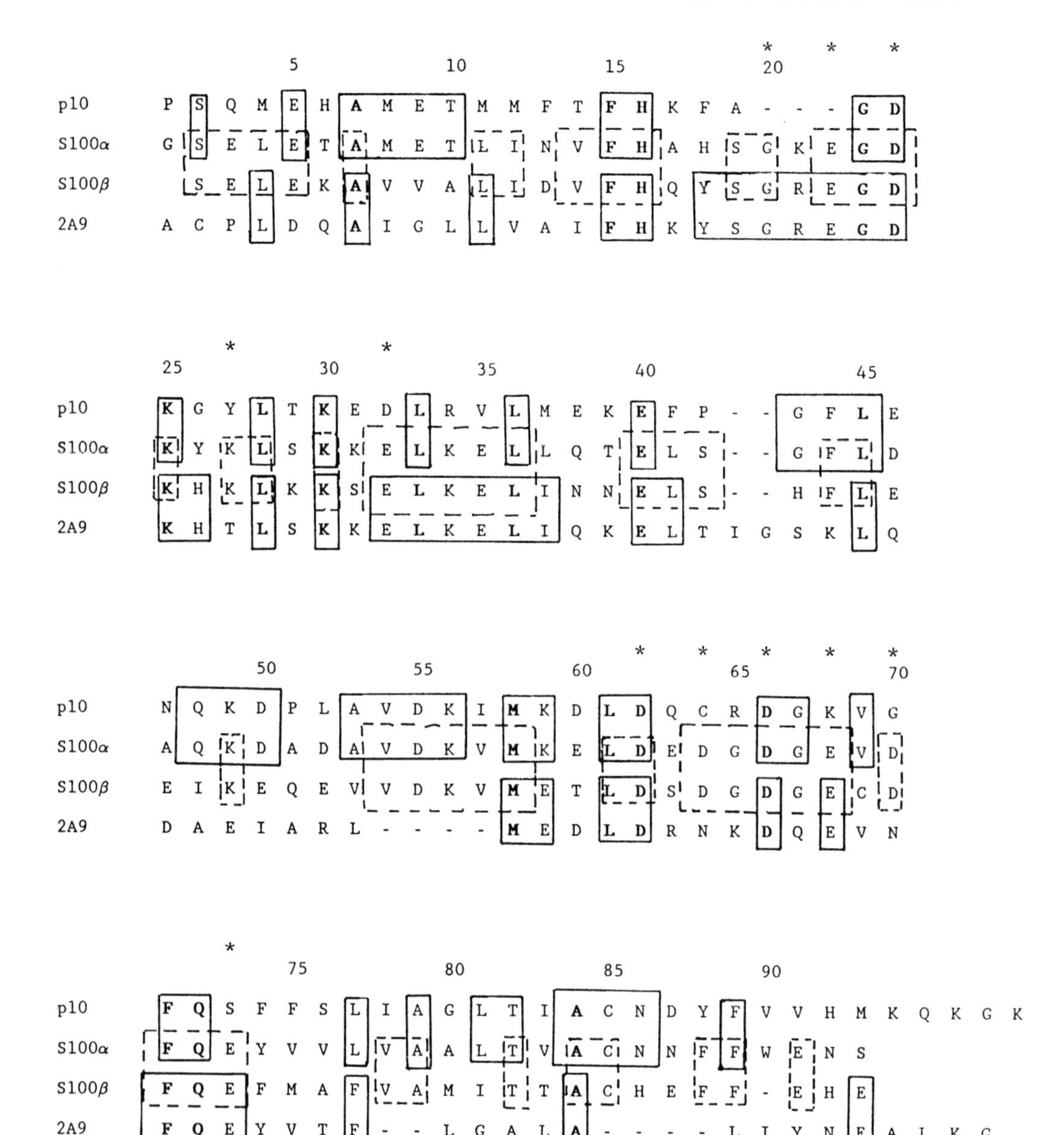

Fig. 1. Comparison of the amino acid sequences of S100α, S100β, p10, and the deduced amino acid sequence of 2A9. Amino acid sequences for bovine brain S100α and S100β (Isobe and Okuyama 1981), pig intestinal p10 (Gerke and Weber 1985), and the amino acid sequence deduced from the nucleotide sequence of 2A9 (Calabretta et al. 1986) are shown. The residue numbers are based on the amino acid sequence of S100α. Gaps or insertions were introduced for the purpose of alignment. *Asterisks* denote amino acid residues postulated to be involved in calcium binding based on homology with the vitamin D-dependent calcium binding protein (Szebenyi et al. 1981; Van Eldik et al. 1982). Residues that are identical between p10 and S100α, and between 2A9 and S100β are indicated by *solid-line boxes*. Residues that are identical between S100α and S100β are indicated by *dashed-line boxes*. Residues that are identical among all four sequences are indicated by *bold letters*

known whether any other calcium/phospholipid binding proteins that have been characterized (Saris et al. 1986) have an associated S100-like protein.

Finally, there has been a very recent report (Ziai et al. 1986) of a protein (PEP-19) isolated from rat cerebellar extracts which shows some sequence homology to S100α. PEP-19 appears to possess the sequence requirements for an EF-hand calcium binding structure and may represent a neuronal-specific calcium modulated protein. PEP-19 is developmentally regulated and increases at least tenfold during postpartum development of rat cerebellum. Further characterization of this protein is required before the significance of its limited sequence homology to S100 can be assessed.

The physiological significance of a family of S100-like proteins is not known at the present time. It is possible that this family of S100-like proteins have functions that are specific to a particular tissue, developmental or differentiation state, or phase of the cell cycle. It will be important to examine the functional similarities or dissimilarities among these structurally related proteins. For example, what are the common and unique molecular targets for S100 proteins, p10, and 2A9? Do p10 and 2A9 bind zinc? What is the distribution of these proteins among tissues and species, and what is the subcellular localization of each? Are there other unidentified members of this family of proteins? These are questions that must be addressed in future studies of S100 and related proteins.

5 Conclusions

Although S100 proteins were originally isolated over 20 years ago, little was known until the past few years about the molecular mechanisms of action of these calcium binding proteins. The early literature on S100 provides suggestive evidence about potential activities, but must be interpreted with caution in light of the more recent reports of multiple S100-like proteins. It is becoming clear that the well-characterized S100α and S100β polypeptides are only two members of a family of structurally similar S100-like proteins. It is possible that some of the activities attributed to S100 may actually reflect the ability of S100 to substitute for an endogenous regulator that is an S100-like protein. Future investigations are required to determine whether S100 proteins are multifunctional like calmodulin, or whether there are multiple S100-like proteins, each having specific molecular target(s). Characterization of S100 and S100-like proteins from various tissues and species, and isolation and identification of their molecular targets will be necessary first steps to understand S100 function. These studies will provide further insight into how the S100 family of proteins are involved in calcium dependent biological responses, and in the longer term, will allow an understanding of how multiple calcium modulated proteins in the same cell coordinate intracellular calcium signals.

References

Baudier J, Briving C, Deinum J, Haglid K, Sorskog L, Wallin M (1982) Effect of S-100 proteins and calmodulin on Ca^{2+}-induced disassembly of brain microtubule proteins *in vitro*. FEBS Lett 147: 165–167

Baudier J, Glasser N, Haglid K, Gerard D (1984) Purification, characterization and ion binding properties of human brain S100β protein. Biochim Biophys Acta 790:164–173

Baudier J, Labourdette G, Gerard D (1985) Rat brain S100b protein: purification, characterization, and ion binding properties. A comparison with bovine S100b protein. J Neurochem 44:76–84

Baudier J, Glasser N, Gerard D (1986) Ions binding to S100 proteins. I. Calcium- and zinc-binding properties of bovine brain S100 αα, S100a (αβ), and S100b (ββ) protein: Zn^{2+} regulates Ca^{2+} binding on S100b protein. J Biol Chem 261:8192–8203

Calabretta B, Battini R, Kaczmarek L, deRiel JK, Baserga R (1986) Molecular cloning of the cDNA for a growth factor-inducible gene with strong homology to S100, a calcium-binding protein. J Biol Chem 261:12628–12632

Cocchia D (1981) Immunocytochemical localization of S-100 protein in the brain of adult rat. An ultrastructural study. Cell Tissue Res 214:529–540

Cocchia D, Michetti F (1981) S-100 antigen in satellite cells of the adrenal medulla and the superior cervical ganglion of the rat. An immunochemical and immunocytochemical study. Cell Tissue Res 215:103–112

Cocchia D, Michetti F, Donato R (1981) Immunochemical and immunocytochemical localization of S100 antigen in normal human skin. Nature (Lond) 294:85–87

Cosgrove JW, Heikkila JJ, Marks A, Brown IR (1983) Synthesis of S100 protein on free and membrane-bound polysomes of the rabbit brain. J Neurochem 40:806–813

Deinum J, Baudier J, Briving C, Rosengren L, Wallin M, Gerard D, Haglid K (1983) The effect of S100α and S100β proteins and Zn^{2+} on the assembly of brain microtubule proteins *in vitro*. FEBS Lett 163:287–291

Donato R (1977) Solubilization and partial characterization of the S-100 protein binding activity of synaptosomal particulate fractions. J Neurochem 28:553–557

Donato R (1984) Mechanism of action of S-100 protein(s) on brain microtubule protein assembly. Biochem Biophys Res Commun 124:850–856

Donato R (1985) Calcium-sensitivity of brain microtubule proteins in the presence of S100 proteins. Cell Calcium 6:343–361

Donato R (1986) S-100 proteins. Cell Calcium 7:123–145

Donato R, Michetti F (1981) Specific binding sites for S-100 protein in isolated brain nuclei. J Neurochem 36:1698–1705

Donato R, Michetti F, Miani N (1975) Soluble and membrane-bound S-100 protein in cerebral cortex synaptosomes. Properties of the S-100 receptor. Brain Res 98:561–573

Donato R, Isobe T, Okuyama T (1985) S-100 proteins and microtubules: analysis of the effects of rat brain S100 (S100β) and ox brain $S100a_0$, S100α and S100β on microtubule assembly-disassembly. FEBS Lett 186:65–69

Donato R, Battaglia F, Cocchia D (1986a) Effects of S100 proteins on assembly of brain microtubule proteins: correlation between kinetic and ultrastructural data. J Neurochem 47:350–354

Donato R, Prestagiovanni B. Zelano G (1986b) Identity between cytoplasmic and membrane-bound S-100 proteins purified from bovine and rat brain. J Neurochem 46:1333–1337

Dunn R, Landry C, O'Hanlon D, Dunn J, Allore R, Brown I, Marks A (1987) Reduction in S100 protein β subunit mRNA in C6 rat glioma cells following treatment with anti-microtubular drugs. J Biol Chem 262:3562–3566

Endo T, Hidaka H (1983) Effect of S100 protein on microtubule assembly-disassembly. FEBS Lett 161:235–238

Fava RA, Cohen S (1984) Isolation of a calcium-dependent 35-kilodalton substrate for the epidermal growth factor receptor/kinase from A-431 cells. J Biol Chem 259:2636–2645

Fujii T, Gocho N, Akabane Y, Kondo Y, Suzuki T, Ohki K (1986) Effect of calcium ions on the interaction of S-100 protein with microtubule proteins. Chem Pharm Bull 34:2261–2264

Gerke V, Weber K (1985) The regulatory chain in the p36-kd substrate complex of viral tyrosine-specific protein kinases is related in sequence to the S100 protein of glial cells. EMBO J 4: 2917–2920

Ghandour MS, Langley OK, Labourdette G, Vincendon G, Gombos G (1981) Specific and artefactual localizations of S100 protein: an astrocyte marker in rat cerebellum. Dev Neurosci 4:66–78

Glenney J Jr (1986) Two related but distinct forms of the Mr 36,000 tyrosine kinase substrate (calpactin) that interact with phospholipid and actin in a Ca^{2+}-dependent manner. Proc Natl Acad Sci USA 83:4258–4262

Glenney J Jr, Tack B (1985) Amino-terminal sequence of p36 and associated p10: identification of the site of tyrosine phosphorylation and homology with S100. Proc Natl Acad Sci USA 82:7884–7888

Glenney J Jr, Boudreau M, Galyean R, Hunter T, Tack B (1986) Association of the S100-related calpactin I light chain with the NH_2-terminal tail of the 36-kDa heavy chain. J Biol Chem 261: 10485–10488

Gopalakrishna R, Barsky S, Anderson W (1985) Isolation of S100 binding proteins from brain by affinity chromatography. Biochem Biophys Res Commun 128:1118–1124

Haglid K, Stavrou D (1973) Water-soluble and pentanol-extractable proteins in human brain normal tissue and human brain tumors, with special reference to S100 protein. J Neurochem 20:1523–1532

Haglid K, Hansson H-A, Ronnback L (1977) S-100 in the central nervous system of rat, rabbit, and guinea pig during postnatal development. Brain Res 123:331–345

Hexham JM, Totty NF, Waterfield MD, Crumpton MJ (1986) Homology between the subunits of S100 and a 10kDa polypeptide associated with p36 of pig lymphocytes. Biochem Biophys Res Commun 134:248–254

Hidaka H, Endo T, Kawamoto S, Yamada E, Umekawa H, Tanabe K, Hara K (1983) Purification and characterization of adipose tissue S100b protein. J Biol Chem 258:2705–2709

Hyden H, McEwen B (1966) A glial protein specific for the nervous system. Proc Natl Acad Sci USA 55:354–358

Hyden H, Ronnback L (1975) Membrane-bound S100 protein on nerve cells and its distribution. Brain Res 100:615–628

Hyden H, Ronnback L (1978) The brain-specific S100 protein on neuronal cell membranes. J Neurobiol 9:489–492

Ishikawa H, Nogami H, Shirasawa N (1983) Novel clonal strains from adult rat anterior pituitary producing S-100 protein. Nature (Lond) 303:711–713

Isobe T, Okuyama T (1981) The amino acid sequence of the α subunit in bovine brain S100a protein. Eur J Biochem 116:79–86

Isobe T, Ishioka N, Okuyama T (1981) Structural relation of two S100 proteins in bovine brain: subunit composition of S100α protein. Eur J Biochem 115:469–474

Isobe T, Ishioka N, Masuda T, Takahashi Y, Ganno S, Okuyama T (1983) A rapid separation of S100 subunits by high performance liquid chromatography: the subunit compositions of S100 proteins. Biochem Intl 6:419–426

Isobe T, Takahashi K, Okuyama T (1984) S100a_0 ($\alpha\alpha$) protein is present in neurons of the central and peripheral nervous system. J Neurochem 43:1494–1496

Jensen R, Marshak D, Anderson C, Lukas T, Watterson D (1985) Characterization of human brain S100 protein fraction: amino acid sequence of S100β. J Neurochem 45:700–705

Kato K, Kimura S (1985) S100a_0 ($\alpha\alpha$) protein is mainly located in the heart and striated muscles. Biochim Biophys Acta 842:146–150

Kato K, Kimura S, Haimoto H, Suzuki F (1986) S100a_0 ($\alpha\alpha$) protein: distribution in muscle tissues of various animals and purification from human pectoral muscle. J Neurochem 46:1555–1560

Kligman D, Marshak D (1985) Purification and characterization of a neurite extension factor from bovine brain. Proc Natl Acad Sci USA 82:1736–1739

Kretsinger R (1980) Structure and evolution of calcium modulated proteins. CRC Crit Rev Biochem 8:119–174

Kuo WN, Dominguez JL, Shabazz KA, White DJ, Nicholson J, Puente K, Shells P, Redding C, Blake T, Sen S (1986a) Modulation of the catalytic subunit of cyclic AMP-dependent protein kinase by calmodulin, S-100 protein, parvalbumin and troponin. Cytobios 46:139–146

Kuo WN, Blake T, Cheema IR, Dominguez J, Nicholson J, Puenta K, Shells P, Lowery J (1986b) Regulatory effects of S-100 protein and parvalbumin on protein kinases and phosphoprotein phosphatases from brain and skeletal muscle. Mol Cell Biochem 71:19–24

Kuwano R, Usui H, Maeda T, Fukui T, Yamanari N, Ohtsuka E, Ikehara M, Takahashi Y (1984) Molecular cloning and the complete nucleotide sequence of cDNA to mRNA for S100 protein of rat brain. Nucl Acids Res 12:7455–7465

Kuwano R, Maeda T, Usui H, Araki K, Yamkuni T, Ohshima Y, Kurihara T, Takahashi Y (1986) Molecular cloning of cDNA of S100α subunit mRNA. FEBS Lett 202:97–101

Mahony J, Brown I (1980) Analysis of messenger RNA coding for S100 protein in the mammalian brain. J Neurochem 35:823–828

Marshak DR, Umekawa H, Watterson DM, Hidaka H (1985) Structural characterization of the calcium binding protein S100 from adipose tissue. Arch Biochem Biophys 240:777–780

Masure H, Head J, Tice H (1984) Studies on the α-subunit of bovine brain S-100 protein. Biochem J 218:691–696

Miani N, Michetti F, DeRenzis G, Caniglia A (1973) Effect of a brain-specific protein (S100 protein) on the nucleolar RNA-polymerase activity in isolated brain nuclei. Experientia (Basel) 29:1499–1501

Michetti F, Donato R (1981) Subnuclear distribution of the S100 protein specific binding sites in rat brain. J Neurochem 36:1706–1711

Michetti F, Miani N, DeRenzis G, Caniglia A, Correr S (1974) Nuclear localization of S100 protein. J Neurochem 22:239–244

Michetti F, DeRenzis G, Donato R, Miani N (1976) Brain-specific effect of the S100 protein on the RNA-polymerase I activity in isolated nuclei. Brain Res 105:372–375

Michetti F, Dell'Anna E, Tiberio G, Cocchia D (1983) Immunochemical and immunocytochemical study of S100 protein in rat adipocytes. Brain Res 262:352–356

Michetti F, Lauriola L, Rende M, Stolfi V, Battaglia F, Cocchia D (1985) S100 protein in the testis. An immunochemical and immunohistochemical study. Cell Tissue Res 240:137–142

Molin S-O, Rosengren L, Haglid K, Baudier J, Hamberger A (1984) Differential localization of "brain-specific" S-100 and its subunits in rat salivary glands. J Histochem Cytochem 32:805–814

Molin S, Rosengren L, Baudier J, Hamberger A, Haglid K (1985) S-100 alpha-like immunoreactivity in tubules of rat kidney. A clue to the function of a "brain-specific" protein. J Histochem Cytochem 33:367–374

Moller M, Ingild A, Bock E (1978) Immunohistochemical demonstration of S-100 protein and GFA protein in interstitial cells of rat pineal gland. Brain Res 140:1–13

Moore BW (1965) A soluble protein characteristic of the nervous system. Biochem Biophys Res Commun 19:739–744

Moore BW (1982) Chemistry and biology of the S100 protein. Scand J Immunol 15:53–74

Nagata Y, Ando M, Miwa M, Kato K (1984) Effects of various forms of stimulation on the content of enolase isozymes and S-100 protein in superior cervical sympathetic ganglia excised from rats. J Neurochem 43:1205–1212

Nakajima T, Yamaguchi H, Takahashi K (1980) S100 protein in folliculostellate cells of the rat pituitary anterior lobe. Brain Res 191:523–531

Patel J, Marangos PJ, Heydorn WE, Chang G, Verma A, Jacobwitz D (1983) S-100-mediated inhibition of brain protein phosphorylation. J Neurochem 41:1040–1045

Perumal AS (1976) Nuclear protein kinase of brain: effect of S100 protein. Res Comm Chem Pathol Pharm 13:489–496

Perumal A, Rapport M (1978) *In vitro* phosphorylation of S100 protein by brain nuclear protein kinases. Life Sci 22:803–808

Qi DF, Kuo JF (1984) S100 modulates Ca^{2+}-independent phosphorylation of an endogenous protein (M_r = 19K) in brain. J Neurochem 43:256–260

Qi DF, Turner RS, Kuo JF (1984) S100 and other acidic proteins promote Ca^{2+}-independent phosphorylation of protamine catalyzed by a new protein kinase from brain. J Neurochem 42: 458–465

Rusca G, Calissano P, Alema S (1972) Identification of a membrane bound fraction of the S100 protein. Brain Res 49:223–227

Saris CJM, Tack BF, Kristensen T, Glenney JR Jr, Hunter T (1986) The cDNA sequence for the protein-tyrosine kinase substrate p36 (calpactin I heavy chain) reveals a multidomain protein with internal repeats. Cell 46:201–212

Shashoua VE, Hesse GW, Moore BW (1984) Proteins of the brain extracellular fluid: evidence for release of S100 protein. J Neurochem 42:1536–1541

Stefansson K, Wollman RL, Moore BW (1982) Distribution of S100 protein outside the central nervous system. Brain Res 234:309–317

Suzuki F, Kato K (1985) Inhibition of adipose S100 protein release by insulin. Biochim Biophys Acta 845:311–316

Suzuki F, Nakajima T, Kato K (1982) Peripheral distribution of nervous system-specific S100 protein in rat. J Biochem 92:835–838

Suzuki F, Kato K, Nakajima T (1983) Enhancement of adipose S100 protein release by catecholamines. J Biochem 94:1707–1710

Suzuki F, Kato K, Nakajima T (1984) Regulation of nervous system-specific S-100 protein and enolase levels in adipose tissue by catecholamines. J Neurochem 42:130–134

Sviridov SM, Korochkin LI, Ivanov VN, Maletskaya EI, Bakhtina TK (1972) Immunohistochemical studies of S-100 protein during postnatal ontogenesis of the brain of two strains of rats. J Neurochem 19:713–718

Szebenyi DME, Obendorf SK, Moffat K (1981) Structure of vitamin D-dependent calcium-binding protein from bovine intestine. Nature (Lond) 294:327–332

Tabuchi K, Kirsch WM (1975) Immunocytochemical localization of S-100 protein in neurons and glia of hamster cerebellum. Brain Res 92:175–180

Takahashi K, Isobe T, Ohtsuki Y, Akagi T, Sonobe H, Okuyama T (1984) Immunohistochemical study on the distribution of α and β subunits of S-100 protein in human neoplasm and normal tissues. Virchows Archiv B 45:385–396

Van Eldik LJ, Zimmer DB (1987) Secretion of S-100 from rat C6 glioma cells. Brain Res 436: 367–370

Van Eldik LJ, Zendegui JG, Marshak DR, Watterson DM (1982) Calcium-binding proteins and the molecular basis of calcium action. Intl Rev Cytol 77:1–61

Ziai R, Pan Y-CE, Hulmes JD, Sangameswaran L, Morgan JI (1986) Isolation, sequence, and developmental profile of a brain-specific polypeptide, PEP-19. Proc Natl Acad Sci USA 83:8420–8423

Zimmer DB, Van Eldik LJ (1986) Identification of a molecular target for the calcium modulated protein S100. Fructose-1,6-bisphosphate aldolase. J Biol Chem 261:11424–11428

Zimmer DB, Van Eldik LJ (1987) Tissue distribution of rat S100α and S100β and S100-binding proteins. Am J Physiol: Cell Physiol 252:C285–C289

Zomzely-Neurath C, Walker W (1980) Nervous system-specific proteins: 14-3-2 protein, neuron-specific enolase, and S100 protein. In: Bradshaw RA, Schneider DM (eds) Proteins of the nervous system. 2nd edn. Raven Press, New York, pp 1–57

Note Added in Proof:

Additional proteins and cDNA structures with similarity to S100 have been described recently: 1) a growth-related cDNA that codes for an S100-like protein [Jackson-Grusby LL, Swiergiel J, Linzer DIH (1987) A growth-related mRNA in cultured mouse cells encodes a placental calcium binding protein. Nucl Acids Res 15:6677–6690]; 2) cDNAs that are induced upon treatment of PC12 cells with nerve growth factor and that code for S100-like proteins [Masiakowski P, Shooter EM (1988) Nerve growth factor induces the genes for two proteins related to a family of calcium-binding proteins in PC12 cells. Proc Natl Acad Sci USA, in press]; and 3) the cystic fibrosis serum antigen [Dorin JR, Novak M, Hill RE, Brock DJH, Secher DS, Hcyningen VV (1987) A clue to the basic defect in cystic fibrosis from cloning the CF antigen gene. Nature 326:614–617].

Comparative Studies on Oncomodulin

J. P. MacManus, L. M. Brewer and M. F. Gillen[1]

1 Introduction

Oncomodulin is the name given to a calcium-binding protein of Mr 11700 kDa first noticed in 1979 in extracts of rat hepatomas (MacManus 1979), and later shown to be related, by comparison of amino acid sequences, to the β-parvalbumin subfamily of the large superfamily of calmodulin proteins (MacManus and Whitfield 1983; MacManus et al. 1983a,b). Many calcium-binding proteins have been described in tumour tissues, some, such as calmodulin or S100, increasing above the normal untransformed level (Watterson et al. 1976; MacManus et al. 1981; Isobe et al. 1984; Van Eldik et al. 1984), and others, such as protein A (Pfyffer et al. 1984) and oncomodulin, appearing de novo following neoplastic transformation. The primary structures of all these proteins are highly conserved. Calmodulin occurs in all eukaryotes with minimal sequence differences between those of protozoans, plants, invertebrates and vertebrates (including human) (Van Eldik et al. 1982; Roberts et al. 1985; Klee and Vanaman 1982). The parvalbumins are also conserved but restricted to vertebrates (Heizmann 1984).

The genes for some of these proteins have been described, and exhibit some conservation of structure (Nabeshima et al. 1984; Hardin et al. 1985; Epstein et al. 1986). This suggests that these genes arose from a common precursor, supporting similar ideas conceived from protein sequence comparison (Kretsinger 1980; Demaille 1982). The nucleic acid probes are highly specific; for example, the rat parvalbumin probe will not react with the chicken genome (Berchtold and Means 1985).

2 Conservation of the Oncomodulin Gene

The rat oncomodulin gene is thought to be single copy because the cDNA probing of genomic digests, obtained with different restriction enzymes, has always yielded only one band of labeled DNA from rat liver (Gillen et al. 1986, 1987). This present availability of nucleic acid probes for oncomodulin has allowed the screening of genomic blots obtained from various species.

---

1 Cellular Oncology Group, Division of Biological Sciences, National Research Council of Canada, Ottawa K1A OR6, Canada. NRC publication no. 28380

Ch. Gerday, R. Gilles, L. Bolis (Eds.)
Calcium and Calcium Binding Proteins

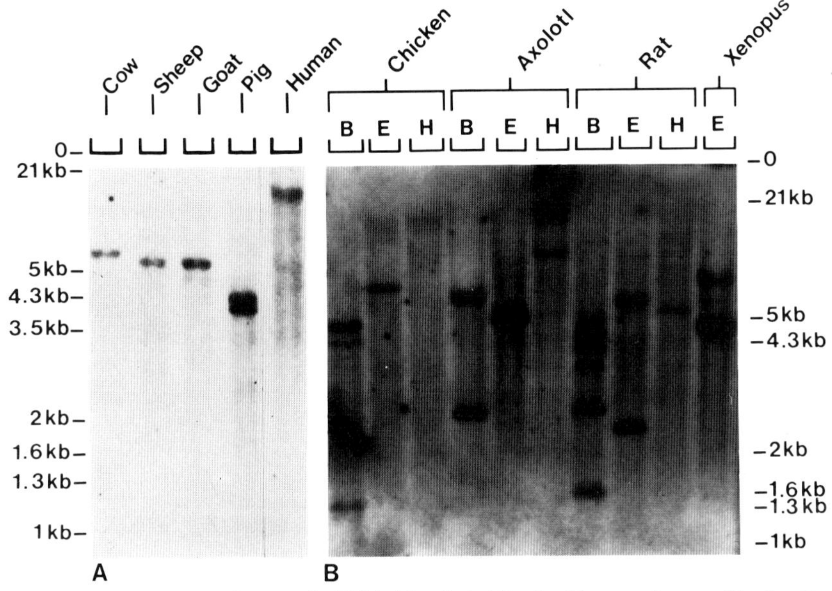

Fig. 1A,B. Restriction digests of genomic DNA blot hybridized with a probe specific for the oncomodulin gene. **A** *Eco* RI digests of mammalian DNAs. Lanes are as indicated. **B** Digests of avian and amphibian DNAs with rat DNA as a control. This blot is hybridized under less stringent conditions than the blot in **A**. Lanes are as indicated. *B Bam* HI; *E Eco* RI; *H Hind* III

The presence of the gene is obvious (Fig. 1) in other mammals (cow, goat, pig, sheep and human). More interesting, and in contrast to parvalbumin, the oncomodulin probe also seems to hybridise to DNA isolated from Aves (*Gallus domesticus:* chicken), and Amphibia (*Amblystoma mexicanum:* axolotl; *Xenopus laevis:* toad). This suggests that the oncomodulin gene is highly conserved, maybe more so than parvalbumin.

3 Messenger RNA for Oncomodulin

The messenger RNA originally reverse transcribed and cloned to obtain complementary nucleic acid probes was isolated from rat Morris hepatoma 5123tc (Gillen et al. 1986, 1987). When Northern blots of RNA isolated from such tumors are probed, a single band of 0.7 kb can be visualised (Fig. 2). This single species is similar to the single species of 0.6 kb found for the mRNA of the 9 kDa calbindin D (Perret et al. 1985), but contrasts with the two mRNAs found for parvalbumin of 0.7 and 1.1 kb (Berchtold and Means 1985) and for calmodulin of 1.6 and 1.9 kb (Lagace et al. 1983).

The size of the oncomodulin mRNA visualised in other tumors, e.g. the virally transformed rat cells, SSV-NRK and B77-ASV-NRK, or the human choriocarcinoma BeWo is also 0.7 kb in all cases (Fig. 2). As will be described below (Sect. 5) the only normal tissues known to express oncomodulin are extraembryonic, for example, placenta. RNA blots from placenta or rat or human also display only one mRNA of size 0.7 kb (Fig. 2).

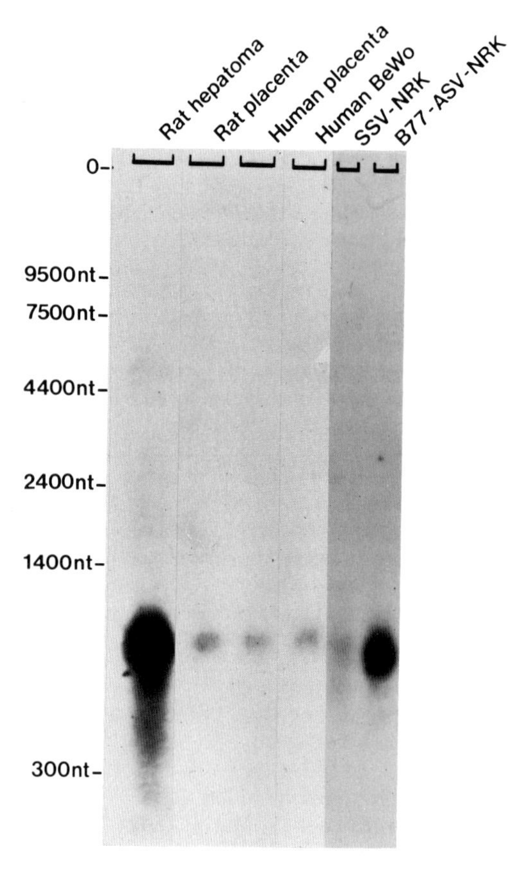

Fig. 2. RNA blot hybridization with a probe specific for oncomodulin mRNA. Lanes are as indicated. The rat placental RNA is from a sample at 20 days of gestation while the human placental RNA is from a 14-week placenta

Besides comparisons of size, the nucleotide sequence of the oncomodulin mRNA can be compared to those for the other related calcium-binding proteins. Since the amino acid sequences are related (MacManus and Whitfield 1983; MacManus et al. 1983a,b, 1987), it is to be expected that some homology would occur within the coding regions of such mRNAs. This is the case for oncomodulin mRNA when compared to calmodulin, S100, or calbindin D, but especially for the closely related parvalbumin where 59% homology for mRNA is reflected in 55% identity in amino acid sequence.

Of more interest is examination of the non-coding regions. When the 5′-non-coding region plus the initiation codon region of the parvalbumin sequence was compared to the calmodulin sequence in the first domain, evidence was found of remnants of a fourth N-terminal domain in the parvalbumin nucleic acid, with 28 bases of 38 identical, i.e. 74% homologous (Epstein et al. 1986). When such a comparison was made between oncomodulin and calmodulin, only 42% homology was found (Gillen et al. 1987). This leads to the provocative conclusion that oncomodulin is much less like calmodulin than is parvalbumin, and that oncomodulin diverged from calmodulin before parvalbumin.

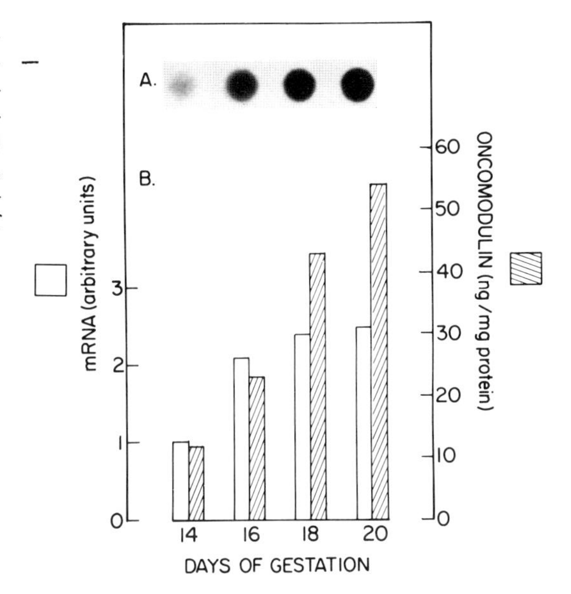

Fig. 3A,B. Levels of oncomodulin message and product in rat placental samples at various times following fertilization. **A** Rat placental RNA dot blotted and hybridized to a radiolabeled oncomodulin mRNA-specific probe. *Dots* correspond to the days of gestation in **B. B** Measured amounts of oncomodulin and its mRNA

The complementary nucleic acid probes have also given an opportunity to quantitate the relative concentration of oncomodulin mRNA. There is no evidence to date of hormonal control of oncomodulin expression. However, the level of this protein does change in placenta during rat development (Brewer and MacManus 1985). The amounts of oncomodulin mRNA and protein were measured in the same samples of rat placenta at various times following fertilisation, and the levels of message and product were shown to increase concomitantly (Fig. 3). One can conclude that transcriptional control dictates oncomodulin concentration.

4 Conservation of Oncomodulin Structure

The suggested conservation of the nucleic acids for oncomodulin described up to this point in various species is reflected by the identity in all 108 residues of the amino acid sequences of both rat and human oncomodulin (MacManus et al. 1983, 1987). Besides this demonstration of conservation, which suggests selective pressure to maintain structure/function, this identity also has a practical outcome. The identity of both rat and human oncomodulin validates the use of antisera to rat oncomodulin for quantitation (MacManus et al. 1984a) and immunolocalisation (MacManus et al. 1985) in human tissue. Evidence of similar tryptic peptide mapping profiles of mouse oncomodulin (MacManus et al. 1982), and HPLC retention time of rabbit oncomodulin (MacManus, unpublished) to those of rat protein, also suggest closely related proteins in these species.

5 Normal Tissues Where Oncomodulin is Found

For several years following the finding of oncomodulin in Morris hepatomas, both calcium-binding assays and quantitative immunological methods, such as radioimmunoassay, failed to detect oncomodulin in any normal adult rat or human tissue (MacManus 1979, 1981a; MacManus et al. 1982, 1984a). This remains the case today, even with the additional screening methods of immunohistochemical staining (Brewer et al. 1984; MacManus and Brewer 1987), immunoradiometric assay (IRMA) or ELISA (MacManus, unpublished; Brousseau et al. 1985).

There are many examples of proteins being oncodevelopmental, e.g. hCG, placental alkaline phosphatase, α-fetoprotein and carcinoembryonic antigen (Fishman 1983). This prompted the screening of developing tissues for oncomodulin, and in 1985 it was shown that indeed oncomodulin is also oncodevelopmental, occurring extraembryonically in both rat and human placenta (Brewer and MacManus 1985; MacManus et al. 1985; MacManus and Brewer 1987). It was also found in rat extraembryonic membranes, i.e. both the parietal and visceral yolk sacs, and the amnion (Brewer and MacManus 1985). All available methods of immunological, or nucleic acid screening to date have not detected oncomodulin expression in any fetal tissue. The proteins from rat and human placenta have been shown to be identical to authentic oncomodulin from hepatoma (MacManus et al. 1985).

The location of oncomodulin in human placenta is very striking, being located in the cytotrophoblastic layer of the villi (MacManus et al. 1985) and the cytotrophoblastic shell (Brewer and MacManus 1987). These recent studies also suggest the presence of oncomodulin in intermediate trophoblast (Brewer and MacManus 1987). The fully differentiated, hormone-secreting, syncytiotrophoblast appears to contain no oncomodulin (MacManus et al. 1985; Brewer and MacManus 1987). Oncomodulin has been quantitated by IRMA during the first and early second trimester of gestation, but the level is very variable, ranging from 2–100 ng/g. By term, the cytotrophoblastic cell population is greatly reduced (Boyd and Hamilton 1970), which would explain the decreased number of cells immunohistochemically stained for oncomodulin (Brewer and MacManus 1987), and the decreased ability to quantitate oncomodulin by 38–40 weeks.

The restriction of oncomodulin to the cytotrophoblastic cell of the human placenta is of considerable theoretical and heuristic interest. This cell type is where several cellular oncogenes (c-*cis*, c-*myc*, c-*fos*) are expressed normally (Pfeifer-Ohlsson et al. 1984, 1985; Goustin et al. 1985; Muller 1986). Transformation by retroviruses containing these and other oncogenes leads to oncomodulin expression (Durkin et al. 1983; Fig. 2). In addition, the human trophoblast is considered to share some tumor-like properties with malignant neoplasms, such as invasiveness, extravasation and lodgement in lung (metastatis), and evasion of immune surveillance (Aitken 1979; Ilgren 1983). Thus the protein oncomodulin is normally expressed in an extraembryonic cell type that may bear some resemblance to a neoplastic cell, and is never expressed in normal fetal or adult tissue, but reappears in many cancers.

The location of oncomodulin in the pure trophoblast of the rat ectoplacental cone as it burrows into the uterine endometrium after 9 days of gestation has been demonstrated (MacManus and Brewer 1987). The rat placenta is not villous, and oncomodulin is not as restricted in its location (as in the human) when the ectoplacental cone differentiates into the complex spongiotrophoblast and labyrinth of the mature placenta by day 14 of gestation. Oncomodulin has been immunohistochemically located and quantitated (Brewer and MacManus 1985) in both labyrinth (the area of fetomaternal exchange) and the spongiotrophoblast (closest to the uterus). It is in the greatest amounts in this outer area of the placenta, suggesting it has nothing to do with calcium transport from mother to fetus.

Immunohistochemical localisation of oncomodulin in trophoblast of species other than human and rat can also be demonstrated. Mouse placenta was stained using rabbit anti-rat hepatoma oncomodulin and an avidin-biotin-peroxidase system (Vectastain ABC), and the resultant positive staining was similar to that observed previously in rat placenta (Brewer and MacManus 1985).

Other species which stained palely using the rabbit anti-oncomodulin were found to stain well with an affinity purified anti-rat hepatoma oncomodulin isolated from yolks of eggs produced by immunized chickens. Using this antibody and a modified avidin-biotin-peroxidase system, positive staining was observed in placentas of sheep, bat, and cat. Due to the small sample sizes obtained (sheep and cat, 2 each; bat placenta, only 1), these results are considered preliminary.

The sheep (*Ovis aries*) placenta is typical of ruminants in that it is composed of cotyledons or tufts of chorionic villi grown into an endometrial elevation or cup of the maternal tissue (Amoroso 1952; Ramsey 1975). In the sheep cotyledon (Fig. 4a,b) positive staining for oncomodulin appeared on the trophoblast of the fetal chorio-allantoic villi, while the interdigitating maternally derived septa lacked positive staining. Characteristic of the sheep trophoblast are large binucleate trophoblast cells, which appeared slightly positive. Control sections stained with pre-immune serum IgG lacked any positive staining.

The bat (*Myotis lucifugus*) has a placenta (at term) more similar to that of the rat and mouse than the sheep, but differs in that the cytotrophoblastic villi (fetal) are interpenetrated by syncytiotrophoblastic (also fetal) elements (Wimsatt 1962). The cytotrophoblast showed positive staining for oncomodulin (Fig. 4c), while only pale staining, slightly greater than background, was observed on the syncytiotrophoblast.

The cat (*Felis domestica*) placenta at term is lacking in cytotrophoblast (Amoroso 1952), similar to the situation in the human. In the cat, the darkest staining elements were large (Fig. 4d) cells termed decidual giant cells, believed to be of maternal origin. These cells are not always readily distinguishable from isolated cytotrophoblast cells, however, and such persisting cells may also be positive, as was cytotrophoblast in the junctional zone at the base of the placenta.

This restriction of oncomodulin to the extraembryonic structures of placenta, yolk sacs and amnion in mammals begs the large question as to possible sites of expression in other vertebrates. As seen in Section 1, some preliminary evidence exists for the presence of the oncomodulin gene in the genome of Aves and Amphibia (Fig. 1). These obviously have no placenta, but Aves is at least amniotic. The amnion would be the prime candidate for screening in this class. However, the Amphibia have no comparable extraembryonic structures during the larval stages as possible sites of expression.

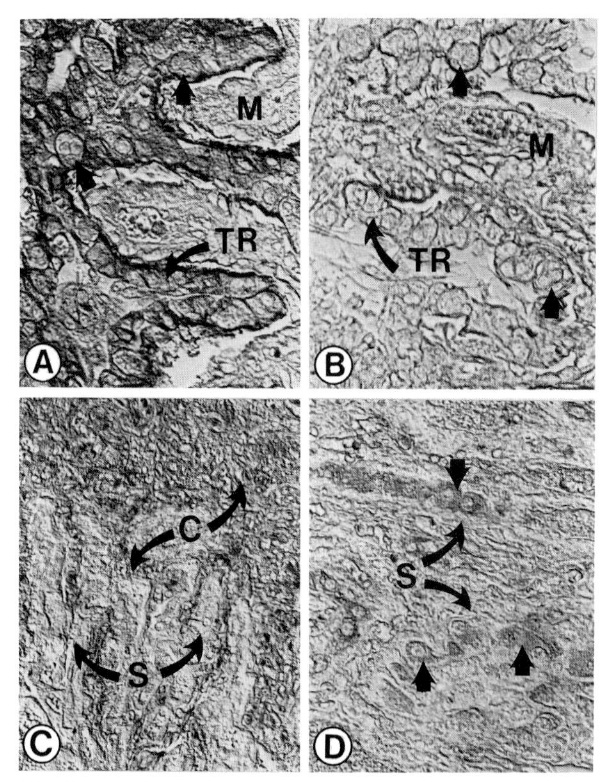

Fig. 4A–D. Immunohistochemical localization of oncomodulin on placental sections (Carnoy's fixed, paraffin embedded) using an affinity purified, chicken egg yolk anti-rat oncomodulin, in an avidin-biotin-peroxidase complex (Vectastain ABC) system. **A** Sheep placenta (125 days of gestation) shows positive staining in the trophoblast (*TR*) of fetal villi, while the interdigitating maternal columns (*M*) lack oncomodulin positive staining. Binucleate trophoblast cells (*arrows*) appear slightly positive. **B** A control section of sheep placenta stained using pre-immune chicken serum IgG shows background staining. **C** Bat placenta (term) shows positive staining for oncomodulin in cytotrophoblast (*C*), while the syncytiotrophoblast (*S*) is very slightly stained. **D** Cat placenta (term) shows positive staining in larbe cells (*arrows*), which appear to be decidual giant cells and possibly include some cytotrophoblast cells. The syncytiotrophoblast (*S*) is only slightly stained. Nomarski optics. *A, B, D* X425, *C* X265

6 Oncomodulin in Tumors

Besides the placenta (and extraembryonic membranes) where oncomodulin is found (less than 1 μg/g for human; 10 μg/g for rat), no other normal tissue seems to contain oncomodulin. These negative findings include rapidly proliferating normal tissues such as fetal liver and regenerating adult liver. Human benign proliferative disorders such as psoriatic skin are also free of oncomodulin. However, the majority of tumors studied show the reappearance of this calcium-binding protein, which initially prompted a portion of the term "oncomodulin".

The chemically induced rat Morris hepatomas were the first tissues to yield the Mr 11700 kDa protein, and the hepatoma 5123tc remains the richest known source of oncomodulin with 100–200 μg/g of tissue for ourselves (MacManus and Whitfield 1983; MacManus and Brewer 1987) and others (Klee and Heppel 1984; Henzl et al. 1986). The presence of oncomodulin has also been demonstrated in hamster tissue transformed by the chemicals aristolochic acid or MNNG (N-methyl-N′-nitro-N-nitroso-guanidine) (Sommer et al. 1985). Virally transformed rat or mouse cells also contain oncomodulin, both in cells transformed by RNA viruses (avian, murine, or simian sarcoma viruses, or murine osteosarcoma virus) or DNA viruses (SV40, polyoma) (MacManus et al. 1982; Durkin et al. 1983; MacManus et al. 1987; Fig. 2). Human tumors also contain oncomodulin, although at apparently lower levels (2 μg/g or less) than rodent tumors (MacManus et al. 1984a; Brousseau et al. 1985; MacManus et al. 1987).

Immunohistochemical localisation has proved of immense value in demonstrating oncomodulin in tissue sections, for example in micrometastases of hepatoma in rat lung (Brewer et al. 1984), and in clinical material (Brousseau et al. 1985). The presence of this oncodevelopmental protein in tumors of a wide variety of tissues in a number of species, induced by a wide variety of carcinogens, suggested the use of oncomodulin detection as a tumor diagnostic (MacManus et al. 1984b). The demonstration of oncomodulin in the blood or tumor-bearing rats supports such a practical outcome of this research (MacManus 1984).

7 Oncomodulin: a Missing Link?

The studies outlined herein using both nucleic acid or immunological probes, in conjunction with protein analyses of various sorts, demonstrate that oncomodulin is a highly conserved calcium-binding protein in mammals, and perhaps in lower vertebrates. It is an oncodevelopmental protein in several species, appearing during prenatal development in extraembryonic structures such as placenta, never in the fetus or normal adult, but reappearing in a majority of cancers in the adult.

The function of oncomodulin during development remains unknown. Test tube experiments have demonstrated the unusual ability of this two-site calcium-binding parvalbumin-like protein to modulate a calmodulin-dependent enzyme, and suggested the second portion of the term "oncomodulin". Thus it was shown that oncomodulin stimulated both rat and bovine heart cyclic-AMP phosphodiesterase in a calcium-dependent manner (MacManus 1981; Mutus et al. 1985). Other calmodulin-dependent enzymes, calcineurin, phosphorylase kinase, myosin light chain kinase, were not stimulated. Thus oncomodulin has a very limited ability to mimic calmodulin. Modifications of calmodulin by protein engineering leads to calmodulin mutants that can no longer modulate most calmodulin-dependent enzymes. However, phosphodiesterase remains stimulable (Roberts et al. 1985, 1986; Putkey et al. 1986). Thus the structural requirements of a phosphodiesterase modulator are not very stringent, and somehow oncomodulin meets these minimal specifications, but not the entire set required to totally mimic calmodulin. Moreover, the skepticism toward the ability of a two-site calcium-binding protein having modulator ability is further allayed by the recent elegant demonstration of the ability of another two-site protein, S-100, to stimulate aldolase (Zimmer and Van Eldik 1986). The unusual modulator behaviour for the parvalbumin-like oncomodulin perhaps arises from other unusual properties in metal binding. Parvalbumins are known to bind two calcium or magnesium ions in the CD and EF sites. Oncomodulin appears to have a CD-site preference for calcium shown by UV, circular dichroic or fluorescence spectroscopy (MacManus et al. 1984c), or at least non-parvalbumin properties by fluorescence Tb or Eu substitution (Henzl et al. 1986), or NMR monitored Lu exchange (Williams et al. 1987). A calcium-specific conformational change is at the heart of calmodulin action. So the oncomodulin primary structure may be β-parvalbumin-like, but somehow the tertiary structure is not exactly so, but exhibits properties of metal binding and calcium-dependent protein-protein interaction which are reminiscent of the archetypic modulator calmodulin. Whether oncomodulin turns out to be a "mini cardiac troponin C" (MacManus et al. 1984c), a "missing link" between parvalbumin and the well-established larger modulator proteins remains to be seen. The least to be hoped is for other workers to be provoked by these ideas, and the function of oncomodulin in placenta to be unearthed.

Acknowledgements. Our thanks are due to R. Ball and L. Joly for excellent technical assistance, D.J. Gillan for the illustration, K. Hamelin for word processing, and J.F. Whitfield for advice and encouragement. Sheep placenta was obtained courtesy of Drs. M. Scott-Dalpé and R. Peterson, Faculty of Medicine, University of Ottawa; bat placenta courtesy of V. Wai-Ting and Dr. B. Fenton, Biology Department, Carleton University; and cat placenta courtesy of S. Prudhomme, Akhetaten Cattery, Gloucester.

References

Aitken RJ, Beaconsfield P, Ginsberg J (1979) Origin and formation of the placenta. In: Beaconsfield P, Villee C (eds) Placenta, a neglected experimental animal. Pergamon Press, Oxford, pp 152–163

Amoroso EC (1952) Placentation. In: Parkes AS (ed) Marshall's physiology of reproduction, 3rd edn, Vol II. Longmans Green, London, pp 127–311

Berchtold MW, Means AR (1985) The Ca^{2+}-binding protein parvalbumin: molecular cloning and developmental regulation of mRNA abundance. Proc Natl Acad Sci USA 82:1414–1418

Boyd JD, Hamilton WJ (1970) The human placenta. Heffer, Cambridge

Brewer LM, MacManus JP (1985) Localisation and synthesis of the tumor protein oncomodulin in extraembryonic tissues of the fetal rat. Dev Biol 112:49–58

Brewer LM, MacManus JP (1987) Detection of oncomodulin, an oncodevelopmental protein in human placenta and choriocarcinoma cell lines. Placenta 8:351–363

Brewer LM, Durkin JP, MacManus JP (1984) Immunocytochemical detection of oncomodulin in tumor tissue. J Histochem Cytochem 32:1009–1016

Brousseau P, Sallafranque P, Losos G (1985) Diagnostic application of oncomodulin in human cancer. CFBS Proc 28:309

Demaille JG (1982) Calmodulin and calcium-binding proteins: evolutionary diversification of structure and function. In: Cheung WY (ed) Calcium and cell function. Vol II. Academic Press, New York, pp 111–144

Durkin JP, Brewer LM, MacManus JP (1983) Occurrence of the tumor-specific calcium-binding protein, oncomodulin, in virally transformed normal rat kidney cells. Cancer Res 43:5390–5394

Epstein P, Means AR, Berchtold MW (1986) Isolation of a rat parvalbumin gene and full length cDNA. J Biol Chem 261:5886–5891
Fishman WH (1983) Oncodevelopmental marker. Biologic, diagnostic and monitoring aspects. Academic Press, New York
Gillen MF, Rutledge R, Narang SA, Seligy VL, Whitfield JF, MacManus JP (1986) The isolation of the nucleic acids coding for the oncodevelopmental calcium-binding protein oncomodulin. CFBS Proc 29:193
Gillen MF, Banville D, Rutledge RG, Narang S, Seligy VL, Whitfield JF, MacManus JP (1987) A complete complementary DNA for the oncodevelopmental calcium-binding protein, oncomodulin. J Biol Chem 262:5308–5312
Goustin AS, Betsholtz C, Pfeifer-Ohlsson S, Persson H, Rydnert J, Bywater M, Holmgren G, Heildin CH, Westermark B, Ohlsson R (1985) Coexpression of the *sis* and *myc* proto-oncogenes in developing human placenta suggests autocrine control of trophoblast growth. Cell 41: 301–312
Hardin SH, Carpenter CD, Hardin PE, Bruskin AM, Klein WH (1985) Structure of the Spec 1 gene encoding a major calcium-binding protein in the embryonic ectoderm of the sea urchin, *Strongylocentrotus purpuratus*. J Mol Biol 186:243–255
Heizmann CW (1984) Parvalbumin, an intracellular calcium-binding protein; distribution, properties and possible roles in mammalian cells. Experientia 40:910–921
Henzl MT, Hapak RC, Birnbaum ER (1986) Lanthanide-binding properties of rat oncomodulin. Biochim Biophys Acta 872:16–23
Ilgren EB (1983) Control of trophoblastic growth. Placenta 4:307–328
Isobe T, Ichimori K, Nakajima T, Okuyama T (1984) The α-subunit of S100 protein is present in tumor cells of human malignant melanoma, but not in schwannoma. Brain Res 294:381–387
Klee CB, Heppel LA (1984) The effect of oncomodulin on cAMP phosphodiesterase activity. Biochem Biophys Res Commun 125:420–424
Klee CB, Vanaman TC (1982) Calmodulin. Adv Prot Chem 35:213–320
Kretsinger RH (1980) Structure and evolution of calcium-modulated proteins. CRC Crit Rev Biochemistry 8:119–174
Lagace L, Chandra T, Woo SLC, Means AR (1983) Identification of multiple species of calmodulin messenger RNA using a full length complementary DNA. J Biol Chem 258:1684–1688
MacManus JP (1979) The occurrence of a low-molecular weight calcium-binding protein in neoplastic liver. Cancer Res 39:3000–3005
MacManus JP (1981a) Development and use of a quantitative immunoassay for the calcium-binding protein (molecular weight 11500) of Morris hepatoma 5123. Cancer Res 41:974–979
MacManus JP (1981b) The stimulation of cyclic nucleotide phosphodiesterase by a Mr 11500 calcium-binding protein from hepatoma. FEBS Lett 126;245–249
MacManus JP (1984) Plasma oncomodulin is proportional to tumor burden in rats bearing Morris hepatomas 5123D, 5123tc, 7288, 7777. Tumor Biol 5:189–197
MacManus JP, Brewer LM (1987) Isolation, localisation and properties of the oncodevelopmental calcium-binding protein oncomodulin. Methods Enzymol 139:156–168
MacManus JP, Whitfield JF (1983) Oncomodulin: a calcium-binding protein from hepatoma. In: Cheung WY (ed) Calcium and cell function. Vol IV. Academic Press, New York, pp 411–440
MacManus JP, Braceland BM, Rixon RH, Whitfield JF, Morris HP (1981) An increase in calmodulin during growth of normal and cancerous liver *in vivo*. FEBS Lett 133:99–103
MacManus JP, Whitfield JF, Boynton AL, Durkin JP, Swierenga SHH (1982) Oncomodulin: a widely distributed tumour-specific calcium-binding protein. Oncodev Biol Med 3:79–90
MacManus JP, Watson DC, Yaguchi M (1983a) The complete amino acid sequence of oncomodulin – a parvalbumin-like calcium-binding protein from Morris hepatoma 5123tc. Eur J Biochem 136:9–17
MacManus JP, Watson DC, Yaguchi M (1983b) A new member of the troponin C superfamily – comparison of the primary structures of rat oncomodulin and rat parvalbumin. Biosci Rep 3: 1071–1075

MacManus JP, Whitfield JF, Stewart DJ (1984a) The presence in human tumours of a Mr 11700 calcium-binding protein similar to rodent oncomodulin. Cancer Lett 21:309–315

MacManus JP, Brewer LM, Whitfield JF (1984b) The potential of the calcium-binding protein, oncomodulin, as a tumor marker. Protides Biol Fluids 31:399–403

MacManus JP, Szabo AG, Williams RE (1984c) Conformational changes induced by binding of bivalent cations to oncomodulin, a parvalbumin-like tumour protein. Biochem J 220:261–268

MacManus JP, Brewer LM, Whitfield JF (1985) The widely distributed tumour-protein, oncomodulin, is a normal constituent of human and rodent placentas. Cancer Lett 27:145–151

MacManus JP, Brewer LM, Gillen MF (1987) Oncomodulin – an oncodevelopmental calcium-binding protein. In: Anghileri LJ (ed) Role of calcium in biological systems. CRC Press, Boca Raton, pp 1–19

Muller R (1986) Proto-oncogenes and differentiation. Trends Biochem Sci 11:129–132

Mutus B, Karuppiah N, Sharma RK, MacManus JP (1985) The differential stimulation of brain and heart cyclic AMP phosphodiesterase by oncomodulin. Biochem Biophys Res Commun 131: 500–506

Nabeshima Y, Kuriyama Y, Muramatus M, Ogata K (1984) Alternative transcription and two modes of splicing result in two myosin light chains from one gene. Nature (Lond) 308:333–338

Perret C, Desplan C, Thomasset M (1985) Cholecalcin (a 9 kDa cholcalciferol-induced calcium-binding protein) messenger RNA. Eur J Biochem 150:211–217

Pfeiffer-Ohlsson S, Goustin AS, Rydnert J, Wahlstrom T, Bjersing L, Stehelin D, Ohlsson R (1984) Spatial and temporal pattern of cellular *myc* oncogene expression in developing human placenta: implications for embryonic cell proliferation. Cell 38:585–596

Pfeiffer-Ohlsson S, Rydnert J, Goustin AS, Larsson E, Betsholtz C, Ohlsson R (1985) Cell-type specific pattern of *myc* protooncogene expression in developing human embryos. Proc Natl Acad Sci USA 82:5050–5054

Pfyffer GE, Haemmerli G, Heizmann CW (1984) Calcium-binding proteins in human carcinoma cell lines. Proc Nat Acad Sci USA 81:6632–6636

Putkey JA, Draetta GF, Slaughter GR, Klee CG, Cohen P, Stull JT, Means AR (1986) Genetically engineered calmodulins differentially activate target enzymes. J Biol Chem 261:9896–9903

Ramsay EM (1975) The placenta of laboratory animals and man. Holt Rinehart & Winston, New York

Roberts DM, Crea R, Malecha M, Alvarado-Urbina G, Chiarello RH, Watterson DM (1985) Chemical synthesis and expression of a calmodulin gene designed for site-specific mutagenesis. Biochemistry 24:5090–5098

Roberts DM, Rowe PM, Siegal FL, Lukas TJ, Watterson DM (1986) Trimethyllysine and protein function. Effect of methylation and mutagenesis of lysine 115 of calmodulin of NAD kinase activation. J Biol Chem 261:1491–1494

Sommer EW, Baer M, Heizmann CW, Maier P, Zbinden G (1985) Tumor-associated Ca^{2+}-binding proteins in chemically transformed rat fibroblasts. J Cell Biol 101:475a

Van Eldik LJ, Zendegui JG, Marshak DR, Watterson DM (1982) Calcium-binding proteins and the molecular basis of calcium action. Internatl Rev Cytol 77:1–61

Van Eldik LJ, Ehrenfied B, Jensen RA (1984) Production and characterisation of monoclonal antibodies with sepcificity for the S100B polypeptide of brain S100 fractions. Proc Natl Acad Sci USA 81:6034–6038

Watterson DM, Van Eldik LJ, Smith R, Vanaman TC (1976) Calcium dependent regulatory protein of cyclic nucleotide metabolism in normal and transformed chicken embryo fibroblasts. Proc Natl Acad Sci USA 73:2711–2715

Williams TC, Corson DC, Sykes BD, MacManus JP (1987) Oncomodulin: ^1H NMR and optical stopped-flow spectroscopic studies of its solution conformation and metal-binding properties. J Biol Chem 262:6248–6256

Wimsatt WA (1962) Some aspects of the comparative anatomy of the mammalian placenta. Am J Ob Gyn 84:1568–1594

Zimmer DB, Van Eldik LJ (1986) Identification of a molecular target for the calcium modulated protein S100. Fructose 16 bisphosphate aldolase. J Biol Chem 261:11424–11428

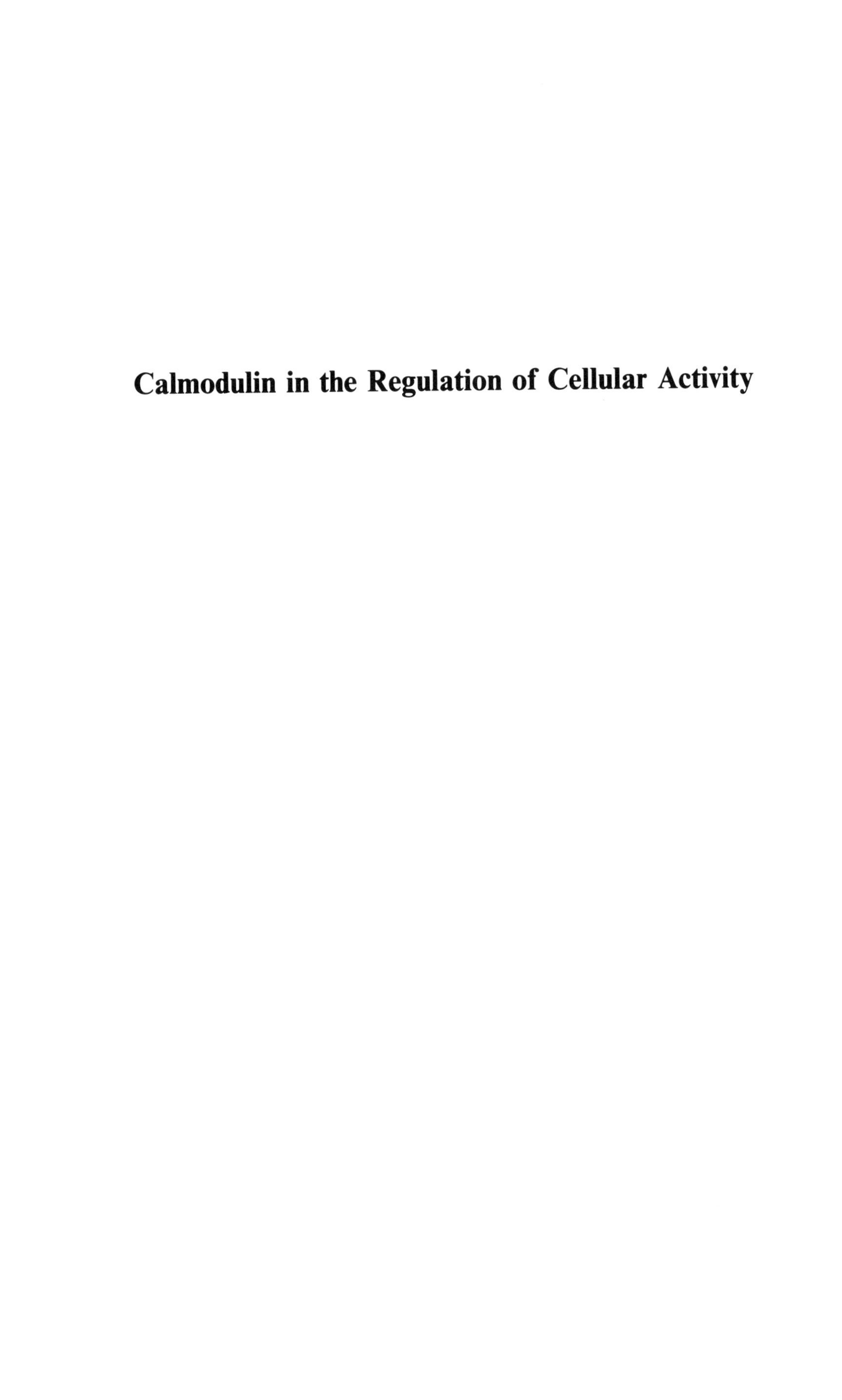

Calmodulin in the Regulation of Cellular Activity

Cation Binding to Calmodulin and Relation to Function

J. A. Cox[1], M. Comte[1], A. Mamar-Bachi[1], M. Milos[2] and J.-J. Schaer[2]

1 Introduction

Calmodulin (CaM) is intimately involved in the stimulus-response coupling in eukaryotic cells since it is the prime sensor of transient increases of the free intracellular [Ca^{2+}] and conveys the signal to multiple target enzymes, especially protein kinases and phosphatases. For a general survey of the properties of CaM, the reader is referred to Cheung (1980), Klee and Vanaman (1982), Manalan and Klee (1984) and Wang et al. (1985) (see also Cheung this Vol.). This paper deals more specifically with the mode of action of CaM. Although in the last decade our understanding of its action has considerably increased, some enigmas remain and, unfortunately, some basic and long-lasting controversies on its mode of action have not been solved yet. In this paper, two of these are examined: (a) cation binding to CaM, and (b) the thermodynamics of the interaction of CaM with its targets.

2 Cation Binding to CaM

2.1 Structural Observations on the Cation Binding Domains of CaM

The elucidation of the primary structure of CaM by Watterson et al. (1980) cleary showed that CaM contains 4 Ca^{2+}-binding domains with an α-helix-loop-α-helix repeat composed of 11,12 and 11 amino acid residues respectively (Table 1). For convenience, these Ca^{2+}-binding domains are numbered I, II, III and IV (starting with the N-terminal side). Nothing in this amino acid sequence permits speculation as to the affinity or selectivity of the different domains. Indeed, sequence homology is very high in each of the Ca^{2+}-binding loops, especially in the positions 1, 3, 5, and 12 (for numerotation, see Table 1) involved in the coordination and those (4 and 6) involved in proper bending of the loop. Close examination of the amino acid sequence reveals two intriguing particularities (Table 1): (a) 3 out of 4 Ca^{2+}-binding loops contain one "acid pair", a definition coined by R. Reid, Ottawa (pers. commun.) to indicate that opposing

1 Department of Biochemistry and
2 Physical Chemistry, University of Geneva, 30, Quai Ernest Ansermet, 1211 Geneva 4, Switzerland

Ch. Gerday, R. Gilles, L. Bolis (Eds.)
Calcium and Calcium Binding Proteins

Table 1. The four metal-binding domains of CaM

		E-helix — Loop — F-helix
		-9 -6 X Y Z −Y −X −Z
		* * 1 2 3 4 5 6 7 8 9 10 11 12
Domain I	(9–42)	I–A–E–F–K–E–A–F–S–L–F–D–K–D–G–D–G–T–I–T–T–K–E–L–G–T–V–M–R–S–L–G–Q–N
		* *
Domain II	(45–78)	E–A–E–L–Q–D–M–I–N–E–V–D–A–D–G–N–G–T–I–D–F–P–E–F–L–T–M–M–A–R–K–M–K–D
		* *
Domain III	(82–105)	E–E–E–I–R–E–A–F–R–V–F–D–K–D–G–N–G–Y–I–S–A–A–E–L–R–H–V–M–T–N–L–G–E–K′
		* *
Domain IV	(118–149)	D–E–E–V–D–E–M–I–R–E–A–D–I–D–G–D–G–Q–V–N–Y–E–E–F–V–Q–M–M–T–A–K

The numbers in parentheses correspond to the positions in the complete sequence of CaM; K′: trimethyllysine.

Ca^{2+}-ligands, i.e. pair-wise X and –X, or Y and –Y, or Z and –Z, are both free carboxylic side chains. Indeed, domain I contains one Asp-Glu pair in the Z direction, domain II one Asp-Asp pair in X, domain III none and domain IV one Asp-Glu pair in Z. (b) All four domains contain in the N-terminal α-helix flanking the Ca^{2+}-binding loop a very well conserved Glu residue in position –9 and a Glu or Asp residue in position –6 (for numeration, see Table 1). Since these positions are situated on the same side of the helix, one predicts a high negative charge density at these particular loci of CaM and possibly weak cation binding activity (see also Sect. 2.5).

The elucidation of the crystal structure of CaM to 2 Å resolution (Babu et al. 1985) confirmed that the 4 Ca^{2+}-binding domains have the right geometry for complexing Ca^{2+}, i.e. they possess the EF-hand structure (Kretsinger 1973). The molecule looks like a dumbbell with the globular (ca. 20 Å in diameter) N- and C-halves linked by a 20 Å-long α-helical stretch. Interestingly in the crystal the two halves of CaM are remarkably similar; the corresponding α-carbon positions can be superimposed with an average difference of ca. 1 Å (Babu et al. 1985). Within each half the two Ca^{2+}-binding domains are in close contact: the distance between the two bound Ca^{2+} ions is 11.3 Å and Ca^{2+}-binding loops are oriented as antiparallel β-sheet strands with one or two hydrogen bonds extending from one Ca^{2+}-binding loop to the other. Thus, from the structural point of view, allostery in metal ion binding might occur within each half of the molecule. A study on the domain organization in CaM (Tsalkova and Privalov 1985) leads to the conclusion that there is no significant allosteric interaction between the two halves of the molecule.

2.2 Ca^{2+}-Binding to CaM

Direct Binding Studies. In 1975, a thorough equilibrium dialysis study of Ca-binding to skeletal muscle troponin C and the effects of Mg^{2+} on these binding properties established unambiguously the affinities of the different Ca-binding sites of this regulatory protein and introduced for the first time the notion of Ca^{2+}-specific and Ca^{2+}-Mg^{2+} mixed sites (Potter and Gergely 1975). In the line of this classification, the metal-binding properties, measured directly by equilibrium dialysis, of parvalbumins (Moeschler et al. 1980) and invertebrate soluble sarcoplasmic Ca^{2+}-binding proteins (Wnuk et al. 1979; Cox and Stein 1981) could easily be interpreted. Furthermore, a wealth of kinetic, conformational and NMR studies confirmed the direct binding data (Wnuk et al. 1982). Unfortunately this was not the case with CaM. Eight different groups mentioned by Burger et al. (1984) (Ogawa and Tanokura 1984; Iida and Potter 1986), who performed direct binding studies in buffers of sufficient ionic strength to eliminate electrostatic effects, experienced that the shape of the binding isotherm is close to the ideal Langmuir isotherm. For instance, in Fig. 1, a Ca^{2+}-saturation curve is depicted, which we have tried to fit to two models: either four independent Ca^{2+}-binding sites of equal affinity, or four interacting Ca-binding sites (Burger et al. 1984). The first model is described by the Langmuir equation:

$$\nu = \frac{4\,K'\,[Ca^{2+}]}{1 + K'\,[Ca^{2+}]}\,, \tag{1}$$

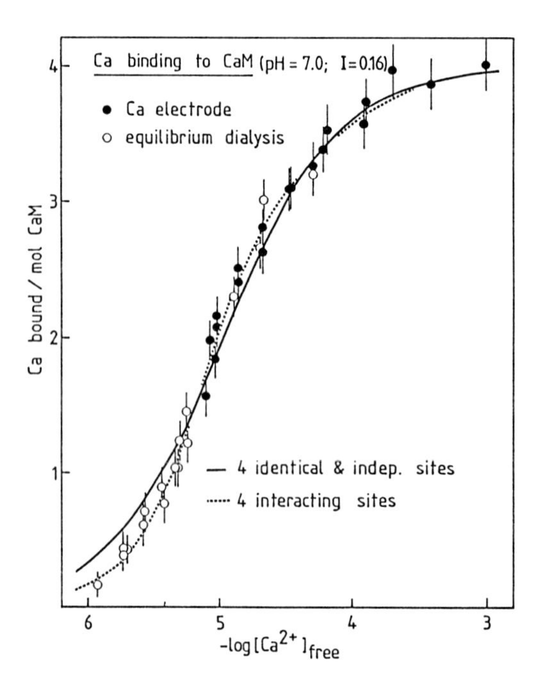

Fig. 1. Ca^{2+}-binding to CaM at pH 7.0 and ionic strength equal to 150 mM. *Solid* and *dotted lines* represent the theoretical curves obtained by iterative fitting according to the Langmuir and Adair equations, respectively. (Burger et al. 1984)

where ν is the amount of bound Ca^{2+} per mole of protein and K′ is the intrinsic binding constant of the four sites. The second model of different or interacting sites is described by the Adair equation:

$$\nu = \frac{K_1[Ca^{2+}] + 2K_1K_2[Ca^{2+}]^2 + 3K_1K_1K_2K_3[Ca^{2+}]^3 + 4K_1K_2K_3K_4[Ca^{2+}]^4}{1 + K_1[Ca^{2+}] + K_1K_2[Ca^{2+}]^2 + K_1K_1K_2K_3[Ca^{2+}]^3 + K_1K_2K_3K_4[Ca^{2+}]^4}, \quad (2)$$

where K_1, K_2 etc. are the stoichiometric (or macroscopic) binding constants of the first, second, etc., site. The experimental accuracy did not allow us to exclude one of these models. It should be noted that the above mentioned studies yielded accurate determination of the stoichiometric affinity constants (K), but cannot yield much information on the site binding constants (k). Nevertheless, if no allosteric interaction occurs between the four binding sites, the following relationship exists between the two sets of constants (Klotz 1985):

$$\begin{aligned} K_1 &= k_I + k_{II} + k_{III} + k_{IV} \\ K_1K_2 &= k_Ik_{II} + k_Ik_{III} + k_Ik_{IV} + k_{II}k_{III} + k_{II}k_{IV} + k_{III}k_{IV} \\ K_1K_2K_3 &= k_Ik_{II}k_{III} + k_Ik_{II}k_{IV} + k_{II}k_{III}k_{IV} + k_Ik_{III}k_V \\ K_1K_2K_3K_4 &= k_Ik_{II}k_{III}k_{IV}. \end{aligned} \quad (3)$$

Hence, particular models of site-binding constants which do not adequately describe the experimental binding isotherm can be rejected.

Kinetic Studies. From studies using $^{43}Ca^{2+}$-NMR (Andersson et al. 1982; Teleman et al. 1986) and stopped-flow fast kinetics (Martin et al. 1985; Suko et al. 1985) it appears that CaM has two pairs of sites, with very different Ca^{2+} off-rates. In the presence of 0.1 M KCl, one pair has a K_{off} of 300 to 500/s and the other a k_{off} of 10 to 40/s. With a diffusion controlled onrate constant of 10^8/M/s (Eigen and Hammes 1963), these off-rate constants point to the existence of two low- (fast exchanging) and two high- (slowly exchaning) affinity pairs of sites with a 10 to 20-fold difference in the binding constants. ^{43}Ca NMR and fast kinetics studies on the trypsin fragments of CaM enabled these authors to propose domains III and IV as the high affinity Ca-binding sites in intact CaM. The model based on these findings predicts that upon raising the free Ca^{2+}-concentration, the ion first binds to the C-terminal half of CaM, which happens to contain the 2 Tyr residues, and later to domains I and II. The model is consistent with studies on the conformational changes of the Tyr residues (see below), but not with the direct binding studies mentioned above.

The Model of Paired Cooperative Sites. Recently, Wang (1985) proposed an elegant theoretical explanation of the discrepancies between the above two lines of studies. He simulated Ca-binding curves on the basis of a model assuming two pairs of binding sites of respectively high and low affinity. With properly chosen values of the binding constants for the two pairs, and assuming relatively strong positive cooperativity between the two sites in each pair, an overall binding isotherm with the appearance of a non-cooperative binding curve of four equivalent sites can be generated. Given the importance of this idea, we have checked this model on the basis of the experimental binding isotherm of Fig. 1. The theoretical alternative of Wang consisted of a tenfold difference of affinity between the N- and C-terminal sites, together with a 20-fold cooperative increase of the affinity of the remaining site after the first Ca^{2+} has bound to a particular pair. In a cooperative system, the site-binding constants have not a fixed value, and the Wang model states that binding of the first Ca^{2+} to the N-terminal half is governed by an affinity constant, $K_N = k_I = k_{II}$ and binding of the second by K'_N, with $K'_N = 20\ K_N$. The same statement can be made for the C-terminal sites with the understanding the $K_C = 10\ K_N$. In this model, the relationship between the site-binding constants and stoichiometric constants is given by the following set of equations:

$$\begin{aligned} K_1 &= 2K_N + 2K_C \\ K_1K_2 &= K_NK_N' + 4K_NK_C + K_CK_C' \\ K_1K_2K_3 &= 2K_NK_N'K_C + 4K_NK_CK_C' \\ K_1K_2K_3K_4 &= K_NK_N'K_CK_C'. \end{aligned} \tag{4}$$

Table 2 shows the most plausible values for the 4 independent parameters K_N. K'_N, K_C and K'_C and for the corresponding stoichiometric Ca^{2+}-binding constants. Figure 2 shows that a moderately good fitting is indeed obtained between the model of Wang and the experimental data. Figure 2 also shows that simulations with a 15-fold (instead of tenfold) difference in affinity between the N- and the C-terminal pairs of sites lead already to an important discrepancy between the model and the experimental isotherm.

As analyzed in Table 2, the model of Wang predicts an off-rate constant of ca. 800/s for dissociation of the first Ca^{2+} from the N-terminal half of CaM and of ca. 80/s for

Table 2. Plausible binding and rate constants of Ca^{2+} interaction with CaM according to the model of Wang

Microscopic binding constants[a] (M^{-1})		Off-rate constants[b] (s^{-1})	Stoichiometric binding constants (M^{-1})	
k_N =	6.15×10^3	ca. 16,000	$K_1 = 1.35 \times 10^5$ [c]	(1.16×10^5) [d]
k'_N =	1.23×10^5	800	$K_2 = 5.76 \times 10^5$	(2.65×10^5)
k_C =	6.15×10^4	1,600	$K_3 = 1.31 \times 10^4$	(8.33×10^4)
k'_C =	1.23×10^6	80	$K_4 = 5.62 \times 10^4$	(1.90×10^4)

[a] Based on the following constraints: $k'_N/k_N = k'_C/k_C = 20$; $k_C/k_N = 10$.
[b] Assuming an on-rate constant for Ca^{2+} binding of 10^8 M^{-1} s^{-1} (Eigen and Hammes 1963).
[c] Calculated from the microscopic constants using Eq. (4).
[d] Values of Burger et al. (1984) (model 2).

dissociation of the first Ca^{2+} from the C-terminal half. Due to the positive cooperative interaction within pairs, the dissociation rate of the second Ca^{2+} is much faster than that of the first, which is thus rate-limiting. Interestingly, the experimentally determined rate constant for the fast exchanging N-terminal pair, as determined by $^{43}Ca^{2+}$-NMR (Teleman et al. 1986), ^{3}H-NMR (Ikura 1986) and stopped-flow (Martin et al. 1985) is in the same range as the predicted one, i.e. ca. 500/s, thus validating the model of Wang for these sites. However, the experimental k_{off} for the high affinity C-terminal pair is between 3 and 10/s (Malencik et al. 1981; Martin et al. 1985; Suko et al. 1985; Teleman et al. 1986; Ikura 1986), i.e. much lower than the one predicted by the model of Wang (1985).

For the purpose of the further development in this chapter, especially for the analyses of the conformational or microcalorimetric changes, it was of interest to calculate the appearance of the different $CaM.Ca_n$ species as a function of free $[Ca^{2+}]$ or, in stoichiometric titrations of metal-free CaM with Ca^{2+}, as function of the ratio of

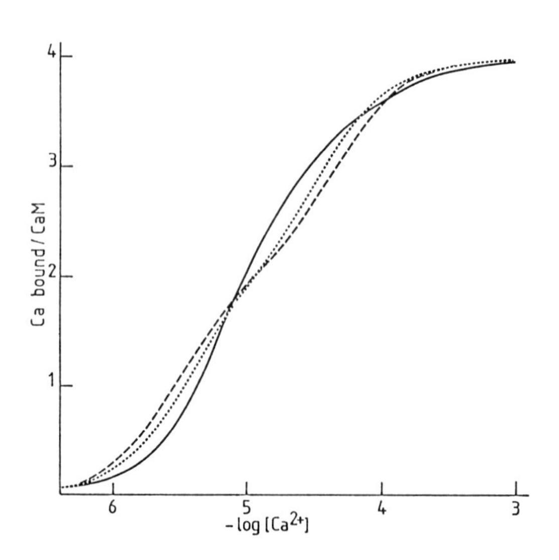

Fig. 2. Ca^{2+}-binding to CaM. *Dotted line* coresponds to the corresponding curve in Fig. 1 (Adair fitting). *Solid line* is a simulated one using the stoichiometric constants of Table 2, assuming the model of paired cooperative sites with $K'_N = 20\,K_N$; $K_C = 10\,K_N$. *Dashed line* represents another simulation of Wang's model with $K'_N = 20\,K_N$; $K_C = 15\,K_N$

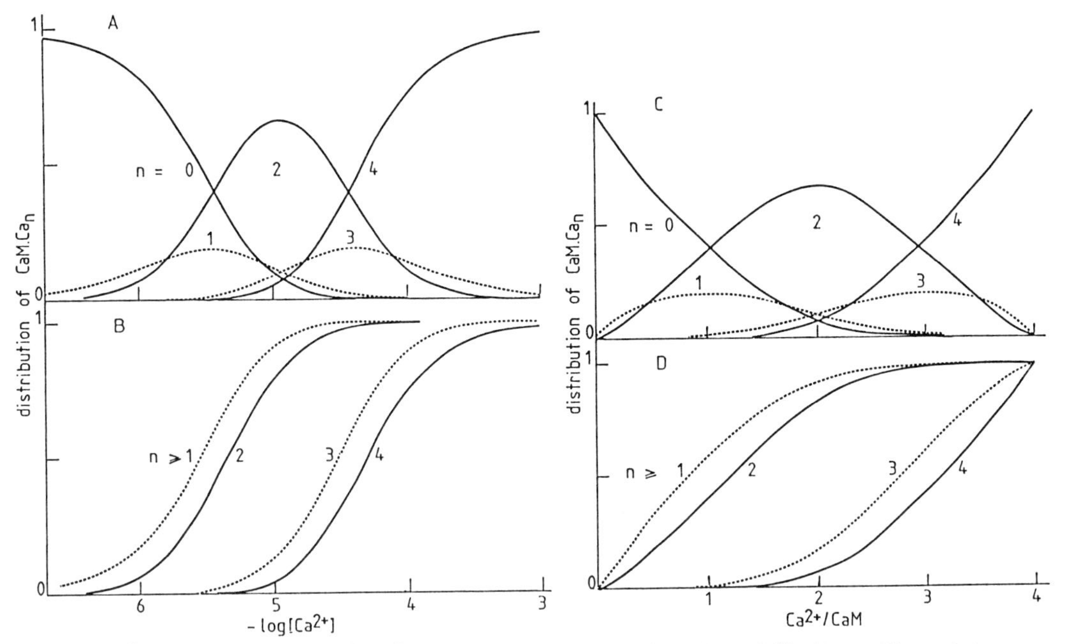

Fig. 3. Distribution of the $CaM.Ca_n$ species as a function of the free $[Ca^{2+}]$ (**A** and **B**) or of the mean saturation (**C** and **D**) using the constants of Table 2 for the model of paired cooperative sites (model of Wang). **A** and **C** are the curves for the individual species, **B** and **D** those for CaM bearing at least n Ca^{2+}

added Ca^{2+} per CaM. Figure 3 is generated as previously described (Cox et al. 1981) using the constants of Table 2. Interestingly, in Wang's model the species $CaM.Ca_1$ and $CaM.Ca_3$ represent at the upmost 20% of the total CaM population at any concentration of free or added Ca^{2+}. In contrast $CaM.Ca_2$ is very abundant (67% of the total population) at mean half maximal saturation of CaM.

Microcalorimetric Studies. In the light of the controversy between the direct binding and the kinetic studies, it was of interest to look to more fundamental parameters of the binding process, namely the enthalpy and entropy changes. Enthalpy titration studies were carried out by Tanokura and Yamada (1984) and by Milos et al. (1986). The former authors observed two categories of enthalpy changes: 1.6 Ca^{2+} binds with ΔH^o = 3.8 kJ/site and 1.4 Ca^{2+} bind with ΔA^o = 7.4 kJ/site. However, at slightly different pH, Milos et al. (1986) observed an identical enthalpy change for each of the four Ca^{2+}-binding domains with ΔH^o = 4.9 kJ/site (after correction for proton release, see Sect. 2.4). Figure 4 shows the linearity of the enthalpy titration at *high* concentration of CaM (where one may assume that all added Ca^{2+} is bound to the protein), at *intermediate* and at *low* concentrations of CaM, where the smoothening of the curve at the equivalence point yields a good estimate of the affinity of the binding sites (Milos et al. 1986). The ΔH values are the same for each binding step (□) and the four sites have the same affinity of 10^5 M^{-1} for Ca^{2+} (● and ×). Figure 4 further shows that the theoretical curve for the enthalpy change, calculated on the assumption of the

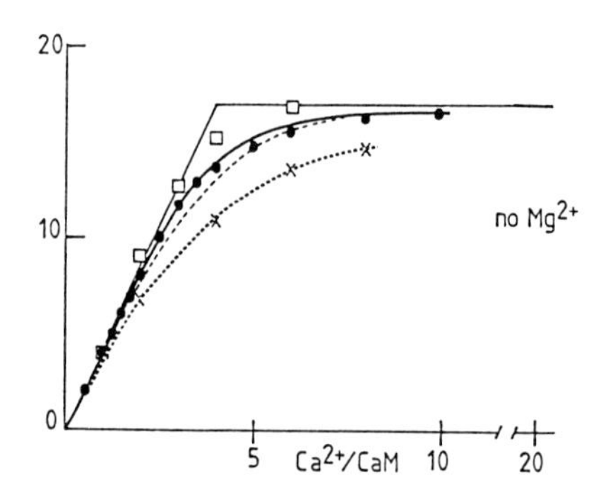

Fig. 4. Enthalpy titrations of Ca^{2+}-binding to CaM at different protein concentrations: (□) 200 mM; (•) 50 mM; (X) 16 mM. *Solid* and *dotted lines* are calculated assuming four identical affinity constants equal to 10^5 M; *dashed line* is calculated with the constants for the paired cooperative model at 50 μM CaM. (After Milos et al. 1986)

model of Wang with the numerical values of Table 2 (thin dashed line), yields a poor fit with the experimental data.

Ca^{2+}-Induced Conformational Studies. The interpretation of the Ca^{2+}-induced conformational changes has been influenced by the model of Ca^{2+}-binding favored by the authors. In stoichiometric titrations, the changes in the environment of the Tyr_{99} and Tyr_{138} are complete when two Ca^{2+} ions are bound (Klee and Vanaman 1982; Krebs 1981). This would favor the model of two high affinity sites in the Ca-terminal half where these aromatic groups are located. In more refined analyses (Crouch and Klee 1980; Burger et al. 1984), it appeared that this conformational change is concomitant with the formation of the $CaM.Ca_2$ species and it is therefore assumed to correspond to the formation of this species and not that of $CaM.Ca_1$. According to Wang's model, this would mean that the whole C-terminal half of CaM has to be saturated with Ca^{2+} before the Tyr conformational change occurs; binding of Ca^{2+} to either domain III or domain IV alone does not perturb the Tyr residues.

Two other conformational changes are very helpful for the understanding of the molecular model of Ca^{2+}-binding to CaM: (a) the change in the α-helical content occurs in two steps (Burger et al. 1984): halfmaximal increase of the far UV ellipticity occurs upon binding of the first Ca^{2+}, whereas the maximal value is reached upon binding of the third Ca^{2+}; (b) the increase in volume of CaM upon binding of Ca^{2+} is not monotonic but suggests a symmetry of pairs with the following incremental volume changes (in ml per mole of CaM):

$$CaM \xrightarrow[37]{} CaM.Ca_1 \xrightarrow[51]{} CaM.Ca_2 \xrightarrow[33]{} CaM.Ca_3 \xrightarrow[49]{} CaM.Ca_4$$

(Kupke and Dorrier 1986). The authors interpreted the data as follows: upon binding of each Ca^{2+} a mean volume increase of ca. 35 ml/mol of CaM is due to dehydration of the Ca^{2+} ion and of the charged groups in the binding loop; the additional volume change in step 2 (14 ml/mol) and in step 4 (16 ml/mol) occurs as a result of conformational changes in CaM. The data of Burger et al. (1984) and of Kupke and Dorrier (1986) thus support the model of Wang rather than the model of four independent sites. However, with its strong positive cooperativity, the orthodox Wang model predicts that upon addition of 1 Ca^{2+} per CaM the whole population of protein would adopt the distribution 0.4 CaM + 0.2 $CaM.Ca_1$ + 0.4 $CaM.Ca_2$ rather than 1.0 $CaM.Ca_1$

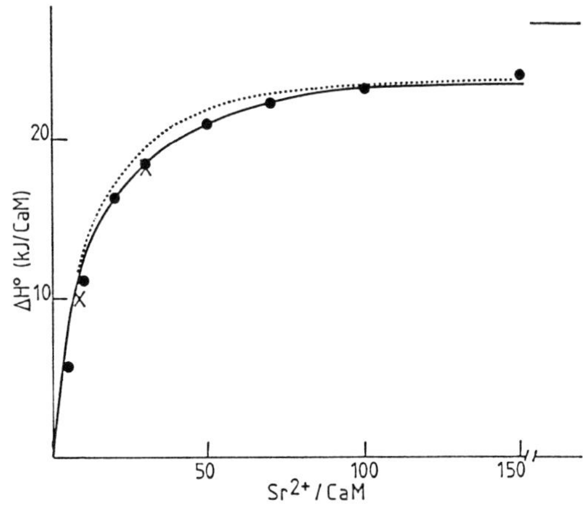

Fig. 5. Enthalpy titration of Sr^{2+}-binding to CaM in 50 mM PIPES-NaOH (•) or Tris HCl (×), pH 7.5, 150 mM NaCl. CaM concentration after mixing was 100 μM. *Solid line* represents the theoretical curve obtained by fitting according to the Langmuir equation with value indicated in Sect. 2.3. *Dotted line* represents a fitting assuming the model of paired cooperative sites with the following affinity constants: $K_N = 1.13 \times 10^2\ M^{-1}$, $K'_N = 2.26 \times 10^3\ M^{-1}$, $K_C = 1.13 \times 10^3\ M^{-1}$, $K'_C = 2.26 \times 10^4\ M^{-1}$

(Fig. 3C). Therefore the question remains why the conformational changes are not more or less linear from 0 to 2 Ca^{2+}. The same holds for the case when 2 to 4 Ca^{2+} are added to CaM.

2.3 Sr^{2+}-Binding to CaM

Although Sr^{2+}-binding to CaM is physiologically irrelevant, this cation is very similar in its coordination and kinetic properties to Ca^{2+} and substitutes for it in different biological processes, including activation of CaM-regulated enzymes (Hoar et al. 1979; Cox et al. 1981). These properties are a strong indication that Sr^{2+} binds to the same sites as Ca^{2+} and therefore it was interesting to find our whether with respect to Sr^{2+} these sites are identical and independent or behave as in the Wang model. A previous study (Cox et al. 1981) indicated that the affinity of Sr^{2+} for CaM is rather low, so that only microcalorimetry is very suitable for studying the binding parameters. Figure 5 shows the enthalpy titration of metal-free CaM (50 μM after mixing) with Sr^{2+}. There is an excellent fitting of these data with a model assuming identical and non-interacting sites with a ΔH^o of 27.2 kJ/mol CaM and a dissociation constant of 545 μM. The 55-fold lower affinity for Sr^{2+} as compared to Ca^{2+} is essentially due to both a more endothermic reaction ($\Delta H^o_{Sr^{2+}}$ = 6.8 kJ/site; $\Delta H^o_{Ca^{2+}}$ = 4.9 kJ/site) and a lower increase of reaction entropy ($\Delta S^o_{Sr^{2+}}$ = 83J/K/site; $\Delta S^o_{Ca^{2+}}$ = 113 J/K/site). Figure 5 also shows that simulation assuming the model of Wang fits less well with the experimental data than the isotherm describing independent and identical affinities. Lastly, it should be noted that the enthalpy change is essentially the same in PIPES and Tris buffer, indicating that at pH 7.5 no protons are released or taken up in the binding process.

2.4 Ca^{2+}-H^+-Antagonism in CaM

Proton antagonism in Ca^{2+}-binding to CaM was suggested from the data of Haiech et al. (1981) and was definitely established by Iida and Potter (1986). According to the data of the latter authors, the affinity decrease of all four binding sites for Ca^{2+} is pronounced below pH 7.9, but not significantly altered above this pH. This suggests that a ionisable group with a pK_i around 6 influences Ca^{2+}-binding and, by virtue of reciprocity, is influenced by Ca^{2+}-binding. By measuring proton release upon binding of Ca^{2+} at different pH values, we confirmed that the proton-Ca^{2+} antagonism is strictly identical for the 4 Ca^{2+}-binding sites (Milos et al. 1986) in the pH region 5.0 to 7.5. Maximal proton release as a function of pH follows a classical Langmuir isotherm with a pK_i of 6.17. Three such chemical groups can be titrated by Ca^{2+} addition to CaM. The most plausible ligand having such an ionization constant is a pair of carboxyl groups involved in a close carboxyl-carboxylate interaction. Such interactions have been found in many protein structures and are attributed to the sharing of a proton by the two negatively charged side chains (Sawyer and James 1982). In Ca^{2+}-free CaM, three pairs of two carboxyl groups would be forced in a restricting, near-neighbor arrangment leading to anomalous high pK_i's. Ca^{2+}-binding then disrupts this interaction and forces one or both carboxyl groups in each pair to participate in other bonds, leading to a lowering of the affinity for protons ($pK_i < 4.0$). Actually we have two hypotheses about the location of the three carboxyl-carboxylate bridges in metal-free CaM: (a) In light of a recent report on structure stabilization in the central α-helix of calcium binding proteins (Sundaralingam et al. 1985), we speculate that the following three pairs would form carboxyl-carboxylate bonds: Asp_{78}-Glu_{82}, Asp_{80}-Glu_{83}, and Glu_{84}-Glu_{87}. Indeed, in each of these pairs, one partner at least can alternatively form a salt bridge (in the Ca^{2+}-saturated state): Asp_{78} with Arg_{74}, Asp_{80} with Lys_{77}, Glu_{82} with Arg_{86}, and Glu_{87} with Arg_{90} (Fig. 6). (b) The three ionic geometries responsible for Ca^{2+}-H^+ antagonism may be situated closer to the Ca^{2+}-binding sites. As mentioned in Section 2.1, all domains except III contain an "acid pair" in one of the coordinating positions. It is therefore quite well possible that in the absence of Ca^{2+} these three pairs, are engaged in a carboxyl-carboxylate bridge which stabilizes the metal-free form of CaM.

It should be noted that our model of Ca-H^+ antagonism does *not* correspond to straight competition between Ca^{2+} and H^+, in other words between Ca^{2+} and the carboxyl-carboxylate bridge, but corresponds to a reciprocal attenuation of the affinities. If excess Ca^{2+} shifts the pK_i of these groups from 6 to 4, excess of H^+ would shift the pK_{Ca} from 5 to 3. At low pH (< 4.5), additional non-specific acid effects, such as massive protonation of all the carboxyl groups and precipitation of the protein, surely abolish completely the affinity of Ca^{2+} for the protein.

2.5 Mg^{2+}-Binding to CaM

Until recently, nearly all the information on the interaction of CaM with Mg^{2+} came from Ca-Mg competition studies (Haiech et al. 1981; Ogawa and Tanokura 1984; Iida and Potter 1986). All these studies indicate that Mg^{2+} interacts with metal-free

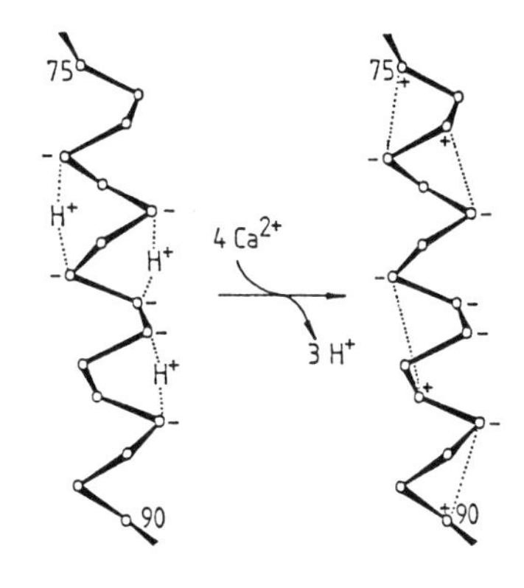

Fig. 6. Possible location of the sites of Ca^{2+}-H^+ antagonism in CaM is depicted. The 3 carboxyl-carboxylate bridges in metal-free CaM (*left*) are disrupted by Ca^{2+} and 4 salt bridges are formed instead (*right*). Possible twisting of the central α-helix is not depicted

CaM with a dissociation constant in the millimolar range. The conformational studies (Richman and Klee 1979; Seamon 1980; Kilhoffer et al. 1981) unambiguously established that the Mg^{2+}-loaded configuration is far from the Ca^{2+}-saturated conformation, thus suggesting, in analogy with the situation in parvalbumin (Cox et al. 1979) and troponin C (Potter and Gergely 1975), that the binding sites are different. However, the interpretation of the Ca-Mg competition studies was less straightforward. Three well-documented studies (Haiech et al. 1981; Ogawa and Tanokura 1984; Iida and Potter 1986) indicate that there is straight competition of Ca^{2+} and Mg^{2+} for the same four sites with affinity constants for Mg^{2+} between 130 and 300 M^{-1} (or 800/M for the last authors). However, other authors did not observe such pronounced competition effects at millimolar free [Mg^{2+}] (Potter et al. 1983; Cox et al. 1981; Burger et al. 1984). The discrepancies can apparently be explained by the low affinity of CaM for Mg^{2+}. Indeed, the interaction is so weak that, at the free [Mg^{2+}] yielding half-maximal saturation, the concentration of bound Mg^{2+} can only be accurately determined provided the equilibrated solution contains above 30 mg of CaM per ml. Recently, we used such high concentrations in a Hummel-Dryer gel filtration method to probe directly the binding of Mg^{2+} to CaM (Milos et al. 1986). The study was further complemented with enthalpy titrations by microcalorimetry, which has the advantage over all other binding methods of measuring accurately the binding process even at low protein- and very high ligand concentrations. The combination of the two methods revealed that 4 Mg^{2+} bind to metal-free CaM with identical affinity: the K_D equals 7 mM (Milos et al. 1986). It should be noted that these experiments were carried out at an ionic strength of 0.2 M and at pH 7.5. At the latter pH, there is no significant proton release as evidenced by enthalpy measurements of Mg^{2+}-binding in two buffers of different protonation enthalpy.

The Mg^{2+}-H^+ antagonism was studied in some detail between pH 7.0 and 5.0 by monitoring proton release upon Mg^{2+} binding to metal-free CaM in a pH stat assembly (Milos et al. 1986). Two interesting informations could thus be obtained: (a) the proton release and hence the Mg^{2+}-H^+ antagonism, is identical for the 4 Mg^{2+}-binding sites; (b) the maximal proton release as a function of the pH, although not very precise, indicates that two protons are released per mole of bound Mg^{2+} from an ionisable group with a pK_i close to 5.0. Hence, the Mg^{2+}-H^+ antagonism is qualitatively very different from the Ca^{2+}-H^+ antagonism, which incites us to assume that the 4 Mg^{2+}-

binding sites are different from the 4 Ca^{2+}-binding sites. The same conclusion is reached from the study of the Ca^{2+}-Mg^{2+} antagonism (Sect. 2.6).

As a very preliminary guess on the location of the Mg^{2+}-binding sites, one would say that two proton ligands with a pK_i of ca. 5.0 are involved in the binding of one Mg^{2+} ion. Likely candidates for binding Mg^{2+} in each of the four domains of CaM are proposed in Table 1. The α-carbon atoms of these two amino acids residues are on the same side of the E helices, and the centers of the β- or δ-carboxyl groups can easily be positioned at about 1.3 Å from each other for the insertion of a Mg^{2+} ion. Such imposed close contact between two carboxyl groups is also present in malonic acid and cyclobutane-1,1-dicarboxylic acid, which display pK_i values of 5.28 and 5.22, respectively, and a K_{Mg} of 129 M^{-1} (Martell and Smith 1977). Interestingly, the putative Mg^{2+}-binding amino acid residues are very well conserved in the four domains of all calmodulins, in some domains of troponin C, and in different regulatory light chains of myosin, where small Mg^{2+} effects upon Ca^{2+} binding have been reported, but not in parvalbumins and invertebrate sarcoplasmic Ca^{2+}-binding proteins, where Mg^{2+} strongly affects the Ca^{2+}-binding properties.

2.6 Ca^{2+}-Mg^{2+} Antagonism in CaM

The rationale for the tentative localization of the Mg^{2+}-binding sites in the N-terminal α-helix of each EF-hand domain stems not only from the arguments developed in Section 2.5, but also from the Ca^{2+}-Mg^{2+} antagonism which we have examined by direct binding methods and by microcalorimetry. Milos et al. (1986) showed results of equilibrium gel filtration experiments in the presence of saturating Ca^{2+} concentrations and varying concentrations of Mg^{2+} and provided direct proof of the existence of distinct Ca^{2+} and Mg^{2+} binding sites. All our data together and especially microcalorimetric determinations of the different affinity constants in the troika CaM-Ca^{2+} lead us to propose the following scheme of antagonism in each of the four metal-binding domains in CaM:

$$\begin{array}{ccc} & (1.05\ 10^5)^4 & \\ P & \rightleftharpoons & P.Ca_4 \\ (140)^4 \updownarrow & & \updownarrow (35)^4 \\ P.Mg_4 & \rightleftharpoons & P.Ca_4.Mg_4 \\ & (2.6\ 10^4)^4 & \end{array}$$

Scheme I

This scheme outlines the negative free energy coupling between the two types of sites: the Ca^{2+} and Mg^{2+} sites are *not* identical; Mg^{2+} is antagonistic to Ca^{2+} at each site and decreases the Ca^{2+}-affinity by a factor of fourfold. Complementary Ca^{2+} decreases the affinity for Mg^{2+} by a factor of fourfold.

Figure 7 dissects the driving forces involved in the interaction of Ca^{2+} and Mg^{2+} with CaM and provides further proof of the relationship between the Ca^{2+} and Mg^{2+} binding processes. The free energy changes and enthalpy changes when Ca^{2+} binds to CaM in the presence of Mg^{2+}, or alternatively Mg^{2+} binds in the presence of Ca^{2+} are neither additive nor mutually exclusive. Interestingly, the positive entropy change is

Fig. 7. Summary of the relative standards free energy, enthalpy and entropy of the Ca^{2+}-, Mg^{2+}- and $Ca^{2+}+Mg^{2+}$-saturated forms of CaM. The metal-free form was taken as reference. The values of the thermodynamic parameters are given for one domain as defined in Table 1

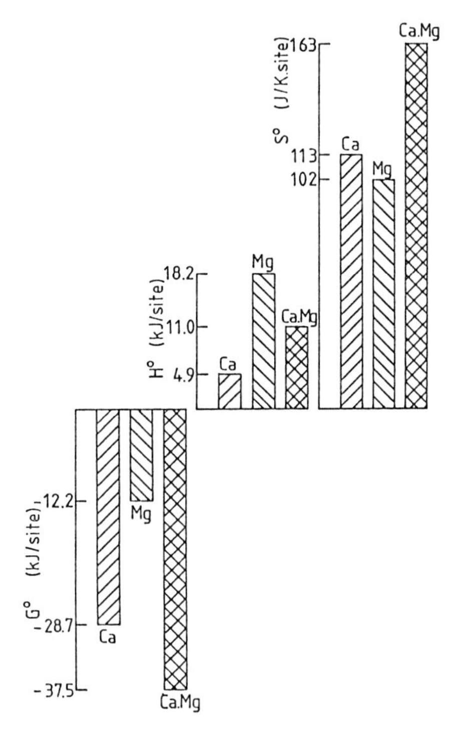

the sole driving force since the three reactions are endothermic. In this respect, the fundamental parameters of Ca^{2+} binding to CaM do not resemble those of other proteins with a rather high affinity for Ca^{2+} ($K > 10^5\ M^{-1}$). In CaM, the reaction is entirely entropy driven with a small endothermic component. In troponin-C, Ca^{2+} binding to the so-called specific sites is entirely enthalpy driven with no change in entropy (Potter et al. 1977). Ca^{2+} binding to the Ca^{2+}-Mg^{2+} sites of the latter protein and also to parvalbumins is both enthalpy and entropy driven (Potter et al. 1977; Moeschler et al. 1980; Tanokura and Yamada 1984). Finally in α-lactalbumin, Ca^{2+} binding is so strongly enthalpy driven that a large negative entropy component accompanies the reaction (Van Ceunebroeck et al. 1985; Schaer et al. 1985). Although there are many structural homologies between the Ca^{2+}-binding domains of these proteins (except that of α-lactalbumin), the fundamental differences shown above may be responsible for their very specific functions in the cell.

3 Ca^{2+} Binding to CaM in the Presence of a Target Enzyme or Peptide

3.1 The Model of Linked Functions and Equation of Huang

Ca^{2+} is needed for the activation of CaM of its target enzymes and quantitative studies on cyclic nucleotide phosphodiesterase (Cox et al. 1981; Huang et al. 1981), cerebellar adenylate cyclase (Malnoë et al. 1982), erythrocyte Ca-Mg ATPase (Cox et al. 1982), skeletal muscle phosphorylase b kinase (Burger et al. 1983), multifunctional protein kinase (Pifl et al. 1984), smooth and skeletal muscle myosin light chain kinase (Blumenthal and Stull 1980; unpublished data from our laboratory) indicate that binding of 3 to 4 Ca^{2+} is required to activate these enzymes. More direct binding studies of CaM to synaptosomal membranes (Malnoë et al. 1982) and to phosphorylase b kinase (Burger et al. 1983) also indicate that the high-affinity complex is only formed if at least 3 Ca^{2+} bind to CaM, in other words that the high affinity E.CaM.Ca_n complex is in direct equilibrium with CaM.Ca_3 and CaM.Ca_4. In the absence of Ca^{2+}, the interaction of CaM with the enzyme displays a low affinity (K_D above μM), but no reliable estimate of this affinity has been published. Wang et al. (1980) were the first to propose a scheme of energy coupling in the interaction between enzyme, CaM and Ca^{2+}.

$$
\begin{array}{ccccccccc}
\text{CaM} & \overset{K_1}{=\!=\!=} & \text{CaM.Ca}_1 & \overset{K_2}{=\!=\!=} & \text{CaM.Ca}_2 & \overset{K_3}{=\!=\!=} & \text{CaM.Ca}_3 & \overset{K_4}{=\!=\!=} & \text{CaM.Ca}_4 \\
\| \; K_e & & \| \; K_d & & \| \; K_c & & \| \; K_b & & \| \; K_a \\
\text{E.CaM} & \underset{K_1'}{=\!=\!=} & \text{E.CaM.Ca}_1 & \underset{K_2'}{=\!=\!=} & \text{E.CaM.Ca}_2 & \underset{K_3'}{=\!=\!=} & \text{E.CaM.Ca}_3 & \underset{K_4'}{=\!=\!=} & \text{E.CaM.Ca}_4
\end{array}
$$

Scheme II

According to this scheme, there are nine independent parameters: the four binding constants of Ca^{2+} to CaM, the four constants that describe binding of Ca^{2+} to enzyme-bound CaM, and one binding constant (arbitrarily we will use K_e in this review) which links the two pathways. Huang and King (1985) have derived a general equation which defines the activation of an enzyme by CaM and Ca^{2+} with the restrictive condition that in the experiments all CaM is in the free form, i.e. $[CaM]_T \gg [Enz]_T$. We used an equivalent equation to describe the binding of CaM to an enzyme at different Ca^{2+} concentrations:

$$
\frac{\text{E.CaM}}{E_T} = \frac{\text{CaM}_F\, \phi_1}{K_e\, \phi_2 + \text{CaM}_F\, \phi_1}\,, \tag{5}
$$

with $\phi_1 = 1 + K_1'|Ca| + K_1'K_2'|Ca|^2 + K_1'K_2'K_3'|Ca|^3 + K_1'K_2'K_3'K_4'|Ca|^4$ and $\phi_2 = 1 + K_1|Ca| + K_1K_2|Ca|^2 + K_1K_2K_3|Ca|^3 + K_1K_2K_3K_4|Ca|^4$.

3.2 Methodology for Determination of the Binding Constants in the Free Energy Coupled Systems

In order to apply the above equation, the concentration of CaM_F, i.e. of all the forms of CaM not bound to the enzyme, has to be known, Therefore experiments in which there is no separation of the two phases (the equilibrated solutions with and without the enzyme) such as optical methods (Malencik and Anderson 1986) will not allow the determination of all the parameters of Scheme II. Similarly equilibrium dialysis experiments in which CaM is excluded from the dialysate (Olwin et al. 1984; Maulet and Cox 1983) do not allow to know CaM_F at each free Ca^{2+} concentration.

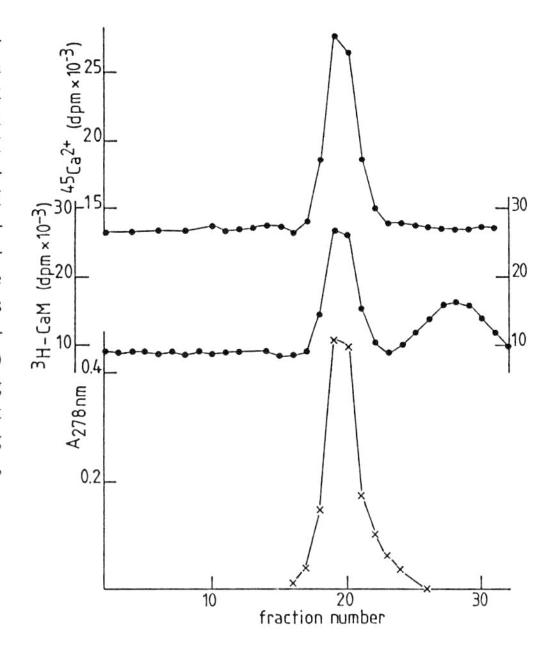

Fig. 8. Typical elution profile of an equilibrium gel filtration experiment of myosin light chain kinase at 4°C on a Sephadex G-75 column (60 × 0.9 cm) equilibrated in 60 mM TES-NaOH, pH 7.0, 150 mM NaCl, 1 mM $MgCl_2$, 15 mM mercaptoethanol containing 0.9 mM CaM (+ 17 nM [^{3}H]-CaM) and 0.91 mM free Ca^{2+} (controlled by means of an EGTA buffer) supplemented with 10 kBg/ml $^{45}CaCl_2$. The concentration of bound Ca^{2+} and CaM was determined in each of the enzyme-containing fractions (about 4 fractions/run) after subtraction of the corresponding mean concentrations in the fractions just before and after the enzyme-containing peak, and reported per mol of enzyme, determined by optical density

How can one evalutate with accuracy the 9 parameters needed to fully describe the system? The K_1, K_2, etc. values are determined independently as indicated in Section 2.2.1. To determine the Ca^{2+}-binding constants to enzyme-bound CaM (K'_1, K'_2, etc.) and to separate this phenomenon from the binding process to free CaM, we recently introduced a method of equilibrium gel filtration: the enzyme migrates and equilibrates in a column containing a Sephadex type of high separation efficiency in the range of the molecular sizes of the target enzyme and CaM. The column is preequilibrated with a fixed concentration of CaM (radiolabelled to facilitate the determination of its concentration) and varying concentrations of Ca^{2+} (maintained at a fixed free level by using EGTA buffers). The amount of enzyme-bound CaM and, in turn, of Ca^{2+} bound to this enzyme-bound CaM can be determined by subtraction (Fig. 8). In this way, the isotherm of Ca^{2+} binding to bound CaM can adequately be constructed and the four K' values estimated by iterative fitting to the Adair equation. Finally in an independent manner one needs to determine K_e. This might be done by equilibrium gel filtration at very low Ca^{2+} levels or, provided the enzyme concentration in the assays is in the same range as the value of $1/K_e$, by optical methods probing conformational changes, and by microcalorimetry.

3.3 Free Energy Coupling in Two Selected Target Enzymes of CaM

Phosphorylase b Kinase. Previously, we studied the free energy coupling in the interactions between Ca^{2+}, CaM and rabbit muscle phosphorylase b kinase by equilibrium gel filtration (Burger et al. 1983). The study was incomplete and the data treatment hampered by the fact that this enzyme contains also many intrinsic Ca^{2+}-binding sites

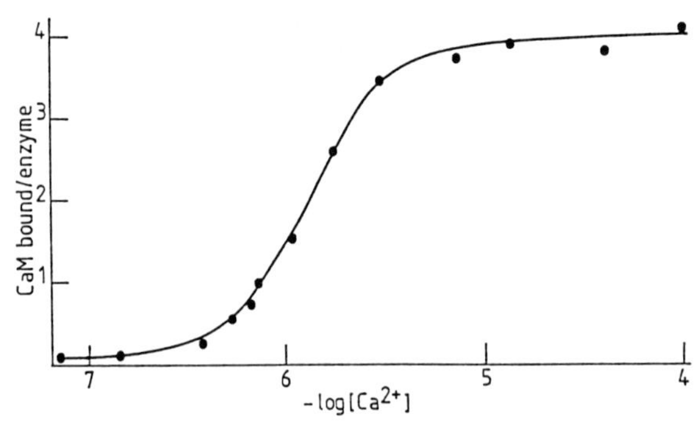

Fig. 9. CaM-binding to phosphorylase b kinase as a function of free $|Ca^{2+}|$ as measured by equilibrium gel filtration in the presence of 0.3 mM free CaM on the column. The experimental data were taken from Burger et al. (1983) with permission. The solid line was the best fitting of the experimental data to the equation of Huang [Eq. (5) in text] with the values for the affinity constants reported on this page

due to the ever-bound δ-subunit (Shenolikar et al.1979). Nevertheless, using model constructions on experimental data, it could be established that the Ca^{2+}-binding isotherm of enzyme-bound CaM (governed by the constants K'_1, K'_2, K'_3 and K'_4) displays strong positive cooperativity in the first three binding steps ($K'_3 = 10^5\ K_3$) followed by a strong leveling off at 3 Ca^{2+} per CaM and a very slow rise up to 4 Ca^{2+} per CaM thereafter (Fig. 7 of Burger et al. 1983). The K_e (= K_c = K_d) value was roughly estimated at $10^4\ M^{-1}$ and $K_a = K_b$ at $10^9\ M^{-1}$. Using the equation of Huang et al., we searched iteratively the constants K_e, K'_1, K'_2, K'_3 and K'_4 involved in the binding of CaM to phosphorylase b kinase. Fig. 9 shows that the experimental data represented in Fig. 4 Burger et al. (1983) fit well into the equation with the following values: $K_e = 3.3 \times 10^4\ M^{-1}$, $K'_1 = 1.16 \times 10^5\ M^{-1}$ (i.e. identical to K_1), $K'_2 = 2.67 \times 10^5\ M^{-1}$ (i.e. identical to K_2), $K'_3 = 1.03 \times 10^9\ M^{-1}$ (i.e 12,600-fold higher than K_3) and $K'_4 = 1.91 \times 10^4\ M^{-1}$ (i.e. identical to K_4).

Myosin Light Chain Kinase (MLCK). Recently, we also studied by equilibrium filtration the energy coupling involved in the interactions between Ca^{2+}, CaM and smooth muscle myosin light chain kinase (Mamar-Bachi and Cox, unpublished observations). This enzyme (130 kDa) binds CaM in a 1 to 1 stoichiometry and has the advantage over phosphorylase b kinase of not binding Ca^{2+} directly to the enzyme. Figure 10 shows the binding of CaM to MLCK as a function of the free Ca^{2+} concentration when the Sephadex G-75 columns are equilibrated in 0.9 μM of free CaM. The position and shape of the curve depends on the concentration of free CaM. Surprisingly, at low free Ca^{2+} concentrations the curve levels off at a plateau value of ca. 0.15 CaM bound per MLCK, indicating that the K_e value is not as low as in the case of phosphorylase b kinase. In order to obtain a good estimate of the K_e value, we measured at $|Ca^{2+}| = 10^{-7}$ M the binding of CaM to MLCK at different concentrations of free CaM up to 7.3 μM. Figure 11 shows that the double reciprocal plot is linear over a one order span of free CaM, crosses the $1/|CaM| = 0$ line at 1 CaM bound per MLCK and allows to extrapolate the K_e value of $1.85 \times 10^5\ M^{-1}$. This corresponds to a dis-

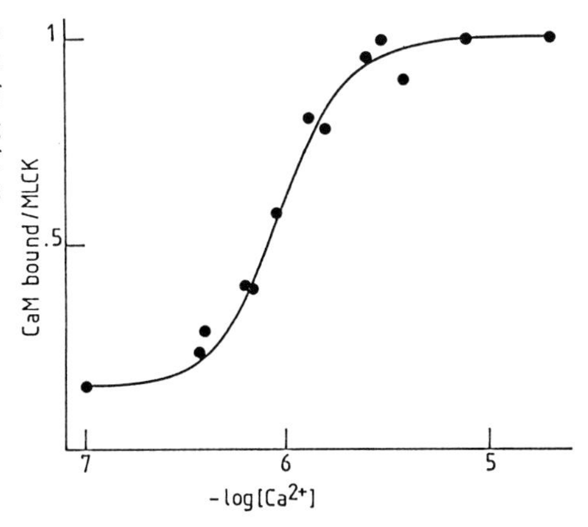

Fig. 10. CaM-binding to myosin light chain kinase as a function of free [Ca^{2+}] as measured by equilibrium gel filtration in the presence of 0.9 mM free CaM. *Solid line* was the best fitting of the experimental data to the equation of Huang with the values for the affinity constants reported on this page

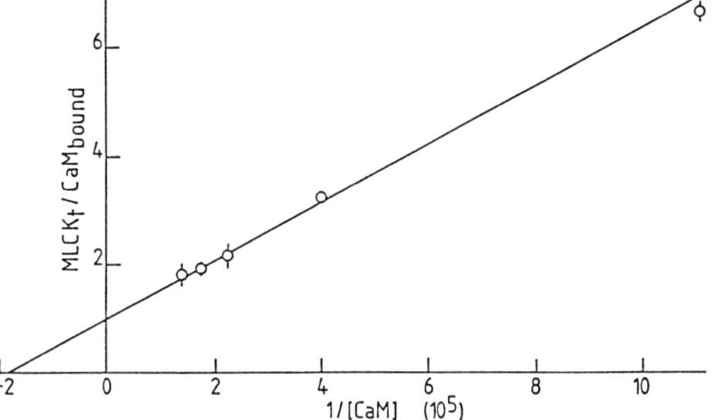

Fig. 11. Double reciprocal plot of CaM binding to myosin light chain kinase as determined by equilibrium gel filtration at a free [Ca^{2+}] of 10^{-7} M. The intersect with the abscissa corresponds to a K_e value of 1.85×10^5 M^{-1}

sociation constant of metal-free CaM to MLCK of 5.4 μM. Figure 12 shows the amount of Ca^{2+} bound per enzyme-bound CaM as a function of the free Ca^{2+} concentration. With this enzyme the binding isotherm could be determined with good precision and the K' constants which best fit the experimental data to the Adair equation are 1.16×10^5 M^{-1} for K'_1, 2.65×10^5 M^{-1} for K'_2, 2.50×10^8 M^{-1} for K'_3 (i.e. 3000-fold higher than K_3) and 1.03×10^5 M^{-1} for K'_4 (5.4-fold higher than K_4). With these values we calculated the theoretical curve of CaM binding as a function of pCa using Eq. (5). The excellent fitting in Fig. 10 guarantees the validity of the model.

3.4 Nature of the Free Energy Coupling in the Interactions Between Ca^{2+}, CaM and its Target

The apparent homotropic cooperativity in Ca^{2+}-binding to the CaM-enzyme complex is due to a progressive stabilisation of the complex from its low- to its high-affinity form. Most of this stabilisation occurs in the third Ca^{2+}-binding step (Burger et al. 1983; see also Fig. 12). The question then arises what kind of forces, i.e. hydrophobic,

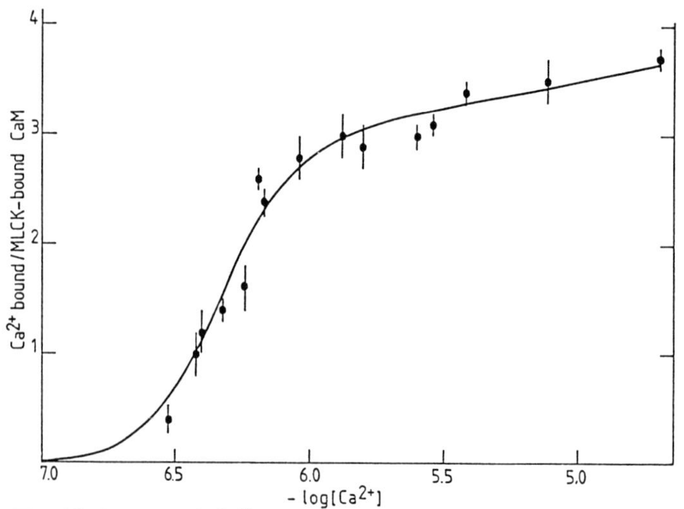

Fig. 12. Binding of Ca^{2+} to enzyme-bound CaM in the experiments described in Fig. 10. The experimental data were obtained as described in Fig. 8. *Solid line* represents the best fitting to the Adair equation [Eq. (2)] with the constants reported on p. 157

electrostatic, van der Waals interactions or hydrogen bonds, are responsible for the observed increase of the Ca^{2+}-affinity. Microcalorimetry is well suited for this purpose, since the relative contributions of enthalpy and entropy changes to the overall free energy change of a reaction are determined by the nature of the force of interaction (Ross and Subramanian 1981). With a natural target enzyme such a microcalorimetric investigation is not yet feasible for technical reasons. Therefore we recently used the model peptide melittin instead of an enzyme and performed an enthalpy titration of Ca^{2+}-binding to the complex (Milos et al. 1987). Melittin is the first (Comte et al. 1983) and most intensively (Maulet and Cox 1983; Steiner et al. 1986; Seeholzer et al. 1986) studied small well known peptide that forms with CaM complexes which resemble those formed with intact target enzymes. Although the melittin model has some draw-backs, the homotropic cooperativity in Ca^{2+}-binding to the CaM-melittin complex resembles the one described for MLCK (p. 157). Table 3 shows the analysis of the thermodynamic parameters: the interpretation of ΔH changes in the first Ca^{2+}-binding step is obscured by an isomerization of the non-specific complex into the high-affinity complex, but binding of the third and fourth Ca^{2+} are 2.5 kJ/site more endothermic than in the absence of melittin. Hence, although the presence of melittin increases the affinity of CaM for Ca^{2+}, Ca^{2+}-binding becomes more endothermic. Thus, the positive free energy coupling in the CaM-Ca^{2+}-melittin system is mostly driven by hydrophobic associations. It is expected that the overall dehydration of non-polar groups in the Ca^{2+}-binding sites, as well as in the two surface-accessible hydrophobic domains of CaM is very pronounced. It remains to be established if in the complexes of CaM with natural targets the standard enthalpy and entropy changes are comparable to those of melittin.

Table 3. Thermodynamic parameters for the interaction of calmodulin, Ca^{2+} and melittin

Interaction	K (M^{-1})	ΔG^0 (kJ/mol)	ΔH^0 (kJ/mol)	ΔS^0 (J/K/mol)
CaM + Ca^{2+}	10^5	− 28.7	4.7	112
$CaM.Mel_1 + Ca^{2+}$	10^5	− 29	0.4	98
$CaM.Mel_1Ca_1 + Ca^{2+}$	10^6	− 34	12.6	156
$CaM.Mel_1Ca_2 + Ca^{2+}$	$\geqslant 10^7$	− 40	7.2	$\leqslant$ 158
$CaM.Mel_1Ca_3 + Ca^{2+}$	$\geqslant 10^7$	− 40	7.2	$\leqslant$ 158
$CaM.Ca_4$ + Mel	3.3×10^8	− 51.4	30.3	275

4 Conclusions

In spite of a great number of investigations along nearly all the possible aspects of the metal-binding properties of CaM, the behavior of the protein still remains very enigmatic. For instance, as for its interaction with Ca^{2+}, two mutually exclusive binding models have been proposed, each of them with its experimental support. The first model of four independent (or nearly independent) binding sites with very similar affinities is favored by direct binding studies, microcalorimetric data and studies on Ca^{2+}-Mg^{2+} and Ca^{2+}-H^+ antagonism. The second model is much more sophisticated with two pairs of sites with pronounced positive cooperativity within each pair and a marked difference in affinity between the pairs. Kinetic data and some conformational studies favor the latter model. At this time, it is difficult to solve the controversy about the two binding models. Specific probes directly involved in Ca^{2+} chelation and specific for each of the four Ca^{2+}-binding domains may become available thanks to two-dimensional resolution of NMR spectra. Crystals of CaM with different contents of Ca^{2+} can constitute an alternative tool for this purpose.

Not only Ca^{2+}-binding, but also its modulation by physiological ions is intriguing. There is indeed strong evidence that CaM contains four additional ion binding sites of low affinity and that the latter modulate the affinity of the strongly binding sites for Ca^{2+}. Under physiological conditions, the low affinity sites are specific for Mg^{2+} and partly occupied by the latter ion. Perhaps the ensuing subtle conformational changes abolish certain hydrophobic interactions of CaM (Milos et al. 1987).

The Ca^{2+}-binding properties of CaM are in a marked way influenced by the presence of the target enzyme with an overall increase of the binding constants and the appearance of quite strong positive cooperativity. In two enzyme systems, the third Ca^{2+}-binding constant was subject to the most dramatic change. The increase in the affinity constants of CaM for Ca^{2+}, the changing affinity of the enzyme for CaM, have been integrated in a coherent system of free energy coupling and could be analyzed quantitatively for the case of phosphorylase b kinase and for myosin light chain kinase with an equation first proposed by C. Huang and collaborators. It can be anticipated that although all the CaM-regulated enzymes are tuned in a similar way by Ca^{2+} and CaM, the variations on the theme are as numerous as in Ravel's Bolero and help to create the complexity in living cells.

Acknowledgements. This study was supported by the Swiss NSF grant Nr. 3.161.0.85 and by the Muscular Dystrophy Association of America, Inc.

References

Andersson T, Drakenberg T, Forsén S, Thulin E (1982) A ^{113}Cd and ^{1}H NMR study of the interaction of calmodulin with D600, trifluoperazine and some other hydrophobic drugs. Eur J Biochem 126:501–505

Babu YS, Sack JS, Greenhough TJ, Brigg CE, Means AR, Cook WJ (1985) Three-dimensional structure of calmodulin. Nature (Lond) 315:37–40

Blumenthal DK, Stull JT (1980) Activation of skeletal muscle myosin light chain kinase by Ca^{2+} and calmodulin. Biochemistry 19:5608–5614

Burger D, Stein EA, Cox JA (1983) Free energy coupling in the interaction between Ca^{2+}, calmodulin and phosphorylase kinase. J Biol Chem 258:14733–14739

Burger D, Cox JA, Comte M, Stein EA (1984) Sequential conformational changes in calmodulin upon binding of calcium. Biochemistry 23:1966–1971

Cheung WY (1980) Calmodulin plays a pivotal role in cellular regulation. Science 207:19–27

Comte M, Maulet Y, Cox JA (1983) Calcium-dependent high-affinity complex formation between calmodulin and melittin. Biochem J 209:269–272

Cox JA, Stein EA (1981) Characterization of a new sarcoplasmic calcium-binding protein with magnesium-induced cooperativity in the binding of calcium. Biochemistry 20:5430–5436

Cox JA, Winge D, Stein EA (1979) Calcium, magnesium and the conformation of parvalbumin during muscular activity. Biochimie (Paris) 61:601–605

Cox JA, Malnoë A, Stein EA (1981) Regulation of brain cyclic nucleotide phosphodiesterase by calmodulin. A quantitative study. J Biol Chem 256:3218–3222

Cox JA, Comte M, Stein EA (1982) Activation of human erythrocyte Ca^{2+}-dependent Mg^{2+}-activated ATPase by calmodulin and calcium. Proc Natl Acad Sci USA 79:4265–4269

Cox JA, Comte M, Malnoë A, Burger D, Stein EA (1984) Mode of action of the regulatory protein calmodulin. In: Sigel H (ed) Metal ion in biological systems. Vol 17. Marcel Dekker Inc, New York Basel, pp 215–273

Crouch TH, Klee CB (1980) Positive cooperative binding of calcium to bovine brain calmodulin. Biochemistry 19:3692–3698

Eigen M, Hammes GG)1063) Elementary steps in enzyme reactions. Adv Enzymol 25:1–38

Haiech J, Klee CB, Demaille J (1981) Effects of cations on affinity of calmodulin for calcium: ordered binding of calcium ions allow the specific activation of calmodulin-stimulated enzymes. Biochemistry 20:3890–3897

Hoar PH, Kerrick WGL, Cassidy PS (1979) Chicken gizzard: relation between calcium-activated phosphorylation and contraction. Science 204:503–506

Huang CY, King MM (1985) Mode of calcium activation of calmodulin-regulated enzymes. Curr Top Cell Regul 27:437–446

Huang CY, Cau V, Chock PB, Wang JH, Sharma RK (1981) Mechanism of activation of cyclic nucleotide phosphodiesterase: requirement of the binding of four Ca^{2+} to calmodulin for activation. Proc Natl Acad Sci USA 78:871–874

Iida S, Potter JD (1986) Calcium binding to calmodulin. Cooperativity of the calcium-binding sites. J Biochem (Tokyo) 99:1765–1772

Ikura M (1986) Protein nuclear magnetic resonance studies on the kinetics of tryptic fragments of calmodulin upon calcium binding. Biochim Biophys Acta 872:195–200

Kilhoffer M-C, Demaille JG, Grard D (1981) Tyrosine fluorescence of ram testis and octopus calmodulins. Effect of calcium, magnesium, and ionic strength. Biochemistry 20:4407–4414

Klee CB, Vanaman TC (1982) Calmodulin. Adv Protein Chem 35:213–303

Klotz IM (1985) Ligand-receptor interactions: facts and fantasies. Quat Rev Biophys 18:227–259

Krebs J (1981) A survey of structural studies on calmodulin. Cell Calcium 2:295-301

Kretsinger RH (1973) Structure and evolution of calcium-modulated proteins. Adv Cyclic Nucleotide Res 11:2–26

Kupke DW, Dorrier TE (1986) Volume changes upon addition of Ca^{2+} to calmodulin: Ca^{2+}-calmodulin conformational states. Biochem Biophys Res Commun 138:199–204

Malencik DA, Anderson SR (1986) Calmodulin-linked equilibria in smooth muscle myosin light chain kinase. Biochemistry 25:709–721

Malencik DA, Anderson SR, Shalitin Y, Schimerlik MI (1981) Rapid kinetic studies on calcium interactions with native and fluorescently labeled calmodulin. Biochem Biophys Res Commun 101:390–395

Malnoë A, Cox JA, Stein EA (1982) Ca^{2+}-dependent regulation of calmodulin binding and adenylate cyclase activation in bovine cerebellar membranes. Biochim Biophys Acta 714:84–92

Manalan SA, Klee CB (1984) Calmodulin. Adv Cyclic Nucleot Protein Phosphorylation Res 18: 227–279

Martell AE, Smith RM (1977) In: Martell AE (ed) Critical stability constants. Vol 3. Plenum, New York London, pp 95–107

Martin SR, Andersson Teleman A, Bayley PM, Drakenberg T, Forsén S (1985) Kinetics of calcium dissociation from calmodulin and its tryptic fragments. A stopped-flow fluorescence study using Quin 2 reveals a 2 domain structure. Eur J Biochem 15:543–550

Maulet Y, Cox JA (1983) Structural changes in melittin and calmodulin upon complex formation and their modulation by calcium. Biochemistry 22:5680–5686

Milos M, Schaer J-J, Comte M, Cox JA (1986) Calcium-proton and calcium-magnesium antagonisms in calmodulin. Biochemistry 25:6279–6287

Milos M, Schaer J-J, Comte M, Cox JA (1987) Microcalorimetric investigation of the interactions in the ternary complex calmodulin-calcium-melittin. J Biol Chem 262:2746–2749

Moeschler H, Schaer J-J, Cox JA (1980) A thermodynamic analysis of the binding of calcium and magnesium to parvalbumin. Eur J Biochem 111:73–78

Ogawa Y, Tanokura M (1984) Calcium binding to calmodulin: effects of ionic strength, Mg^{2+}, pH and temperature. J Biochem (Tokyo) 95:12–28

Olwin BB, Edelman AM, Krebs EG, Storm DR (1984) Quantitation of energy coupling between Ca^{2+}, calmodulin, skeletal muscle myosin light chain kinase, and kinase substrates. J Biol Chem 259:10949–10955

Pifl C, Plank B, Wyskovsky W, Bertel O, Hellmann G, Suko J (1984) Calmodulin-Ca_4 is the active calmodulin-calcium species activating the calcium-, calmodulin-dependent protein kinase of cardiac sarcoplasmic reticulum in the regulation of the calcium pump. Biochim Biophys Acta 773:197–206

Potter JD, Gergely J (1975) The calcium and magnesium binding sites on troponin and their role in the regulation of myofibrillar adenosine triphosphatase. J Biol Chem 250:4628–4633

Potter JD, Hsu FJ, Pownall JH (1977) Thermodynamics of Ca^{2+}-binding to troponin C. J Biol Chem 252:2452–2454

Potter JD, Strang-Brown P, Walker PL, Iida S (1983) Ca^{2+}-binding to calmodulin. Methods Enzymol 102:135–143

Richman PG, Klee CB (1979) Specific perturbation by Ca^{2+} of tyrosyl residue 138 of calmodulin. J Biol Chem 254:5372–5376

Ross PD, Subramanian S (1981) Thermodynamics of protein association reactions: force contribution to stability. Biochemistry 20:3096–3102

Sawyer L, James MNG (1982) Carboxyl-carboxylate interactions in proteins. Nature (Lond) 295:79–80

Schaer J-J, Milos M, Cox JA (1985) Thermodynamics of the binding of calcium and strontium to bovine alpha-lactalbumin. FEBS Lett 190:77–80

Seamon KB (1980) Calcium- and magnesium-dependent conformational states of calmodulin as determined by nuclear magnetic resonance. Biochemistry 19:207–215

Seeholzer SH, Cohn M, Putkey JA, Means AR, and Crespi HL (1986) NMR studies of a complex of deuterated calmodulin with melittin. Proc Natl Acad Sci USA 83:3634–3638

Shenolikar S, Cohen PTW, Cohen P, Nairn AC, Perry SV (1979) The role of calmodulin in the structure and regulation of phosphorylase kinse from rabbit skeletal muscle. Eur J Biochem 100:329–337

Steiner RF, Marshall L, Needleman D (1986) The interaction of melittin with calmodulin and its tryptic fragments. Arch Biochem Biophys 246:286–300

Suko J, Pidlich J, Bertel O (1985) Calcium release from intact calmodulin and calmodulin fragment 78-14 measured by stopped-flow fluorescence with 2-p-toluidinyl naphatalen sulfonate. Eur J Biochem 153:451–457

Sundaralingam M, Drendel W, Greaser M (1985) Stabilization of the long central helix of troponin C by intrahelical salt bridges between charged amino acid side chains. Proc Natl Acad Sci USA 82:7944–7947

Tanokura M, Yamada K (1984) Heat capacity and entropy changes of calmodulin induced by calcium binding. J Biochem (Tokyo) 95:643–649

Teleman A, Drakenberg T, Forsén S (1986) Kinetics of Ca^{2+} binding to calmodulin and its tryptic fragments studied by ^{43}Ca-NMR. Biochim Biophys Acta 873:204–213

Tsalkova TN, Privalov PL (1985) Thermodynamic study of domain organization in troponin C and calmodulin. J Mol Biol 181:533–544

Van Ceunebroeck JC, Hanssens I, Joniau M, Van Cauwelaert F (1985) Thermodynamics of the Ca^{2+} binding to bovine alpha-lactalbumin. J Biol Chem 260:10944–10947

Wang CL (1985) A note on Ca^{2+} binding to calmodulin. Biochem Biophys Res Commun 130: 426–430

Wang JH, Sharma RK, Huang CY, Chau V, Chock PB (1980) On the mechanism of activation of cyclic nucleotide phosphodiesteras by calmodulin. Ann NY Acad Sci 356:190–204

Wang JH, Pallen C, Sharma RK, Adachi AM, Adachi K (1985) The calmodulin regulatory system. Current Topics Cell Regul 27:419–436

Watterson DM, Sharief F, Vanaman TC (1980) The complete amino acid sequence of the Ca^{2+}-dependent modulator protein (calmodulin) of bovine brain. J Biol Chem 255:962–975

Wnuk W, Cox JA, Kohler LG, Stein EA (1979) Calcium and magnesium binding properties of a high affinity calcium-binding protein from crayfish sarcoplasm. J Biol Chem 254:5284–5289

Wnuk W, Cox JA, Stein EA (1982) Parvalbumin and other soluble sarcoplasmic Ca^{2+}-binding proteins. In: Cheung WY (ed) Calcium and cell function. Vol II. Academic Press, New York, pp 243–278

Regulatory Properties of Bovine Brain Calmodulin-Dependent Phosphatase

W. Y. CHEUNG[1]

1 Introduction and Perspective

Phosphorylation is an important cellular regulatory mechanism, controlling and coordinating various enzymic activities. The relative concentration of phosphorylated and nonphosphorylated forms of the appropriate substrates must be properly maintained if they are to function in a regulatory capacity. The level of phosphorylation of a substrate is thus dependent upon a balance between the activities of kinases and those of phosphatases. Although considerable emphasis has been placed on the regulation of protein kinases, it is now evident that protein phosphatases are also subject to an intricate system of control.

Calmodulin mediates many of the functions of Ca^{2+}, including the regulation of protein phosphorylation, catalyzed by calmodulin-dependent protein kinases, and regulation of protein dephosphorylation, catalyzed by a calmodulin-dependent phosphatase (Cheung 1984; Stull et al. 1986). The phosphatase was originally discovered as an "inhibitor" of calmodulin-dependent phosphodiesterase (Wang and Desai 1976; Klee and Krinks 1978), the inhibition resulting from competition with phosphodiesterase for calmodulin (Wang and Desai 1976; Klee and Krinks 1978; Wallace et al. 1978). Prior to identification of its intrinsic enzymic activity, calmodulin-dependent phosphatase has been referred to as an inhibitor protein (Wang and Desai 1976; Klee and Krinks 1978; Wallace et al. 1978), modulator-binding protein (Sharma et al. 1979), calmodulin-binding protein (Sharma et al. 1979), CaM-BP80 (Wallace et al. 1980; Wang et al. 1980), and calcineurin (Klee et al. 1979).

In 1982, Stewart et al. reported that a highly purified skeletal muscle phosphatase displayed a subunit structure strikingly similar to that of bovine brain calcineurin, and that calcineurin displayed protein phosphatase activity that was dependent upon Ca^{2+} and calmodulin. They proposed that calcineurin was most likely a Ca^{2+}/calmodulin-dependent protein phosphatase. We had previously prepared a polyclonal antibody against bovine brain calcineurin (Wallace et al. 1980). By correlating calcineurin activity using radioimmunoassay with phosphatase activity, we provided added evidence that calcineurin is indeed a calmodulin-dependent phosphatase (Yang et al. 1982). Further evidence substantiating the identification of calcineurin as a protein phosphatase, regulated by Ca^{2+}/calmodulin, was subsequently provided by Tonks and Cohen (1983), Manalan and Klee (1983), Pallen and Wang (1983) and Cernoff et al.

1 Department of Biochemistry, St. Jude Children's Research Hospital, Memphis, TN 38101, USA

Ch. Gerday, R. Gilles, L. Bolis (Eds.)
Calcium and Calcium Binding Proteins

(1984). Thus, a protein originally isolated on the basis of its ability to inhibit phosphodiesterase was finally identified as a calmodulin-dependent phosphatase. Although this article focuses on the regulatory properties of the enzyme from bovine brain, those from other tissues are often mentioned for comparison. Several reviews on the phosphatase have appeared (Pallen and Wang 1986; Tallant and Cheung 1986; Balloue and Fischer 1986).

2 Distribution

Calmodulin-dependent phosphatase has been found in many tissues, including bovine brain (Stewart et al. 1982; Yang et al. 1982), bovine cardiac muscle (Wolf and Hofmann 1980; Krinks et al. 1984; Blumenthal et al. 1986), rabbit brain, skeletal muscle, heart muscle, adipose tissue, rat and rabbit liver (Ingebritsen et al. 1983), rabbit reticulocyte lysates (Foulkes et al. 1982), human platelets (Tallant and Wallace 1985), certain forms of human brain tumors (Goto et al. 1986), human erythrocytes (Brissette et al. 1983), human lymphoblastoid cells, pig lymphocytes (Chantler 1985), Xenopus oocytes (Foulkes and Maller 1982), and *Paramecium* (Klumpp et al. 1983). The highest level detected is in rat brain, which contains about 150 mg/kg (Tallant and Cheung 1983).

The level of the phosphatase in various bovine tissues has been determined by radioimmunoassay (Wallace et al. 1980); high levels are found in the neural tissue, predominantly in the cerebrum, olfactory bulb, and cerebellum. Within the cerebrum, the level is especially high in the caudate nucleus and putamen (Table 1). The levels of the enzyme in non-nervous tissues detected by radioimmunoassay are considerably lower than those found by enzyme assay (Ingebritsen et al. 1983). However, Stewart et al. (1983) have estimated that the level of the phosphatase in skeletal muscle is 12.5 to 25 mg/kg, which is comparable to that found in bovine brain by radioimmunoassay. One explanation for this apparent discrepancy is that the phosphatases in non-nervous tissues may be slightly different antigenically and may cross-react poorly with the brain antibody. Alternatively, the non-nervous tissues may contain higher levels of proteases. The phosphatase is highly susceptible to proteolysis and may be degraded to various degrees in the non-nervous tissues, and escape recognition by the antibody.

On a subcellular basis, approximately 50% of the calmodulin-dependent phosphatase is associated with a particulate fraction in both rat cerebrum (Tallant and Cheung 1983) and bovine cerebrum (Tallant 1983). Other workers have detected the protein in postsynaptic densities isolated from dog brain (Carlin et al. 1981) and in synaptic plasma membranes and postsynaptic densities from chick retina (Cooper et al. 1985). In contrast, in liver and skeletal muscle the enzyme is found exclusively in the cytosol (Ingebritsen et al. 1983; Kuret et al. 1986).

In the caudate-putamen of rodent brain, phosphatase is found in association with neuronal elements – but not with oligodendroglia or astrocytes –, with postsynaptic sites within neuronal somata and dendrites and, within the dendrites, with postsynaptic densities and dendritic microtubules (Wood et al. 1980a,b). In chick retina, the

Table 1. Level of calmodulin-dependent phosphatase in various bovine tissues[a]

Tissue	Phosphatase (mg protein per kg tissue)[b]
Adrenal	3.3 ± 0.2
Cerebellum	29.4 ± 12.0
Cerebrum	
Gray matter	36.0 ± 5.0
White matter	15.9 ± 2.5
Caudate nucleus	62.9 ± 10.9
Hippocampus	36.4 ± 7.3
Hypothalamus	31.3 ± 19.7
Putamen	84.8 ± 20.0
Thalamus	19.8 ± 7.0
Heart	0.7 ± 0.3
Kidney	
Cortex	2.3 ± 0.7
Medulla	3.4 ± 1.8
Liver	1.8 ± 0.5
Lung	3.2 ± 1.9
Medulla oblongata	2.5 ± 1.5
Olfactory bulb	17.2 ± 11.8
Pons	6.3 ± 3.9
Skeletal muscle	1.9 ± 0.6
Spleen	1.8 ± 0.6
Testis	3.1 ± 0.9
Tongue	2.8 ± 0.9
Thyroid	2.1 ± 1.5

[a] Adapted from Wallace et al. (1980).
[b] The level of the phosphatase was determined by radioimmunoassay in a 100,000 g supernatant fluid from each tissue.

phosphatase is associated with photoreceptor synaptic terminals in the outer plexiform layer, and in the inner plexiform layer, with dendrites of ganglion cells and presynaptic terminals, some of which belong to bipolar cells (Cooper et al. 1985). In chick forebrain, Anthony et al. (1987) show that the phosphatase is found in the cytoplasm and microsomes, while little is detected in nucleus, myelin, oligodendrocytes, synaptic vesicles, and mitochondria. A considerable amount is found in synaptosomes, where it is localized exclusively in the synaptoplasm, accounting for 0.3% of the synaptoplasmic proteins.

The ontogenic development of the phosphatase has been studied in nervous tissues with radioimmunoassay (Tallant and Cheung 1983) and immunocytochemistry (Cooper et al. 1985). The levels of the protein increase significantly in rat cerebrum and cerebellum and in chick retina and brain during the periods corresponding to major synapse formation (Tallant and Cheung 1983). In a parallel study on developing chick retina, Cooper et al. (1985) have shown a correlation between the appearance of synaptic vesicles and synaptic densities and that of phosphatase at these sites. Collectively, these observations suggest that the phosphatase is developmentally linked

Table 2. Physiochemical properties of bovine brain phosphatase

Property	Value	Reference[a]
Gel filtration (M_r)	80,000	1
Amino acid composition (M_r)	80,000	2
Sedimentation equilibrium (M_r)	84,000	3
Sedimentation and gel filtration (M_r)	86,000	3
Subunit A (M_r)		
SDS-gel electrophoresis	60,000	1
Amino acid composition	61,000	2
Subunit B (M_r)		
SDS-gel electrophoresis (Ca^{2+})	15,000	4
SDS-gel electrophoresis (EGTA)	16,500	5
Amino acid composition	19,000	2
Stokes radius (A)	39–40.5	1, 3
$s_{20\,w}$(S)	4.96	3
Isolectric point		
Holophosphatase	4.5–6.0	1, 2
Subunit A	5.5	6
Subunit B	4.8	6
$E^{1\%}$	9.6–9.7	1, 3, 7

[a] (1) Wallace et al. (1979); (2) Klee et al. (1983a); (3) Sharma et al. (1979); (4) Klee et al. (1979); (5) Tallant and Cheung 1984b); (6) Klee and Haiech (1980); (7) Klee and Krinks (1978).

to the formation of synaptic membranes and densities during the course of synaptogenesis.

3 Molecular Properties

Calmodulin-dependent phosphatase from bovine brain is a globular protein with a molecular weight of approximately 80,000 (Klee and Krinks 1978; Wallace et al. 1979; Sharma et al. 1979; Klee et al. 1983a). Various molecular properties of the phosphatase are summarized in Table 2. The phosphatase displays a typical ultraviolet spectrum with maximum absorbance at 278 nm (Klee and Krinks 1978; Wallace et al. 1979; Sharma et al. 1979), which is characteristic of a protein rich in tyrosine and poor in tryptophan (Klee et al. 1983a).

The phosphatase is a heterodimer, with subunit A of M_r 60,000 and subunit B of M_r 19,000, present in a 1:1 ratio (Wallace et al. 1979; Klee and Krinks 1978; Klee et al. 1979). We have resolved the heterodimer into its subunits with retention of biologcal activity, and have reconstituted an active enzyme from the subunits (Merat et al. 1985). The later process is dependent upon the presence of divalent cations, preferentially Ca^{2+} (Merat and Cheung, in preparation).

Sharma et al. (1979) showed that subunit A separated from subunit B retains its ability to form a Ca^{2+}-dependent complex with calmodulin and to inhibit calmodulin-supported phosphodiesterase activity. We have identified subunit A as the catalytic subunit, which was stimulated by calmodulin, by subunit B, and synergistically by both (Winkler et al. 1984; Merat et al. 1985). Subunit B, even though similar to calmodulin in many respects, does not stimulate the activity of the native enzyme, suggesting that it does not substitute for calmodulin. Limited trypsinization of subunit A increases its catalytic activity to the level observed with calmodulin, and this activity is further stimulated by subunit B but not by calmodulin (Merat et al. 1984). These results suggest that subunit A contains two regulatory domains that are distinct and are specific for each of the regulatory proteins.

Phosphatase subunit B is a Ca^{2+}-binding protein, capable of binding four calcium ions, with a K_d estimated to be less than 10^{-6} M in the presence of physiological concentrations of Mg^{2+} (Klee et al. 1979). The electrophoretic mobility of subunit B undergoes a shift in SDS-polyacrylamide gels, depending upon the presence of EGTA or Ca^{2+}; its molecular weight corresponds to 15,000 in the presence of Ca^{2+} and 16,500 in the presence of EGTA (Table 2). Several other Ca^{2+}-binding proteins, including calmodulin and troponin C, undergo similar changes in their electrophoretic mobility (Burgess et al. 1980; Klee et al. 1979; Wallace et al. 1982). The difference in the apparent molecular weight of these proteins is probably caused by a Ca^{2+}-induced conformational change that makes the protein more compact.

Upon binding Ca^{2+}, the phosphatase undergoes a shift in its UV absorption spectrum that indicates a change in the environment of both tyrosine and tryptophan residues (Klee et al. 1979). In addition, the interaction of the enzyme with its substrate may be dependent upon Ca^{2+}; the phosphatase was retained by a substrate-affinity column (thiophosphorylated myosin light chain) in the presence of Ca^{2+}, but not in the presence of EGTA (Tonks and Cohen 1983).

Aitken et al. (1984) have deduced the primary sequence of subunit B, which is composed of 168 amino acids and has a molecular weight of 19,200. The sequence shows 35% identity with calmodulin and 29% with troponin C; the greatest homology is in the region of the four putative Ca^{2+}-binding loops. The overall sequence is compatible with that of a structure having four Ca^{2+}-binding "EF-hands", as described by Kretsinger (1980), suggesting that subunit B is a member of a homologous family of Ca^{2+}-binding proteins (Aitken et al. 1984). As do other members of this class of proteins, subunit B contains a large number of acidic residues and no tryptophan or cysteine (Klee and Haiech 1980; Aitken et al. 1984).

The amino-terminus of phosphatase subunit B is blocked by acylation with myristic acid, a 14-carbon saturated fatty acid (Aitken et al. 1982). The myristoyl moeity is linked to the N-terminal glycine through an amide bond. The presence of this fatty acid undoubtedly contributes to the hydrophobicity of subunit B, and may play a role in substrate recognition, interaction with membranes, or in maintaining subunit-subunit interactions.

Calmodulin-dependent phosphatase can be phosphorylated with 0.5 to 1 mol of phosphate by protein kinase C, and phosphorylation appears to exert no detected difference in its enzyme activity (Tung 1986). On the other hand, Patel et al. (1986) and Singh and Wang (1986) noted that phosphorylation of phosphatase by protein kinase C increasess phosphatase activity and its affinity for calmodulin.

The phosphatase is an excellent substrate in vitro for carboxylmethylation (Gagnon 1983). In fact, it is a better substrate than calmodulin (Gagnon et al. 1981). Carboxylation of phosphatase reduces the calmodulin-stimulated but not basal activity (Billingsley et al. 1985).

4 Stimulation by Calcium and Calmodulin and by Limited Trypsinization

Calmodulin, in the presence of Ca^{2+}, stimulates the activity of the phosphatase (Yang et al. 1982; Tallant and Cheung 1984b; Stewart et al. 1982, 1983). Half-maximal activation of the brain enzyme is observed at 20–30 nM calmodulin in the presence of a saturating concentration of Ca^{2+} (Tallant and Cheung 1984b; Li 1984). At a saturating level of calmodulin (290 nM), half-maximal activation of the brain enzyme is observed at 0.35 uM Ca^{2+} (Tallant and Cheung 1984b). A Hill coefficient of 4.9 indicates a high degree of cooperativity. Kincaid and Vaughan (1986) have made a direct comparison of Ca^{2+} requirements for calmodulin interaction with and activation of protein phosphate. They found that filling of two Ca^{2+} sites permits calmodulin to form a complex with the enzyme; this complex is inactive – full activation requires additional filling of 1 to 2 sites.

Calmodulin stimulation of phosphatase activity is inhibited by trifluoperazine, a calmodulin antagonist (Yang et al. 1982; Stewart et al. 1982, 1983; Tallant and Cheung 1984b). Trifluoperazine (50 μM) abolishes calmodulin stimulation of the phosphatase without affecting the basal activity (Yang et al. 1982; Tallant and Cheung 1984b). The brain phosphatase interacts with a norchlorpromazine-Affi-Gel column in the presence or absence of Ca^{2+} (Klee et al. 1983b); it is not known whether this interaction is through subunit A or subunit B.

The stoichiometry of the calmodulin phosphatase complex is 1 to 1. Formation of the complex depends upon the level of Ca^{2+}. The rate of formation or dissociation is rapid, and the phosphatase activity can be readily reversed by changes in the Ca^{2+} concentration (Tallant and Cheung 1984b).

The formation of a Ca^{2+}-dependent complex between calmodulin and the phosphatase is compatible with the proposed mechanism for calmodulin-stimulated enzymes. The mode of action involves the binding of Ca^{2+} to calmodulin; this binding forms an active Ca^{2+}-calmodulin complex, which then interacts with the enzyme to form the activated holoenzyme (Cheung 1980).

Calmodulin exerts its effect by increasing the maximal velocity of the phosphatase without altering its affinity for the substrate, either with casein (Table 3, see also Tallant and Cheung 1984b), p-nitrophenyl phosphate (Pallen and Wang 1984; Li 1984), inhibitor-1, phosphorylase kinase, or myosin light chain (Stewart et al. 1983). Although the kinetic parameters are altered by various divalent cations, the changes induced by calmodulin appear independent of the divalent cation (Pallen and Wang 1984; Li 1984).

Limited trypsinization stimulates calmodulin-dependent phosphatase and renders it calmodulin-independent (Manalan and Klee 1983; Tallant and Cheung 1984a). Stimulation of a calmodulin-dependent enzyme by limited proteolysis was first noted with cyclic nucleotide phosphodiesterase (Cheung 1971). Subsequently, other calmodulin-dependent enzymes have been shown to be stimulated by limited proteolysis, rendering

Table 3. Kinetic parameters of phosphatase

Enzme	Substrate	Calmodulin	K_m (uM)	V_{max} (nmol/mg/min)
Bovine brain	Casein[a]	–	2.7	1.7
		+	2.1	41
	Histone[b]	+	4.2	33.3
	p-Nitrophenyl phosphate[c]	–	12,500	1087
	4-Methylumbelliferyl phosphate[d]	+	1,300	100
Rabbit skeletal muscle	Inhibitor-1[e]	–	2.5	170
		+	2.5	2080
	Phosphorylase kinase[e]	–	5.9	1040
	Myosin light chain[e]	+	3.7	790
Bovine heart	Regulatory subunit of type II cAMP-dependent protein kinase[f]	+	5	1000

[a] Assayed with Ca^{2+}; Tallant and Cheung (1984b).
[b] Assayed with Ca^{2+}; King and Huang (1983).
[c] Assayed with Mn^{2+}; Li (1984).
[d] Assayed with Ca^{2+}; Anthony et al. (1986).
[e] Assayed with Mn^{2+}; Stewart et al. (1983).
[f] Assayed with Mn^{2+}; Blumenthal and Krebs (1983).

them insensitive to calmodulin. The trypsinized phosphatase, when fully activated, has a molecular weight of 60,000 and is a heterodimer of subunit B and a 43,000-Da fragment of subunit A (Tallant and Cheung 1984a; Manalan and Klee 1983). The trypsinized phosphatase no longer binds to calmodulin and is insensitive to calmodulin.

The trypsinized enzyme, although insensitive to stimulation by calmodulin, is stimulated by Ca^{2+} (Tallant and Cheung 1984a). The mechanism by which Ca^{2+} affects the proteolyzed phosphatase is not understood, but may be due to the Ca^{2+}-binding properties of subunit B.

Subunit A appears to have at least two regulatory domains. One domain, relatively resistant to trypsin, interacts with subunit B; the other domain, highly susceptible to tryptic attack, interacts with calmodulin. This calmodulin-binding domain inhibits the enzyme activity. The inhibition can be relieved by a conformation change upon binding calmodulin or by limited proteolysis (Manalan and Klee 1983; Tallant and Cheung 1984a).

5 Regulation by Divalent Cations

At a neutral pH, the activity of the phosphatase is significantly increased by Ni^{2+}, Mn^{2+}, and other divalent cations; the order of effectiveness in stimulating the phosphatase, in the presence or absence of calmodulin, is $Ni^{2+} > Mn^{2+} > Co^{2+} > Ca^{2+}$, Sr^{2+}, or Ba^{2+} using p-nitrophenyl phosphate as a substrate (King and Huang 1983; Pallen and Wang 1984; Li 1984). Maximal activation is achieved at 100 uM Mn^{2+}, with a K_a

of 40 uM (Pallen and Wang 1984; Li 1984). Mg^{2+}, at a neutral pH, has no effect on phosphatase activity (Pallen and Wang 1984; Li 1984); however, at an alkaline pH (8.6), it also stimulates phosphatase activity (Li 1984). The order of effectiveness of the metal ions at pH 8.6 is $Mg^{2+} > Ni^{2+} > Mn^{2+} > Ca^{2+}$, using p-nitrophenyl phosphate as a substrate. The stimulatory concentration of Mg^{2+} was high (20 mM), the K_a being 5 mM. Since the cellular concentration of Ni^{2+} and Mn^{2+} is extremely low, while that of Mg^{2+} is in the millimolar levels, Li (1984) has suggested that Mg^{2+} may represent the physiological regulator.

The pH optimum of phosphatase is dependent upon the metal ion. In the presence of Ca^{2+}, Mn^{2+}, or Ni^{2+}, the phosphatase shows optimal activity at a neutral pH (Tallant and Cheung 1984b; Pallen and Wang 1983; Li 1984), whereas with Mg^{2+}, phosphatase activity reaches a maximum at pH 8.6 (Li 1984; Li and Chan 1984).

The kinetic parameters of the phosphatase have been determined in the presence of Mn^{2+} or Ni^{2+} at a neutral pH, and of Mg^{2+} at an alkaline pH (Pallen and Wang 1984; Li 1984). The K_m of the Mg^{2+}-stimulated activity is significantly increased over that found with Mn^{2+}, which is higher than that of Ni^{2+}. The maximal velocity of the enzyme is higher with Mg^{2+} than with either Mn^{2+} or Ni^{2+}; the V_{max} if the Mn^{2+}- and Ni^{2+}-stimulated activities are almost identical. Calmodulin increases the V_{max} of the Ni^{2+}-, Mn^{2+}-, or Mg^{2+}-supported activities without affecting the K_m, and the effect of calmodulin is independent of the divalent cation.

Stimulation of phosphatase activity by Ni^{2+} is time-dependent and invariably preceded by a pronounced lag phase (King and Huang 1983; Pallen and Wang 1984). The addition of Ni^{2+} quenches the tryptophan fluorescence of the phosphatase approximately 30%, and the quenching occurs at a faster rate than does the enzyme activation (King and Huang 1983). In addition, subunit A, separated from subunit B, displays a similar Ni^{2+}-induced quenching of its tryptophan fluorescence; this finding suggests that the primary Ni^{2+} binding site is on subunit A (King et al. 1984). Pallen and Wang (1984) have noted that the rate of initial activation of the phosphatase as a function of Ni^{2+} concentration is hyperbolic; the rate-limiting step in the activation process may be a conformational change induced by the metal binding, rather than the binding of the metal itself.

The Ni^{2+}-induced activation of the phosphatase is not reversed with EDTA, although the chelator readily reverses the Mn^{2+}-stimulated activity (King and Huang 1983; Pallen and Wang 1984). Thus, the activated enzyme appears to have a much higher affinity for Ni^{2+} than for Mn^{2+}, even though the nonactivated phosphatase has a higher affinity for Mn^{2+} than for Ni^{2+}. Neither the Mn^{2+}- nor the Ni^{2+}-stimulated activity of the phosphatase is reversed by dilution or by extensive dialysis against metal-free buffer; both metal ions may be very tightly bound to the activated phosphatase (Pallen and Wang 1984).

Mn^{2+} and Ni^{2+} appear to compete in stimulating phosphatase activity (Pallen and Wang 1984). Since Ni^{2+} activates the enzyme to a higher level than does Mn^{2+}, the two activities may be differentiated. When the two metal ions are added simultaneously, an intermediate activity is obtained, suggesting that the two ions compete for the same binding site. Moreover, the Ni^{2+}-activated enzyme does not bind Mn^{2+}, and the Mn^{2+}-activated enzyme does not bind Ni^{2+}. More recent experiments, however, show that the enzyme either binds 2 mol of Mn^{2+} and 1 mol of Ni, or 1 mol of Mn^{2+} and 2 mol of Ni^{2+} (Pallen and Wang 1986). The implication is that the enzyme has one site specific for Mn^{2+}, a second site specific for Ni^{2+}, and a third site for either Mn^{2+} or Ni^{2+}. Once one metal is bound to this site, it cannot easily be displaced by the metal. Stimulation by Ni^{2+} appears to require a certain cysteine residue or subunit A of phosphatase (King 1986).

Subunit A of phosphatase requires divalent metal ions for activity (Winkler et al. 1984; Merat et al. 1984; Gupta et al. 1984). Mn^{2+} is most effective; Ni^{2+} or Co^{2+} are less effective (Merat et al. 1984). The Mn^{2+}-stimulated activity of subunit A is not easily reversed by EGTA (Merat et al. 1984), in agreement with the ineffectiveness of EDTA to fully reverse the activity of the Mn^{2+}-stimulated holoenzyme (Pallen and Wang 1984). Subunit A which is stimulated by limited proteolysis is also dependent upon Mn^{2+} for activity (Merat et al. 1984).

Calmodulin-stimulated phosphatase activity is not affected by low concentrations of Cu^{2+}, Be^{2+}, Cd^{2+}, Fe^{2+}, Al^{3+}, Fe^{3+}, or Pb^{2+} (Pallen and Wang 1984; King and Huang 1983; Gupta et al. 1984). However, millimolar concentrations of Fe^{2+}, Cu^{2+}, Pb^{2+}, or Cd^{2+}, and micromolar concentrations of Zn^{2+} inhibits phosphatase activity (Gupta et al. 1984); enzyme activity is also inhibited by F^- (Tallant and Cheung 1984b; Klee et al. 1983a) and vanadate ion (Chan et al. 1986).

Atomic absorption spectrophotometric determinations on the phosphatase reveals almost stoichiometric amounts of bound Fe and Zn (King and Huang 1984). These metals are associated with subunit A (Merat and Cheung, in preparation). No significant amounts of Mn^{2+}, Ni^{2+}, or Co^{2+} are detected. The phosphatase thus appears to be an Fe- and Zn-containing metalloenzyme, but one which requires an addition divalent cation for maximal activity. Activation by the metal ions could result from formation of a more stable conformation of the enzyme, or by facilitating the interaction of the substrate with the enzyme. Although evidence in vitro suggests that Mn^{2+} or Mg^{2+} satisfies the divalent cation requirement, the physiological metal cation has not been established.

6 Substrate Specificity

Calmodulin-dependent phosphatase from bovine brain has been shown to dephosphorylate various substrates, including the α-subunit of phosphorylase kinase, inhibitor-1 (Stewart et al. 1982, 1983), casein, histone (Yang et al. 1982), protamine, phosvitin, troponin I (Tallant and Cheung 1984b), myelin basic protein (Gupta et al. 1985; Wolff and Sved 1985), myosin light chain from both smooth and skeletal muscle (Manalan and Klee 1983), the regulatory subunit of Type II cAMP-dependent protein kinase (Blumenthal et al. 1986), MAP-2 (a microtubule-associated protein) (Murthy and Flavin 1983), tubulin, tau factor (Goto et al. 1985), bovine brain phosphodiesterase (Sharma and Wang 1985), EGF-receptor kinase (Pallen et al. 1984) and four mammalian brain phosphoproteins – DARPP-32, G-substrate, Protein K.-F., and synapsin I (King et al. 1984). However, comparison of the catalytic efficiency of the phosphatase toward some of these substrates indicates that dephosphorylation of only DARPP-32, G-substrate, and Protein K.-F. may occur at a significant rate in vivo.

DARPP-32, a soluble neuronal phosphoprotein highly concentrated in the neostriatum, is found predominantly in regions that are highly innervated by dopaminergic neurons (Hemmings et al. 1984b; Walaas et al. 1983; Walaas and Greengard 1984; Ouimet et al. 1984; Hemmings and Greengard 1986). In its phosphorylated state, it inhibits type 1 protein phosphatase at nanomolar concentrations (Hemmings et al. 1984a), and is dephosphorylated by calmodulin-dependent phosphatase at a significant rate in vitro (King et al. 1984). Both calmodulin-dependent phosphatase and DARPP-32 are located at identical sites within the basal ganglia of the brain (Wallace et al. 1980; Walaas et al. 1983; Walaas and Greengard 1984) and within the cell (Wood et al. 1980a; Ouimet et al. 1984).

G-substrate, a soluble, 23,000-Da protein which is phophorylated on two threonine residues by cGMP-dependent protein kinase, is also dephosphorylated by calmodulin-dependent phosphatase (King et al. 1984). It is relatively enriched in the cerebellum, where it is located almost exclusively within the Purkinje cells (Schlicter et al. 1978; Aswad and Greengard 1981; Detre et al. 1984). G-substrate has also recently been identified as an inhibitor of phosphatase 1 in its phosphorylated state (Nestler and Greengard 1984).

Protein K.-F. is dephosphorylated by the calmodulin-dependent phosphatase at significant rates in vitro (King et al. 1984). It is a membrane-bound phosphoprotein (M_r = 18,000) that is phosphorylated on a serine residue by a cyclic nucleotide/Ca^{2+}-independent protein kinase present in the particulate fraction (Greengard and Chan 1983). Protein K.-F. and myelin basic protein have many properties in common and may be related proteins (Greengard and Chan 1983). Myelin basic protein appears to be a good substrate for the phosphatase (Gupta et al. 1985), but phosphatase is not found in myelin, raising the question of whether myelin basic protein is a physiological substrate (Anthony et al. 1987).

Synapsin I (as a M_r = 80,000 and 86,000 doublet in non-avian tissues) is phosphorylated at various sites by cAMP- and by Ca^{2+}/calmodulin-dependent protein kinases; it is present throughout the brain and is localized on synaptic vesicles (Camilli and Greengard 1986). It is dephosphorylated by calmodulin-dependent phosphatase in vitro, albeit not at a significant rate (King et al. 1984). MAP-2, a microtubule-associated protein of molecular weight 270,000 (Murthy and Flavin 1983), is dephosphorylated in a calmodulin-dependent manner also at a rate probably not physiologically significant.

Although calmodulin-dependent phosphatase from bovine brain dephosphorylates casein, histone, troponin I, protamine, and phosvitin (Yang et al. 1982; Tallant and Cheung 1984b), the rates of dephosphorylation are fairly low, and their physiological relevance appears uncertain.

DARPP-32, G-substrate, and inhibitor-1 appear to share certain common properties (Nimmo and Cohen 1978; Aswad and Greengard 1981; Hemmings et al. 1984b). They are phosphorylated on threonine residues by cyclic nucleotide-dependent protein kinases, and the sequences around their phosphorylation sites are similar (Cohen et al. 1977; Aitken et al. 1981; Hemmings et al. 1984c). In addition, their phosphorylated forms function as inhibitors of phosphatase 1. The ability of calmodulin-dependent phosphatase to dephosphorylate the three proteins suggests a mechanism by which Ca^{2+} antagonizes the effects of cyclic nucleotides. It also suggests the presence of a potential protein phosphatase cascade, in which calmodulin-dependent phosphatase regulates the activity of type 1 phosphatases by modulating the state of phosphorylation of various phosphatase inhibitors.

Calmodulin-dependent phosphatase dephosphorylates various low-molecular-weight, nonprotein phosphodiesters. Some examples are p-nitrophenyl phosphate (Pallen and Wang 1983), β-naphthyl phosphate, α-naphthyl phosphate, phenylphthalein mono- and diphosphate, phenyl dihydrogen phosphate, methylumbelliferyl phosphate (Li 1984; Wang et al. 1984; Anthony et al. 1986) 3-fluoro-DL-tyrosine 6-phosphate and tetrafluoro-DL-tyrosine 0-phosphate (Martin et al. 1985). The order of effectiveness of dephosphorylation is β-naphthyl phosphate > p-nitrophenyl phosphate > α-naphthyl phosphate. It seems that a hydrophobic group bulkier than the phenyl group may fit better near the active site, because β-naphthyl phosphate is a better substrate than p-nitrophenyl phosphate (Li 1984). The phosphatase also dephosphorylates phosphatyrosine in the presence of either Mn^{2+}, Co^{2+}, or Ni^{2+}; no significant activity is found toward phosphoserine, phosphothreonine, NADP, glucose 6-phosphate, 2′-AMP, 3′-AMP, 5′-AMP, 5′-IMP, 5′-GMP, ADP, ATP, or β-glycerophosphate (Pallen and Wang 1984; Li 1984).

Although the brain phosphatase dephosphorylates a wide range of substrates in vitro, its catalytic efficiency toward these proteins suggests that it may have a more limited specificity in vivo, and no physiological substrates for the brain phosphatase have yet been established. Based on sequence analysis, Blumenthal et al. (1986) noted that an amphipathic B-sheet structure in the substrate may be an important structural determinant. However, the structural determinants may differ for phosphotyrosyl versus phosphoryl substrates. Moreover, metal cations may also affect substrate specificity. For example, Ni^{2+} appears to stimulate preferentially phosphotyrosyl substrates in the presence of Ca^{2+} and calmodulin (Chan et al. 1986).

7 Concluding Remarks

Phosphorylation of proteins by kinases and dephosphorylation by phosphatases constitute major cellular regulatory mechanisms. The activity of the kinases and that of the phosphatases is in turn controlled and coordinated by various external physiological stimuli. The identification of a calmodulin-dependent phosphatase, activated by micromolar concentrations of Ca^{2+}, suggests that this enzyme may be regulated by cellular flux of Ca^{2+}, such as after neural or hormonal stimulation. The steady state concentration of Ca^{2+} in the cytosol of mammalian cells is in the range of 10^{-8} to 10^{-7} M, and upon stimulation of the intracellular Ca^{2+} reaches the micromolar level, which is sufficient to stimulate the phosphatase.

Calmodulin-dependent protein phosphatase appears to be an important regulatory enzyme in brain, skeletal muscle, and possibly other tissues. Our understanding of its exact role in these tissues is dependent, to a large extent, upon identification of its endogenous substrates.

Acknowledgements. I am grateful to Drs. Frank Anthony and Dennis Merat for their many helpful comments on the manuscript and to Ann Suttle for manuscript preparation. The work in our laboratory is supported by GM 36734 and Ca 21765 from the National Institutes of Health and by American-Lebanese-Syrian Associated Charities.

References

Aitken A, Bilham T, Cohen P, Aswad D, Greengard P (1981) A specific substrate from rabbit cerebellum for guanosine 3':5'-monophosphate-dependent protein kinase. III. Amino acid sequences at the two phosphorylation sites. J Biol Chem 256:3501–3506

Aitken A, Cohen P, Santikarn S, Williams DH, Calder AG, Smith A, Klee CB (1982) Identification of the NH_2-terminal blocking group of calcineurin B as myristic acid. FEBS Lett 150:314–318

Aitken A, Klee CB, Cohen P (1984) The structure of subunit B of calcineurin. Eur J Biochem 139: 663–671

Anthony FA, Merat DL, Cheung WY (1986) A spectrofluorimetric assay of calmodulin-dependent protein phosphatase using 4-methylumbelliferyl phosphate. Anal Biochem 155:103–107

Anthony FA, Winkler MA, Edwards HH, Cheung WY (1987) A quantitative subcellular localization of calmodulin-dependent phosphatase in chick forebrain. J Neuroscience (in press)

Aswad DW, Greengard P (1981) A specific substrate from rabbit cerebellum for guanosine 3':5'-monophosphate-dependent protein kinase. I. Purification and characterization. J Biol Chem 256:3487–3493

Ballou LM, Fischer EH (1986) Phosphoprotein phosphatases. The Enzymes 16:311–361

Billingsley ML, Kincaid RL, Lovenberg W (1985) Stoichiometric methylation of calcineurin by protein carboxyl 0-methyltransferase and its effects on calmodulin-stimulated phosphatase activity. Proc Natl Acad Sci 82:5612–5616

Blumenthal DK, Takio K, Hansen RS, Krebs EG (1986) Dephosphorylation of cAMP-dependent protein kinase regulatory subunit (Type II) by calmodulin-dependent protein phosphatase. J Biol Chem 261:8140–8145

Brissette RE, Cunningham EB, Swislocki NI (1983) A Ca^{2+}-dependent phosphoprotein phosphatase of the erythrocyte membrane. Fed Proc, Fed Am Soc Exp Biol 42:2030 (abstract)

Burgess WH, Jemiolo DK, Kretsinger RH (1980) Interaction of calcium and calmodulin in the presence of sodium dodecyl sulfate. Biochim Biophys Acta 623:257–270

Camilli PD, Greengard P (1986) Synapsin I; A synaptic vesicle-associated neuronal phosphoprotein. Biochem Pharm 35:4349–4357

Carlin RK, Grab DJ, Siekevitz P (1981) Function of calmodulin in postsynaptic densities. III. Calmodulin-binding proteins of the postsynaptic density. J Cell Biol 89:449–455

Chan CP, Gallis B, Blumenthal DK, Pallen CJ, Wang JH, Drebs EG (1986) Characterization of the phosphotyrosyl protein phosphatase activity of calmodulin-dependent protein phosphatase. J Biol Chem 261:9890–9895

Chantler PD (1985) Calcium-dependent association of a protein complex with the lymphocyte plasma membrane – probable identity with calmodulin-calcineurin. J Cell Biol 101:207–216

Chernoff J, Sells MA, Li H-C (1984) Characterization of phosphotyrosyl-protein phosphatase activity associated with calcineurin. Biochem Biophys Res Commun 121:141–148

Cheung WY (1971) Cyclic 3',5'-nucleotide phosphodiesterase. Evidence for and properties of a protein activator. J Biol Chem 246:2859–2869

Cheung WY (1980) Calmodulin plays a pivotal role in cellular regulation. Science 207:19–27

Cheung WY (1984) Biological functions of calmodulin. Harvey Lec 79:173–216

Cohen P, Rylatt DB, Nimmo GA (1977) The hormonal control of glycogen metabolism: The amino acid sequence at the phosphorylation site of protein phosphatase inhibitor-1. FEBS Lett 76:182–186

Cooper NGF, McLaughlin BJ, Tallant EA, Cheung WY (1985) Calmodulin-dependent protein phosphatase: immunocytochemical localization in chick retina. J Cell Biol 101:1212–1218

Detre JA, Nairn AC, Aswad DW, Greengard P (1984) Localization in mammalian brain of G-substrate, a specific substrate for guanosine 3′:5′-cyclic monophosphate-dependent protein kinase. J Neurosci 4:2843–2849

Foulkes JG, Maller JL (1982) In vivo actions of protein phosphatase inhibitor-2 in Xenopus oocytes. FEBS Lett 150:155–160

Foulkes G, Ernst V, Levin D (1982) Separation and identification of Type 1 and Type 2 protein phosphatases from rabbit reticulocyte lysates. Fed Proc, Fed Am Soc Exp Biol 41:648 (abstract)

Gagnon C (1983) Enzymatic carboxyl methylation of calcium-binding proteins. Can J Cell Biol 61:921–926

Gagnon C, Kelley S, Manganiello V, Vaughan M, Odya C, Strittmatter W, Hoffman A, Hirata F (1981) Modification of calmodulin function by enzymatic carboxyl methylation. Nature (Lond) 291:515–516

Goto S, Yamamoto H, Fukunaga K, Iwasa T, Matsukado Y, Miyamoto E (1985) Dephosphorylation of microtubule-associated protein 2, tau factor, and tubulin by calcineurin. J Neurochem 45:276–283

Goto S, Matsukado Y, Mihara Y, Inoue N, Miyamoto E (1986) Calcineurin as a neuronal marker of human brain tumors. Brain Res 371:237–243

Greengard P, Chan K-FJ (1983) Identification, purification, and partial characterization of a membrane-bound phosphoprotein from bovine brain. Fed Proc, Fed Am Soc Exp Biol 42: 2048 (abstract)

Gupta RC, Khandelwal RL, Sulakhe PV (1984) Intrinsic phosphatase activity of bovine brain calcineurin requires a tightly bound trace metal. FEBS Lett 169:251–255

Gupta RC, Khandelwal RL, Sulakhe PV (1985) Resolution of bovine brain calcineurin subunits: stimulatory effect of subunit B on subunit A phosphatase activity. FEBS Lett 190:104–108

Hemmings HC Jr, Greengard P (1986) DARPP-32, a dopamine- and adenosine 3′:5′-monophosphate-regulated phosphoprotein: Regional, tissue, and phylogenetic distribution. J Neurosci 6: 1469–1481

Hemmings HC Jr, Greengard P, Tung HYL, Cohen P (1984a) DARPP-32, a dopamine-regulated neuronal phosphoprotein, is a potent inhibitor of protein phosphtase-1. Nature (Lond) 310: 503–505

Hemmings HC Jr, Nairn AD, Aswad DW, Greengard P (1984b) DARPP-32, a dopamine and adenosine 3′:5′-monophosphate-regulated phosphoprotein enriched in dopamine-innervated brain regions. II. Purification and characterization of the phosphoprotein from bovine caudate nucleus. J Neurosci 4:99–110

Hemmings HC Jr, Williams KR, Konigsberg WH, Greengard P (1984c) DARPP-32, a dopamine- and adenosine 3′:5′-monophosphate-regulated neuronal phosphoprotein. I. Amino acid sequence around the phosphorylated threonine. J Biol Chem 259:14486–14490

Ingebritsen TS, Stewart AA, Cohen P (1983) The protein phosphatases involved in cellular regulation. 6. Measurement of type-1 and type-2 protein phosphatases in extracts of mammalian tissues: An assessment of their physiological roles. Eur J Biochem 132:297–307

Kincaid RL, Vaughan M (1986) Direct comparison of Ca^{2+} requirement for calmodulin interaction with and activation of protein phosphatase. Proc Natl Acad Sci USA 83:1193–1197

King MM (1986) Modification of the calmodulin-stimulated phosphatase, calcineurin, by sulfhydryl reagents. J Biol Chem 261:4081–4084

King MM, Huang CY (1983) Activation of calcineurin by nickel ions. Biochem Biophys Res Commun 114:955–961

King MM, Huang CY (1984) The calmodulin-dependent activation and deactivation of the phosphoprotein phosphatase, calcineurin, and the effect of nucleotides, pyrophosphate and divalent metal ions. J Biol Chem 259:8847–8856

King MM, Huang CY, Chock PB, Nairn AC, Hemmings HC Jr, Chan K-FJ, Greengard P (1984) Mammalian brain phosphoproteins as substrates for calcineurin. J Biol Chem 259:8080–8083

Klee CB, Haiech J (1980) Concerted role of calmodulin and calcineurin in calcium regulation. Ann NY Acad Sci 356:43–54

Klee CB, Krinks MH (1978) Purification of cyclic 3′,5′-nucleotide phosphodiesterase inhibitory protein by affinity chromatography on activator protein coupled to Sepharose. Biochemistry 17:120–126

Klee CB, Crouch TH, Krinks MH (1979) Calcineurin: A calcium- and calmodulin-binding protein of the nervous system. Proc Natl Acad Sci USA 76:6270–6273

Klee CB, Krinks MH, Manalan AS, Cohen P, Stewart AA (1983a) Isolation and characterization of bovine brain calcineurin: A calmodulin-stimulated protein phosphatase. In: Means AR, O'Malley BW (eds) Methods in enzymology. Vol 102. Academic Press, New York, pp 227–244

Klee CB, Newton DL, Krinks M (1983b) Versatility of calmodulin as a cytosolic regulator of cellular function. In: Chaiken IM, Wilchek M, Parikh I (eds) Affinity chromatography and biological recognition. Academic Press, New York, pp 55–67

Klumpp S, Steiner AL, Schultz JE (1983) Immunocytochemical localization of cyclic GMP, cGMP-dependent protein kinase, calmodulin and calcineurin in *Paramecium* tetraurelia. Eur J Cell Biol 32:164–170

Kretsinger RH (1980) Structure and evolution of calcium-moudulated proteins. CRC Crit Rev Biochem 8:119–174

Krinks MH, Haiech J, Rhoads A, Klee CB (1984) Reversible and irreversible activation of cyclic nucleotide phosphodiesterase: Separation of the regulatory and catalytic domains by limited proteolysis. Adv Cyclic Nucleotide Protein Phosphorylation Res 16:31–47

Kuret J, Bell H, Cohen P (1986) Identification of high levels of protein phosphatase-1 in rat liver nuclei. FEBS Lett 203:197–202

Li H-C (1984) Activation of brain calcineurin phosphatase towards nonprotein phosphoesters by Ca^{2+}, calmodulin and Mg^{2+}. J Biol Chem 259:8801–8807

Li H-C, Chan WS (1984) Activation of brain calcineurin towards proteins containing Thr(P) and Ser(P) by Ca^{2+}, calmodulin, Mg^{2+} and transition metal ions. Eur J Biochem 144:447–452

Lin YM, Cheung WY (1980) Ca^{2+}-dependent cyclic nucleotide phosphodiesterase. In: Cheung WY (ed) Calcium and cell function. Vol 1. Academic Press, New York, pp 79–104

Manalan AS, Klee CB (1983) Activation of calcineurin by limited proteolysis. Proc Natl Acad Sci USA 80:4291–4295

Martin B, Pallen CJ, Wang JH, Graves DJ (1985) Use of fluorinated tyrosine phosphates to probe the substrate specificity of the low molecular weight phosphatase activity of calcineurin. J biol Chem 260:14932–14937

Merat DL, Hu ZY, Carter TE, Cheung WY (1984) Subunit A of calmodulin-dependent phosphatase requires Mn^{2+} for activity. Biochem Biophys Res Commun 122:1389–1397

Merat DL, Hu ZY, Carter TE, Cheung WY (1985) Bovine brain calmodulin-dependent protein phosphatase: regulation of subunit A activity by calmodulin and subunit B. J Biol Chem 260: 11053–11059

Murthy ASN, Flavin M (1983) Microtubule assembly using the microtubule-associated protein MAP-2 prepared in defined states of phosphorylation with protein kinase and phosphatase. Eur J Biochem 137:37–46

Nestler EJ, Greengard P (1984) Protein phosphorylation in nervous tissue. In: Usdin E, Carlsson A, Dahlstrom A, Engel J (eds) Catecholamines. Part A: Basic and peripheral mechanisms. Liss Inc, New York, pp 9–22

Nimmo GA, Cohen P (1978) The regulation of glycogen metabolism. Phosphorylation of inhibitor-1 from rabbit skeletal muscle, and its interaction with protein phosphatase-III and -II. Eur J Biochem 87:353–367

Ouimet CC, Miller PE, Hemmings HC Jr, Walaas SI, Greengard P (1984) DARPP-32, a dopamine and adenosine 3′:5′-monophosphate regulated phosphoprotein enriched in dopamine-innervated brain regions. III. Immunocytochemical localization. J Neurosci 4:111–124

Pallen CJ, Wang JH (1983) Calmodulin-stimulated dephosphorylation of p-nitrophenyl phosphate and free phosphotyrosine by calcineurin. J Biol Chem 258:8550–8553

Pallen CJ, Wang JH (1984) Regulation of calcineurin by metal ions. Mechanism of activation by Ni^{2+} and enhanced response to Ca^{2+}/calmodulin. J Biol Chem 259:6134–6141

Pallen CJ, Wang JH (1986) Stoichiometry and dynamic interaction of metal ion activators with calcineurin phosphatase. J Biol Chem 261:16115–16120

Pallen CJ, Valentine KA, Wang JH, Hollenberg MD (1985) Calcineurin-mediated dephosphorylation of human placental membrane receptor for epidermal growth factor urogastrone. Biochemistry 24:4724–4730

Patel J, Lanciotti M, Huang CY (1986) Phosphorylation of calmodulin-dependent protein phosphatase by protein kinase C. Fed Proc, Fed Am Soc Exp Biol 45:1884 (abstract)

Schlicter DJ, Casnellie JE, Greengard P (1978) An endogenous substrate for cGMP-dependent protein kinase in mammalian cerebellum. Nature (Lond) 273:61–62

Sharma RK, Wang JH (1985) Differential regulation of bovine brain calmodulin-dependent cyclic nucleotide phosphodiesterase isozymes by cyclic AMP-dependent protein kinase and calmodulin-dependent phosphatase. Proc Natl Acad Sci USA 82:2603–2607

Sharma RK, Desai R, Waisman DM, Wang JH (1979) Purification and subunit structure of bovine brain modulator binding protein. J Biol Chem 254:4276–4282

Singh TJ, Wang JH (1986) Phosphorylation and activation of calcineurin by glycogen synthase (casein) kinase-1 and cyclic AMP-dependent protein kinase. Fed Proc, Fed Am Soc Exp Biol 45:1803 (abstract)

Stewart AA, Ingebritsen TS, Manalan A, Klee CB, Cohen P (1982) Discovery of a Ca^{2+}- and calmodulin-dependent protein phosphatase. Probable identity with calcineurin ($CaM\text{-}BP_{80}$). FEBS Lett 137:80–84

Stewart AA, Ingebritsen TS, Cohen P (1983) The protein phosphatases involved in cellular regulation. 5. Purification and properties of a Ca^{2+}/calmodulin-dependent protein phosphatase (2B) from rabbit skeletal muscle. Eur J Biochem 132:289–295

Stull GT, Nunnally MH, Michnoff CH (1986) Calmodulin-dependent protein kinases. The Enzymes 17:113–166

Tallant EA (1983) Purification and characterization of calmodulin-dependent protein phosphatase from bovine brain. PhD Dessertation Univ Tennessee Center for the Health Sciences, Memphis

Tallant EA, Cheung WY (1983) Calmodulin-dependent protein phosphatase: A developmental study. Biochemistry 22:3630–3635

Tallant EA, Cheung WY (1984a) Activation of bovine brain calmodulin-dependent protein phosphatase by limited trypsinization. Biochemistry 23:260–279

Tallant EA, Cheung WY (1984b) Characterization of bovine brain calmodulin-dependent protein phosphatase. Arch Biochem Biophys 232:260–279

Tallant EA, Cheung WY (1986) Calmodulin-dependent protein phosphatase. In: Cheung WY (ed) Calcium and cell function. Vol 6. Academic Press, New York, pp 71–112

Tallant EA, Wallace RW (1985) Characterization of a calmodulin-dependent phosphatase from human platelets. J Biol Chem 260:7744–7751

Tonks NK, Cohen P (1983) Calcineurin is a calcium ion-dependent, calmodulin-stimulated protein phosphatase. Biochim Biophys Acta 747:191–193

Tung HYL (1986) Phosphorylation of the calmodulin-dependent protein phosphatase by protein kinase C. Biochem Biophys Res Commun 138:783–788

Walaas SI, Greengard P (1984) DARPP-32, a dopamine and adenosine 3′:5′-monophosphate-regulated phosphoprotein enriched in dopamine innervated brain regions. I. Regional and cellular distribution in the rat brain. J Neurosci 4:84–98

Walaas SI, Aswad DW, Greengard P (1983) A dopamine- and cyclic AMP-regulated phosphoprotein enriched in dopamine-innervated brain regions. Nature (Lond) 301:69–71

Wallace RW, Lynch TJ, Tallant EA, Cheung WY (1978) An endogenous inhibitor protein of brain adenylate cyclase and cyclic nucleotide phosphodiesterase. Arch Biochem Biophys 187:328–334

Wallace RW, Lynch TJ, Tallant EA, Cheung WY (1979) Purification and characterization of an inhibitor protein of brain adenylate cyclase and cyclic nucleotide phosphodiesterase. J Biol Chem 254:377–382

Wallace RW, Tallant EA, Cheung WY (1980) High levels of a heat-labile calmodulin-binding protein (CaM-BP_{80}) in bovine neostriatum. Biochemistry 19:1831–1837

Wallace RW, Tallant EA, Dockter ME, Cheung WY (1982) Calcium binding domains of calmodulin. J Biol Chem 257:1845–1854

Wang JH, Desai R (1976) A brain protein and its effect on the Ca^{2+}- and protein modulator-activated cyclic nucleotide phosphodiesterase. Biochem Biophys Res Commun 72:926–932

Wang JH, Sharma RK, Tam SW (1980) Calmodulin-binding proteins. In: Cheung WY (ed) Calcium and cell function. Vol 1. Academic Press, New York, pp 305–328

Wang JH, Pallen CJ, Brown ML, Mitchell KJ (1984) A survey of calcineurin activity toward non-protein substrates. Fed Proc, Fed Am Soc Exp Biol 43:1897 (abstract)

Winkler MA, Merat DL, Tallant EA, Hawkins S, Cheung WY (1984) Catalytic site of calmodulin-dependent protein phosphatase from bovine brain resides in subunit A. Proc Natl Acad Sci USA 81:3054–3058

Wolf H, Hofmann F (1980) Purification of myosin light chain kinase from bovine cardiac muscle. Proc Natl Acad Sci USA 77:5852–5855

Wolff DJ, Sved DW (1985) The divalent cation dependence of bovine brain calmodulin-dependent phosphatase. J Biol Chem 260:4195–4202

Wood JG, Wallace RW, Whitaker JN, Cheung WY (1980a) Immunocytochemical localization of calmodulin and a heat-labile calmodulin-binding protein (CaM-BP_{80}) in basal ganglia of mouse brain. J Cell Biol 84:66–76

Wood JG, Wallace RW, Whitaker JN, Cheung WY (1980b) Immunocytochemical localization of calmodulin in regions of rodent brain. Ann NY Acad Sci 356:75–82

Yang S-D, Tallant EA, Cheung WY (1982) Calcineurin is a calmodulin-dependent protein phosphatase. Biochem Biophys Res Commun 106:1419–1425

Calmodulin and Calmodulin Binding Proteins During Differentiation of Human Intestinal Brush Borders

C. ROCHETTE-EGLY, B. LACROIX, M. KEDINGER and K. HAFFEN [1]

1 Introduction

Calmodulin is a ubiquitous calcium binding protein (Cheung 1980; Means and Dedman 1980), which in the adult intestinal epithelial cells has been shown to be localized preferentially at the brush border level (Howe et al. 1980; Glenney et al. 1980). In the intestinal epithelial cells, calmodulin functions mainly as a calcium buffer (Glenney and Glenney 1985), activates the actomyosin based contractility system (Mooseker et al. 1983) and interacts with a number of actin binding proteins. These proteins with both actin and calmodulin binding capacities are mainly the 110 kDa protein (Coudrier et al. 1981; Glenney and Glenney 1984; Howe and Mooseker 1983), caldesmon (Bretscher and Lynch 1985) and fodrin (Glenney et al. 1982; Hirokawa et al. 1983). They have been extensively studied and characterized in the intestinal brush borders of avians (Glenney and Weber 1980; Mooseker 1985) but not in humans. The aim of the present chapter is to provide some current views of the human intestinal development and to attempt a correlation between the developmental pattern of calmodulin and calmodulin binding proteins, and the epithelial differentiation accompanying intestinal ontogenic maturation.

2 Current Views on the Organization and the Ontogenesis of Intestinal Epithelial Cells

2.1 Calmodulin and Calmodulin Binding Proteins in the Brush Border Cytoskeleton

In the adult small intestine, the epithelium is a highly differentiated structure composed predominantly of absorptive cells named enterocytes (Haffen et al. 1986). Enterocytes are highly polarized columnar cells, characterized by the presence at their luminal side of brush borders (Louvard et al. 1986; Mooseker 1985). Figure 1 summarizes the architecture of a mature enterocytic brush border. Such a brush border consists of a densely packed array of thousands of microvilli, connected by an underlying terminal web. Each microvillus contains an axial bundle of actin microfilaments

1 Unité INSERM 61, Biologie Cellulaire et Physiopathologie Digestives, 3, Avenue Molière, 67200 Strasbourg, France

Ch. Gerday, R. Gilles, L. Bolis (Eds.)
Calcium and Calcium Binding Proteins

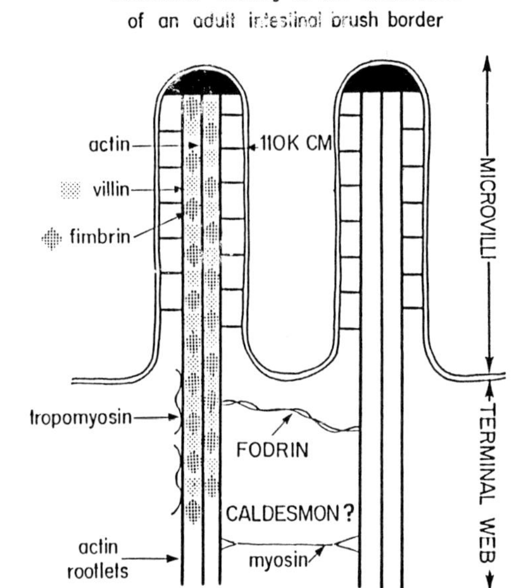

Fig. 1. Schematic drawing depicting the organization and the major proteins of the brush border cytoskeleton

anchored at the tip of the microvilli and linked to the plasma membrane by spirally arranged lateral arms. These actin microfilaments extend as rootlets through the underlying terminal web. The interrootlets zone consists of a dense meshwork of fine, non-actin filamentous material which interdigitates between and cross-links adjacent core rootlets.

The polypeptides which participate in the organization of the brush border cytoskeleton have been recently characterized (Mooseker 1985). Among them are calmodulin and three major calmodulin binding proteins (110 kDa protein, caldesmon and fodrin. In microvilli, calmodulin is associated to the 110 kDa protein as a complex which forms the lateral arms. This complex consists of a dimer of 110 kDa protein with variable amounts of associated calmodulin. Contrariwise, the core bundle of actin microfilaments contains no calmodulin, but two actin-binding proteins, villin and fimbrin. In the terminal web, the actin rootlets (composed of actin, villin, fimbrin and also of tropomyosin) are cross-linked by myosin and a calmodulin-binding protein which has been identified as the commonly distributed non erythroid spectrin; fodrin. Like other spectrins, fodrin consists of a β subunit of 235 kDa, and of a α subunit of 240 kDA which binds calmodulin (Glenney and Glenney 1983). The terminal web also contains a third calmodulin binding protein, named caldesmon (Ishimura et al 1984), but its exact localization as its role are still under investigation.

2.2 Ontogenesis of Intestinal Epithelial Cells

The ontogenic maturation of the human intestinal epithelium is associated mainly with morphogenetic events and ultrastructural modifications (Colony 1983). Using

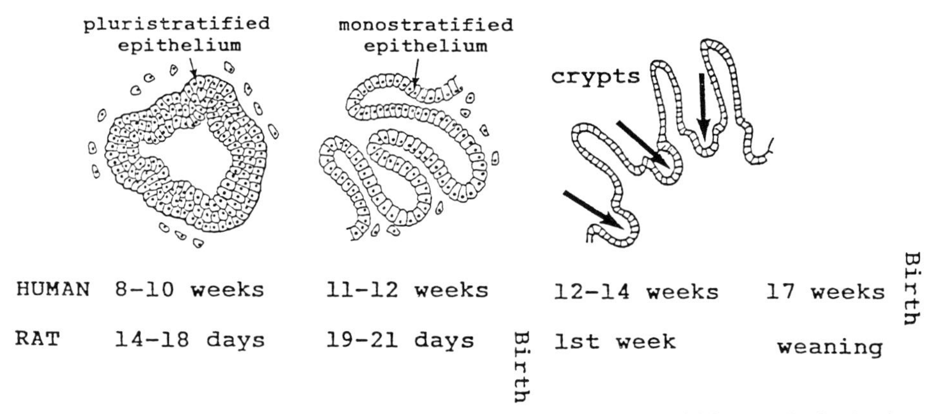

Fig. 2. Schematic representation of the major morphological features which occur in the developing intestinal epithelium

standardized criteria for estimating fetal age (Lacroix et al. 1984a) the exact timing of the early morphogenetic features has been performed (Lacroix et al. 1984b). These processes are closely similar in almost all species, only the timing of the events varying (Vollrath 1969; Mathan et al. 1976; Kedinger et al. 1986), and can be summarized as follows and in Fig. 2.

At early fetal stages (during the first 8–10 weeks of human development), the intestinal anlage consists of a simple tube composed of radially arranged undifferentiated mesenchymal cells (future muscular layers and connective tissue). Up to 8 weeks the luminal side of the jejunal tube is still flat. At 8 weeks, villi start to protrude and the apical layer of cells exhibit short and irregular microvilli. Thereafter, between 10 and 12 weeks of human gestation, jejunal villi increase progressively in height. The cells of the stratified epithelium are progressively replaced by a single monolayer of columnar absorptive cells with well defined microvilli at the apical border. Between 12 and 14 weeks of gestation, villi still extend, crypts start to invaginate at their base and microvilli increase in length, density and regularity. Around 17 weeks the cytodifferentiation of the crypt-villus unit is very similar to that of the adult.

3 Ontogenesis of Calmodulin and Calmodulin Binding Proteins in Human Intestinal Epithelial Cells

3.1 Materials and Methods Used for the Characterization of Human Calmodulin and Calmodulin Binding Proteins

Human intestines were obtained between 6 and 14 weeks of gestation after legal abortions with the consent of the mothers. The specimens were dissected out from abortion material and only healthy and undamaged intestinal tracts were used. Fetal age was determined according to the developmental pattern of hand morphology

(Lacroix et al. 1984a). The specimens analyzed were from 6, 8, 9, 10, 11, 12 and 14 week-old fetuses. Fragments from human adult jejunum were obtained by oral biopsy.

At each stage, a fragment of proximal small intestine (jejunum) was fixed (2% paraformaldehyde in 0.1 M Pipes buffer at pH 7.0 containing 3% sucrose) and processed for immunofluorescence as previously described (Rochette-Egly et al. 1986).

Jejunum fragments were homogenized in calcium, magnesium-free (CMF) Hank's buffer containing 1 mM EGTA and a cocktail of protease inhibitors (pepstatin, 1 μg/ml), antipaïn 1 μg/ml, benzamidin 15 μg/ml, leupeptin 10 μg/ml and aprotinin 10 μg/ml). An aliquot of the homogenates was centrifuged and the supernatant was treated at 80°C and assayed for calmodulin content using a commercial (^{125}I) calmodulin radioimmunoassay kit (New England Nuclear).

At 8 weeks of fetal life, the endoderm was separated from the mesenchyme by incubation of the intestine in 0.03% collagenase containing medium, as previously described (Kedinger et al. 1981). For older fetuses, brush borders were prepared according to Mackensie et al. (1983). Endoderms and brush borders were homogenized in Hanks buffer. Then total jejunums, endoderms or brush borders were processed for electrophoresis in 5–15% linear gradient polyacrylamide slab gels and electrophoretically transferred to nitrocellulose (NC) membranes as already described (Rochette-Egly and Daviaud 1985). The calmodulin binding proteins bound to NC membranes were visualized by the (^{125}I) calmodulin overlay technique described by Flanagan and Yost (1984). Calmodulin binding proteins were also identified immunologically on NC membranes with specific rabbit antibodies by the method of Ngai and Walsh (1985).

Antibodies to chicken gizzard caldesmon (Ngai and Walsh 1985) were a generous gift from Dr. M.P. Walsh (Calgary, Canada). Brain fodrin antibodies (Glenney et al. 1982) were generously provided by Dr. J.R. Glenney (San Diego, USA). Chicken 110 K antibodies (Shibayama et al. 1987) were a generous gift from Dr. M.S. Mooseker (New Haven, USA). Calmodulin antibodies were kindly provided by Dr. J. De Mey (Beerse, Belgium).

3.2 Calmodulin Levels and Localization in the Developing Human Jejunum

Calmodulin levels were evaluated quantitatively by radioimmunoassay on homogenates of human jejunum fragments at different stages of fetal development (Fig. 3). In the jejunum of 8-week-old human fetuses, calmodulin was present at very low levels ranging around 17.7 ± 3.8 ng/mg protein. Thereafter, calmodulin levels increased progressively until 12–14 weeks (four- and tenfold at 10 and 14 weeks respectively). At that stage of development, the amounts present in the jejunum were not significantly different from those in the adult jejunum.

However, immunofluorescence labeling (Fig. 4) led to results contrasting in a certain way with the quantitative evaluation of calmodulin. At 8 weeks of gestation (Fig. 4a), although detectable by radioimmunoassay, calmodulin showed no specific fluorescence in the pluristratified epithelium which is devoid of differentiated brush borders. Thereafter, between 10 and 12 weeks of gestation (Fig. 4b), a weak linear fluorescence appeared at the luminal surface of the epithelial cells parallelling the

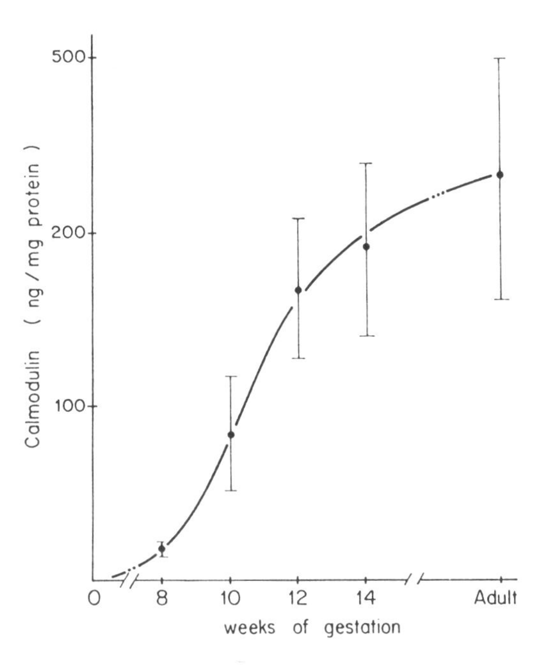

Fig. 3. Calmodulin levels (ng/mg protein) as a function of age. For each fetal stage, 2 to 5 jejunums were assayed in triplicate. Concerning the adults, the values are the mean ± SE of 10 biopsies assayed in triplicate

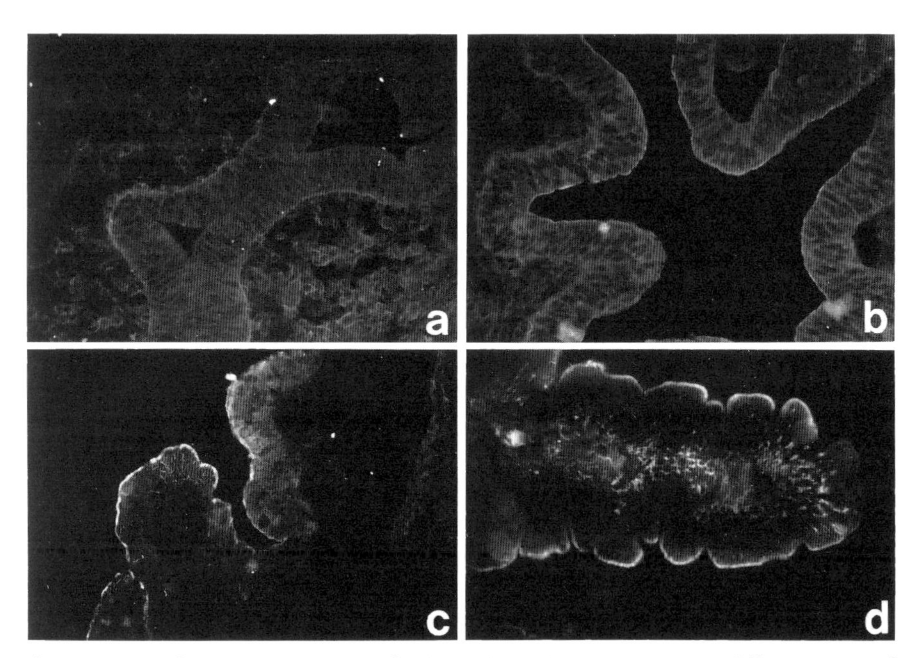

Fig. 4a–d. Immunofluorescence staining of calmodulin in human jejunum at different stages of development. **a** 8-week-old fetus (× 150); no fluorescence was visible for calmodulin; **b** 12-week-old fetus (× 150); **c** 14-week-old fetus (× 150); **d** adult (× 300). In the lamina propria, numerous mast cells show autofluorescence of histamine which is easily distinguished from the greenish specific immunoreactions by its yellowish color (Ishimura et al. 1984; Rochette-Egly et al. 1986; Robine et al. 1985)

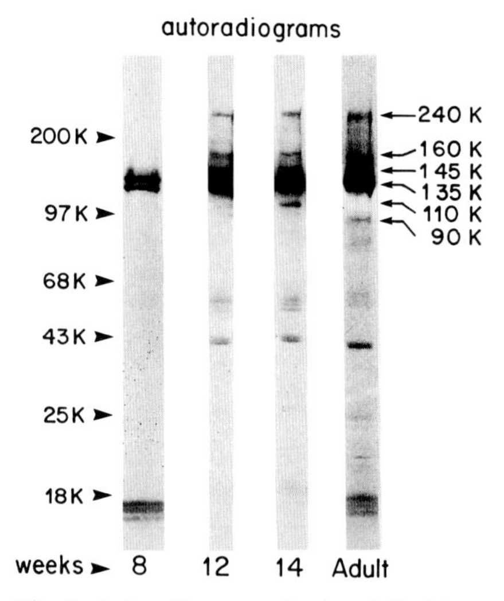

Fig. 5. Autoradiograms of calmodulin binding proteins in total jejunum homogenates at different stages of development. Samples (200 μg proteins) were run on a 5–15% linear gradient acrylamide gel in the presence of 0.1% SDS and electrophoretically transferred to nitrocellulose membranes. The membranes were overlaid with (^{125}I) calmodulin (4 μCi/ml) for 5 h as described by Flanagan and Yost (1984). After washing the membranes were air-dried and (^{125}I) calmodulin binding was visualized by autoradiography on Fuji X-ray film. Molecular weights were calculated by using prestained standards (Bethesda Research Laboratories) in the blots used to produce the autoradiograms

formation of still sparse and irregular microvilli. At 14 weeks (Fig. 4c) the intensity of the apical fluorescence had increased markedly in parallel to the maturation of the apical brush borders. Figure 4d illustrates the adult stage with higher villi lined by regularly labeled brush borders.

3.3 Calmodulin Binding Proteins in the Developing Human Jejunum

Characterization of Calmodulin Binding Proteins. The occurrence of calmodulin binding proteins in the human fetal jejunum was investigated by electrophoresis of total jejunum homogenates and incubation of the nitrocellulose replica with (^{125}I) calmodulin in the presence of calcium (Fig. 5).

At 8 weeks of gestation and up to 11 weeks, only one (^{125}I) calmodulin binding *doublet* was detectable, migrating at a position corresponding to 145–135 kDa. At 12 weeks of gestation this doublet had increased in intensity and two less intense bands had appeared at positions corresponding to 160 kDa and 240 kDa. At 14 weeks of gestation the autoradiograms revealed the appearance of a new major calmodulin binding band migrating at a position corresponding to approximately 110 kDa. At that developmental stage, the number and the intensity of the calmodulin binding bands were similar to those obtained in the adult jejunum. However, it must be stressed that in adults, the 110 kDa calmodulin-binding band was no longer visible, but a 90 kDa one was present.

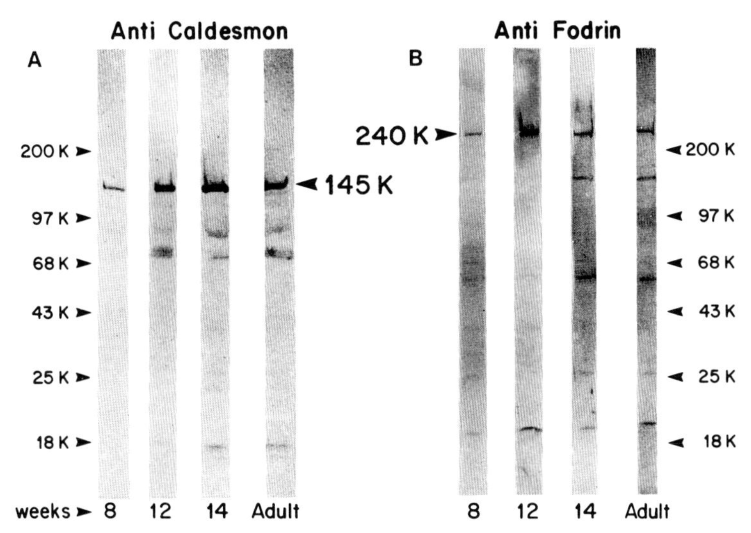

Fig. 6A,B. Immunoblots of fetal and adult human jejunum homogenates with caldesmon (**A**) and fodrin (**B**) antibodies

Identification of the Calmodulin Binding Proteins. The nature of the (^{125}I) calmodulin binding bands was identified by immunoblotting experiments with specific antibodies.

From 8 weeks of gestation up to the adult stage, caldesmon antibodies labeled *one band* with an apparent molecular weight of 145 kDa (Fig. 6A) corresponding to the 145 kDa band of the (^{125}I) calmodulin labeled doublet (Fig. 5). From 12 weeks of fetal life, some lower molecular weight bands (90 kDa and 70 kDa) were also immunoreactive on immunoblots; they may represent proteolytic breakdown products generated during sample preparation as described by others (Ngai and Walsh 1985; Bretscher 1984) in spite of efforts to minimize proteolysis (presence of a cocktail of protease inhibitors in all media).

In the immunoblotting experiments, fodrin antibodies labeled the 240 kDa band (Fig. 6B). By this technique the 240 kDa band was visible as soon as 8 weeks of gestation, whereas it was still not detectable by the (^{125}I) calmodulin overlay technique (Fig. 5). As for caldesmon, some lower molecular weight bands (mainly 160 and 60 kDa) probably resulting from the proteolytic degradation of the 240 kDa protein, were also immunoreactive (Hirokawa et al. 1983).

The results described with total jejunum homogenates were also reproduced with pure endoderms of 8 weeks fetuses and with brush borders of 10–11 weeks fetuses (data not shown).

Concerning the 110 kDa protein, immunoblotting experiments with purified and specific 110 kDa protein antibodies gave different results (Fig. 7). The 110 kDa protein was undetectable by this technique in 8-week-old fetal intestinal homogenates (not shown). However, by increasing the epithelium protein concentration (use of pure endoderm) a faint labeled band could be visualized at that early stage. Surprisingly this faint immunoreactive band was localized at a position corresponding to 135 kDa

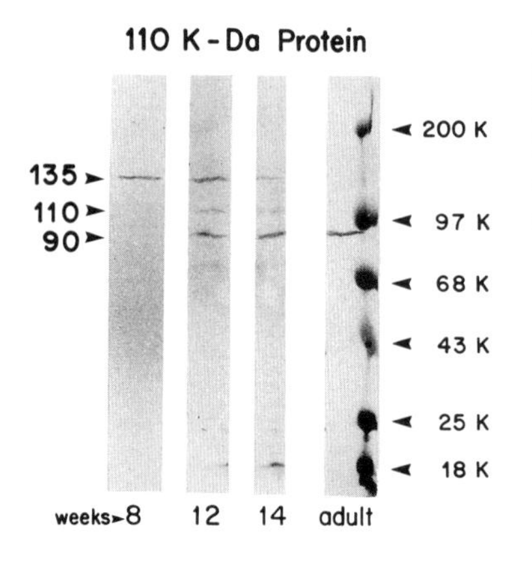

Fig. 7. Immunoblots of fetal and adult human jejunum with 110 kDa antibodies. 8-week-old human fetus = endoderm (200 μg); 12-week-old human fetus = brush borders (200 μg); 14-week-old human fetus and adult = total jejunum homogenates (150 μg)

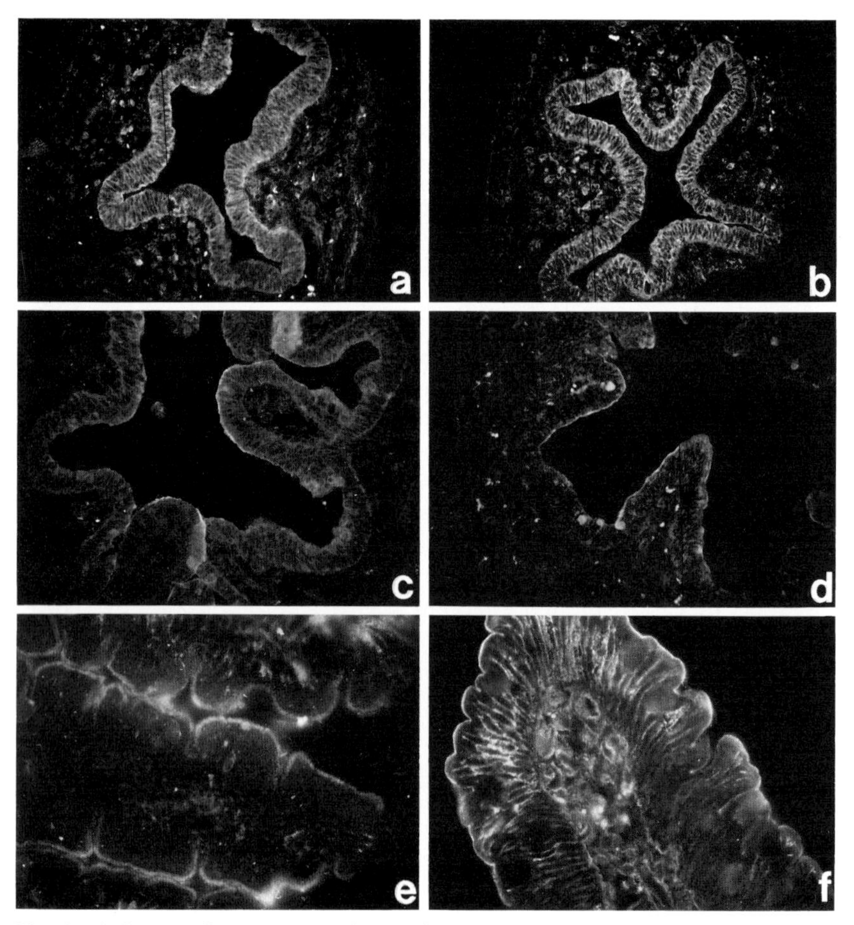

Fig. 8a–f. Immunofluorescence staining of caldesmon (**a**, **c**, **e**) and fodrin (**b**, **d**, **f**) in human jejunum at different stages of development. **a**, **b** 8-week-old fetus (× 150); **c**, **d** 12-week-old fetus (× 150); **e**, **f** adult (× 300)

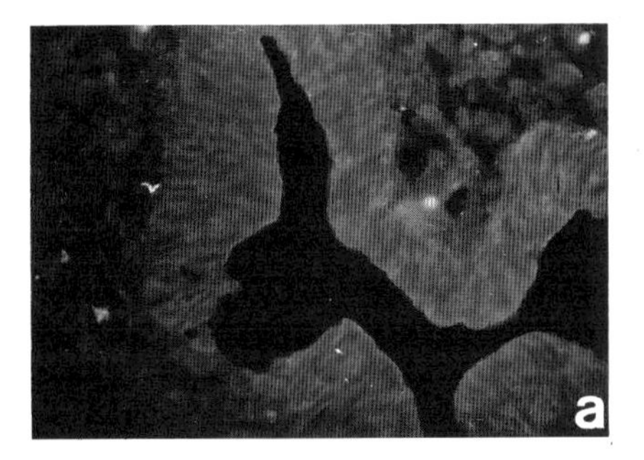

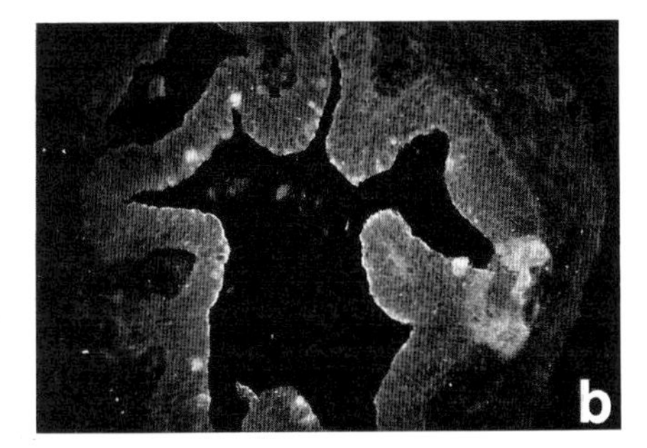

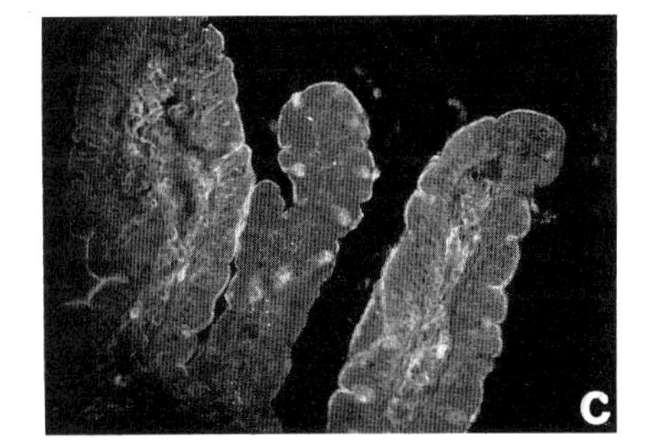

Fig. 9a–c. Immunofluorescence staining of 110 kDa protein in developing human jejunum. **a** 8-week-old fetus (× 300); **b** 12-week-old fetus (× 150); **c** adult (× 150)

[corresponding to the lower band of the (^{125}I) calmodulin labeled doublet] instead of 110 kDa.

At 12 weeks of fetal life, the amount of this protein had increased since it was detectable in total homogenates. Moreover, two other immunoreactive bands with lower molecular weights (110 and 90 kDa) appeared in both total homogenates and brush border preparations. From the 14 week of fetal life, a progressive shift of the two high molecular weight bands occurred towards the lower one which became the unique labeled band in the adult.

Localization of Calmodulin Binding Proteins by Immunofluorescence in Developing Human Fetal Intestine. At 8 weeks of human gestation, immunofluorescence of caldesmon and fodrin revealed an intense labeling uniformely localized throughout the whole endoderm (Fig. 8a,b). However an intensification of the labeling was visible at the apical side of the epithelial cells and included the circumference of individual cells. On the other hand, the 110 kDa protein was difficult to detect at that early stage (Fig. 9a).

At 12 weeks, with the three types of antibodies, the fluorescence was quite restricted to the apical part of the epithelial cells (Figs. 8c,d and 9b). From that stage onwards, the immunofluorescence labeling of either caldesmon, fodrin or 110 kDa protein, increased progressively in parallel to the maturation of brush borders (illustrated in the adult in Figs. 8e,f, 9c). It must be stressed that at the adult stage, the fodrin labeling (Fig. 8f) was obvious not only at the level of brush borders but also along the lateral sides of the cells, as already described in chicken (Glenney et al. 1982) and rat (Rochette-Egly and Haffen 1987) intestinal epithelium.

4 Conclusions

The present chapter emphasizes that in human intestine both the identity and the localization of calmodulin binding proteins are the same as in avians (Mooseker 1985) and rodents (Rochette-Egly and Haffen 1987). Furthermore, it demonstrates the existence of a parallelism between the developmental pattern of calmodulin and calmodulin binding proteins, and the morphological differentiation of human fetal intestine. As soon as the 8th week of fetal life, calmodulin and the three major calmodulin binding proteins, caldesmon, fodrin and the 110 kDa protein, are detectable in undifferentiated human endoderms although the latter one is expressed at very low levels. At these early stages, the apparent discrepancy between the biochemical and cytochemical observations, would rather reflect the low amount of these proteins compared to the others and also the presence of these proteins in the whole epithelial cells rather than associated to recognizable brush borders. Then up to the 14th week of fetal life, calmodulin and calmodulin binding proteins increase markedly and segregate to the brush borders of epithelial cells at the time of their differentiation; in parallel, calmodulin and 110 kDa protein become detectable by immunocytochemistry. Such a phenomenon was also observed by ourselves in rodents for either calmodulin (Rochette-Egly et al. 1986) or calmodulin binding proteins (Rochette-Egly and Haffen 1987). Similarly, other intestinal cytoskeleton proteins are known to be synthetized in undifferentiated epithelial cells and to segregate to brush borders when they are formed (Coudrier et al. 1984; Robine et al. 1985; Mooseker 1985; Shibayama et al. 1986, 1987). Contrariwise, Carboni et al. (1986) showed that in chick intestine, the 110 kDa protein segregates to the brush borders only late in development (the day before hatch). This discrepancy might, in fact, reflect the more gradual and complex brush border assembly in chick intestine (Chambers and Grey 1979; Mooseker 1985).

Another interesting finding arising from the present study was the presence at early development stages (8 weeks), of an immunoreactive form of the 110 kDa protein, with an apparent molecular weight of 135 kDa. This observation was afforded not only in human fetuses but also in 14-day-old rat fetal endoderms (Rochette-Egly and Haffen 1987). This 135 kDa band is able to bind (^{125}I) calmodulin (the lower band of the 145–135 kDa doublet) and is distinct from caldesmon. The possibility that the detection of this protein was due to the presence of contaminating antibodies is excluded since affinity purified 110 kDa protein antibodies were used (Shibayama et al. 1987). A possible explanation is that the 110 kDa protein might be synthesized in undifferentiated cells as a 135 kDa form which cleaves progressively into a 110 kDa fragment when it segregates to differentiating brush borders. A similar phenomenon has already been described for brush border intestinal hydrolases which are synthetized as high molecular weight precursors and cleaved by the pancreatic enzymes (Hauri et al. 1979; Skovberg 1982; Triadou and Zweibaum 1985). Concerning the 90 kDa band observed in the adults, it might represent the common proteolytic fragment of the 110 kDa protein in spite of the presence of a cocktail of protease inhibitors. Finally, the exact basis for the antigenic relationship between the 135 kDa and the 110 kDa protein is unclear, but these results suggest that both proteins share some degree of homology in their molecular structures.

In conclusion, most of the human intestinal maturation occurs very early, between 8 and 14 weeks of gestation, and although the evolution of the 110 kDa protein is unclear, the overall developmental pattern of calmodulin, caldesmon, and fodrin parallels the morphological and functional maturation of brush borders.

Acknowledgements. The authors wish to thank Dr. Glenney for the generous gift of fodrin antibodies, Dr. Mooseker for 110 kDa antibodies and Dr. Walsh for caldesmon antibodies. They are also endebted to Drs. Doffoel and Kern for providing the biopsies and the abortion material. They acknowledge E. Alexandre, C. Arnold and D. Daviaud for their skillful technical assistance, B. Lafleuriel for the realization of graphs, and Mrs C. Haffen for photography.

References

Bretscher A (1984) Smooth muscle caldesmon. Rapid purification and F-actin cross-linking properties. J Biol Chem 259:12873–12880

Bretscher A, Lynch W (1985) Identification and localization of immunoreactive forms of caldesmon in smooth and non muscle cells; a comparison with the distribution of tropomyosin and actinin. J Cell Biol 150:1656–1663

Carboni JM, Shibayama T, Mooseker MS (1986) Brush border assembly: redistribution of the microvillar core protein, 110K during embryogenesis. J Cell Biol 103:112a (abstract)

Chambers C, Grey RD (1979) Development of the structural components of the brush border in absorptive cells of the chick intestine. Cell Tissue Res 204:387–405

Cheung WY (1980) Calmodulin plays a pivotal role in cellular regulation. Science 207:19–27

Colony PC (1983) Successive phases of human fetal intestinal development. In: Kretchmer N, Ninkowski A (eds) Nutritional adaptation of the gastrointestinal tract of the newborn. Raven Press, New York, pp 3–28

Coudrier E, Reggio H, Louvard D (1981) Immunolocalization of the 110 000 molecular weight cytoskeleton protein of intestinal microvilli. J Mol Biol 152:49–66

Coudrier E, Robine S, Huet C, Arpin M, Sahuquillo C, Louvard D (1984) Expression of two structural markers of brush border intestinal mucosa: villin and a membrane glycoprotein (140 Kd), in a human colon carcinoma cell line HT29. J Submicrosc Cytol 16:159–160

Flanagan SD, Yost B (1984) Calmodulin-binding proteins: visualization by (^{125}I) calmodulin overlay on blots quenched with Tween 20 or bovine serum albumin and poly (Ethylene oxide). Anal Biochem 140:510–519

Glenney JR, Glenney P (1983) Spectrin, fodrin and TW 260/240: a family of related proteins lining the plasma membrane. Cell Motil 3:671–682

Glenney JR Jr, Glenney P (1984) The microvillus 110K Da cytoskeletal protein is an integral membrane protein. Cell 37:743–751

Glenney JR Jr, Glenney P (1985) Comparison of Ca^{++} regulated events in the intestinal brush border. J Cell Biol 100:754–763

Glenney JR Jr, Weber K (1980) Calmodulin binding proteins of the microfilaments present in isolated brush borders and microvilli of intestinal epithelial cells. J Biol Chem 255:10551–10554

Glenney JR Jr, Bretscher A, Weber K (1980) Calcium control of the intestinal microvillus cytoskeleton: its implication for the regulation of microfilament organization. Proc Natl Acad Sci USA 77:6458–6462

Glenney JR Jr, Glenney P, Osborn M, Weber K (1982) An F-actin and calmodulin-binding protein from isolated intestinal brush borders has a morphology related to spectrin. Cell 28:843–854

Haffen K, Kedinger M, Lacroix B (1986) Cytodifferentiation of the intestinal villus epithelium. In: Desnuelle P, Sjöström H, Nören O (eds) Molecular and cellular basis of digestion. Elsevier, Amsterdam, pp 311–322

Hauri HP, Quaroni A, Isselbacher KJ (1979) Biogenesis of intestinal plasma membrane: posttranslational route and cleavage of sucrase-isomaltase. Proc Natl Acad Sci USA 76:5183–5186

Hirokawa N, Cheney RE, Willard M (1983) Location of a protein of the fodrin-spectrin-TW260/240 family in the mouse intestinal brush border. Cell 32:953–965

Howe CL, Mooseker MS (1983) Characterization of the 110K dalton actin-calmodulin and membrane-binding protein from microvilli of intestinal epithelial cells. J Cell Biol 97:974–985

Howe CL, Mooseker MS, Graves TA (1980) Brush-border calmodulin: a major component of the isolated microvillus core. J Cell Biol 86:916–923

Ishimura K, Jufita H, Ban T, Matsuda H, Sobue K, Kakiuchi S (1984) Immunocytochemical demonstration of caldesmon (a calmodulin, F-actin interacting protein) in smooth muscle fibers and absorptive intestinal cells in the small intestine of the rat. Cell Tissue Res 235:207–209

Kedinger M, Simon PM, Grenier JF, Haffen K (1981) Role of epithelial-mesenchymal interactions in the ontogenesis of intestinal brush border enzymes. Develop Biol 86:339–347

Kedinger M, Haffen K, Simon-Assmann P (1986) Control mechanisms in the ontogenesis of villus cells. In: Desnuelle P, Sjöström H, Nören O (eds) Molecular and cellular basis of digestion. Elsevier, Amsterdam, pp 323–334

Lacroix B, Wolff-Quenot MJ, Haffen K (1984a) Early human hand morphology: an estimation of fetal age. Early Hum Dev 9:127–136

Lacroix B, Kedinger M, Simon-Assmann P, Haffen K (1984b) Early organogenesis of human small intestine: scanning electron microscopy and brush-border enzymology. Gut 25:925–930

Louvard D, Reggio H, Coudrier E (1986) Cell surface asymmetry is a prerequisite for the function of transporting and secreting epithelia. In: Desnuelle P, Sjöström H, Nören O (eds) Molecular and cellular basis of digestion. Elsevier, Amsterdam, pp 25–42

Mackensie NR, Morris B, Morris R (1983) Protein binding to brush borders of enterocytes from the jejunum of the neonatal rat. Biochim Biophys Acta 755:204–209

Mathan M, Moxey PC, Trier JS (1976) Morphogenesis of fetal rat duodenal villi. Am J Anat 146: 73–92

Means AR, Dedman JR (1980) Calmodulin – an intracellular calcium receptor. Nature (Lond) 285:73–77

Mooseker MS (1985) Organization, chemistry and assembly of the cytoskeletal apparatus of the intestinal brush border. Ann Rev Cell Biol 1:209–241

Mooseker MS, Keller TCS, Hirokawa N (1983) In: Brush border membranes – Ciba Foundation Symposium 95. Pitman books Ltd, London, p 195

Ngai PK, Walsh MP (1985) Detection of caldesmon in muscle and non-muscle tissues of the chicken using polyclonal antibodies. Biochem Biophys Res Commun 127:533–539

Robine S, Huet C, Moll R, Sahuquillo-Nerino C, Coudrier E, Zweibaum A, Louvard D (1985) Can villin be used to identify malignant and undifferentiated normal digestive epithelial cells? Proc Natl Acad Sci USA 82:8488–8492

Rochette-Egly C, Daviaud D (1985) Calmodulin binding to nitrocellulose and zetapore membranes during electrophoretic transfer from polyacrylamide gels. Electrophoresis 6:235–238

Rochette-Egly C, Garaud JC, Kedinger M, Haffen K (1986) Calmodulin in epithelial intestinal cells during rat development. Experientia 42:1043–1046

Rochette-Egly C, Haffen K (1987) Developmental pattern of calmodulin-binding proteins in rat jejunal epithelial cells. Differentiation (in press)

Shibayama T, Carboni T, Mooseker MS (1986) Brush border assembly: distribution of microvillar core proteins during embryogenesis. J Cell Biol 103:112a (abstract)

Shibayama T, Carboni JM, Mooseker MS (1987) Assembly of the intestinal brush border: appearance and redistribution of microvillar core proteins in developing chick enterocytes. J Cell Biol 105:335–344

Skovbjerg H (1982) High molecular weight pro sucrase-isomaltase in human fetal intestine. Pediatr Res 16:948–949

Triadou N, Zweibaum A (1985) Maturation of sucrase-isomaltase complex in human fetal small and large intestine during the gestation. Pediatr Res 19:136–138

Vollrath L (1969) Über die Entwicklung des Dünndarms der Ratte. In: Brodal A, Hild W, Ortman R, Schiebler TH, Töndury G, Wolff E (eds) Ergebnisse der Anatomie und Entwicklungsgeschichte. Bd 4. Springer, Berlin Heidelberg New York, pp 1–66

Site-Specific Mutagenesis and Protein Engineering Approach to the Molecular Mechanism of Calcium Signal Transduction by Calmodulin

J. Haiech[1] and D. M. Watterson[2]

1 Introduction

Calmodulin is an ubiquitous, small and acidic protein involved in calcium signal management in all eukaryotic cells (Van Eldik et al. 1982; Cox et al. 1984). The cell, when stimulated, transiently increases its cytosolic calcium concentrations from approximately 0.1 μM to 10 μM. In this range, calmodulin binds calcium ions, modifies its conformation and thereby modifies the function of calmodulin binding proteins. Therefore, calmodulin is a key element in the transduction of a quantitative calcium signal into a cellular response (Klee et al. 1980; Kilhoffer et al. 1983).

This transduction has at least four components (Haiech and Demaille 1981):

1. The molecular mechanisms which determine the calcium dependent interactions of calmodulin with its target proteins (kinematic regulation).
2. The localization of calmodulin and target proteins in the cell (topological regulation).
3. The ratio between the concentration of calmodulin and the different target proteins (genomic regulation).
4. The post translational modification of the target proteins and calmodulin (concerted regulation).

Clearly, these are not exclusive or all-encompassing classifications and are intended only to simplify the complex nature of calmodulin regulation. A variety of approaches are being utilized to elucidate how each of these components contributes to the involvement of calmodulin functioning as a signal transducer. In this paper, we will briefly discuss some examples of how site-directed mutagenesis and protein engineering are being used as one approach in attempts to gain a better understanding of the molecular mechanisms of calmodulin action.

2 Calmodulin as a Transducer of Calcium Signals

As indicated above, in order for calmodulin to serve as a transducer of calcium signals, it must bind calcium, undergo conformational changes and properly transduce these

1 LP8402 CNRS-INSERM, Route de Mende, BP5051, 34033 Montpellier Cedex, France
2 Department of Pharmacology, Vanderbilt University Medical Center, and Laboratory of Cellular and Molecular Physiology, Howard Hughes Medical Institute, Nashville, TN 37232, USA

Ch. Gerday, R. Gilles, L. Bolis (Eds.)
Calcium and Calcium Binding Proteins

alterations in calmodulin structure to the calmodulin binding protein. How calmodulin functions in this role has been studied over the past 10 years by using two approaches: examination of the interaction of calcium with calmodulin and the interaction of calcium/calmodulin with a specific enzyme or binding protein. The first approach provides some understanding of calcium binding and calcium induced conformational changes, and the second approach provides information on the interaction of calmodulin, upon binding calcium, with a given calmodulin binding protein. A third approach, which has not been initiated until recently due to technical limitations, is the examination of the interaction of calcium with the calmodulin-calmodulin binding protein complex. This latter approach requires the availability of relatively large amounts of chemically homogeneous calmodulin binding proteins or functional fragments of these proteins. The purification of various calmodulin binding proteins, the identification of calmodulin binding sites in these binding proteins, and the production of analogs by the use of peptide synthesis and expression cloning of DNA have now made this a feasible approach. In addition, this latter approach can be expanded to include an analysis of the chemical complementarities in the multiple points of interaction between calmodulin and its binding proteins that must be involved in transducing a change in calcium concentration into a catalytic event.

2.1 Calcium Binding to Calmodulin

Calcium binding has been studied by gel filtration, ultrafiltration, flow dialysis, equilibrium dialysis, calcium titration using calcium sensitive dyes or a calcium electrode. Calmodulin possesses four calcium sites and two to four other low affinity cationic sites (Milos et al. 1986). The binding properties are dependent on pH, ionic strength and cation concentration, e.g., magnesium concentration (Kilhoffer et al. 1983).

As pointed out by Klotz in a recent review (Klotz 1985), the binding mechanism of a protein with four sites is described by 15 independent binding constants. From binding studies, it is only possible to obtain four macroscopic binding constants. Therefore, to determine the 15 independent binding constants, it is necessary either to obtain data from other experiments and/or to propose a binding mechanism which can be experimentally verified. Under most of the conditions used by researchers, the Scatchard representation of data from binding experiments was a straight line. Therefore, calmodulin was believed to have four independent calcium sites characterized by four intrinsic constants of equal magnitude (all the coupling factors between the sites are set to 1). However, a more detailed study of calcium binding to calmodulin by Crouch and Klee (1980) indicated a positive cooperativity between at least the first two sites occupied. In this case, at least one coupling factor is necessary to describe the binding mechanisms. A study of calcium binding under different conditions (different magnesium and potassium concentrations) allowed a description of the system by using three coupling constants (Haiech et al. 1981). It was proposed that the calcium binding to calmodulin was sequential and ordered. Other proposals (Wang 1985) suggest that calmodulin is composed of two independent pairs of calcium binding sites and each pair cooperatively binds calcium. In this latter case, two coupling factors are necessary to describe the system.

2.2 Calcium Induced Conformational Change

When calmodulin binds 1 to 4 calcium ions per mole of protein, its conformation is modified. If we consider only one site, we have the following scheme:

$$CaM + Ca \underset{k_{-1}}{\overset{k_1}{\rightleftharpoons}} CaMCa \underset{k_{-2}}{\overset{k_2}{\rightleftharpoons}} CaMCa^* .$$

The first step is associated with the true binding (or ligation) of calcium, and the second one with the conformational change. When followed by spectroscopic techniques such as fluorescence, NMR, or UV spectroscopy, only CaMCa* is monitored. Therefore, the signal could be associated with any of the 15 independent binding constants or linear combinations of such constants. For instance, a spectroscopic signal could monitor the occupancy of one specific site, two sites together and so on. From NMR and fluorescence studies (Forsen et al. 1986), it was concluded that calmodulin possesses two high affinity sites and two low affinity sites. This is based on the observations that the environment of residues located in the carboxy terminal part of the molecule, as followed by NMR resonance or the change in fluorescence intensity of tyrosine-138, was altered under conditions where two calcium ions should be bound per mole of protein. This conclusion appeared not to be in complete agreement with data obtained from calcium binding studies, but the assumption that binding between the different sites is coupled allowed full agreement between the two sets of data (Haiech et al. 1981; Kilhoffer et al. 1983; Wang 1985). However, the exact nature of this coupling is not yet fully understood and has not been described in mathematical terms.

Finally, it should be noted that the calcium-dependent parameter being monitored in an analysis of CaMCa* is not necessarily the same parameter whose change results in activity. For example, the calcium dependent increase in overall helix content of calmodulin that is estimated from circular dichroism spectroscopy is not well-correlated with the calcium dependent conformational change that results in enzyme stimulation. This was first indicated in studies by Stevens and co-workers (Walsh et al. 1978), who showed that limited chemical modifications of calmodulin can result in inactivation of calmodulin but not result in changes in the CD spectrum in the presence of calcium. Later studies (Craig et al. 1987), using site-specific mutagenesis and protein engineering, confirmed that the increase in helix content in the presence of calcium could occur in calmodulins in which activity with at least two enzymes had been lost. Thus, caution should be used in evaluating the results of experiments in which the calcium dependence of a structural parameter change is correlated with both calcium binding, per se, and calcium dependent activation of an enzyme.

2.3 Interaction of Calmodulin with Target Proteins

In addition to the necessity of analyzing conformational change and calcium binding to calmodulin, it is necessary to study both the interaction of calmodulin with calcium and calmodulin binding proteins, and the activation of the calmodulin binding protein.

Although activation of an enzyme by calmodulin clearly requires the binding of calmodulin to the enzyme, the binding of calmodulin to an enzyme does not necessarily result in activation. Based on precedents in protein chemistry, it is reasonable to suppose that the binding of calmodulin to an enzyme is the result of numerous contact points, or pairs of chemical complementarities, each of which is being utilized at some point in time but not necessarily at the same time as all of the other contact points. That is, there is a set of dynamic equilibria that result in binding between calmodulin and the enzyme, and another set of dynamic equilibria that result in enzyme activation. The set of contact points that are utilized in protein-protein binding is distinct from the set used for activation, but many of the individual contact points are probably identical between the two sets. Thus, functional conformational changes are those which result in a shift to the set of equilibria found in the activated complex.

Site-specific mutagenesis and protein engineering, as well as peptide synthesis under some circumstances, help in simplifying the study of calmodulin *binding* to its target protein vs. *activating* its target protein. Through the use of calmodulin binding peptides or calmodulin binding fragments (domains) based on the calmodulin binding site of a target enzyme, the basic features and energetics of binding can be partially dissociated from the activation step. Analogous to the interaction of calcium with calmodulin, calmodulin interaction with a calmodulin binding protein can be written as involving at least two steps:

$$\text{Ca·CaM} + \text{CaMBP} \underset{k_{-1}}{\overset{k_1}{\rightleftharpoons}} \text{Ca·CaM·CaMBP} \underset{k_{-2}}{\overset{k_2}{\rightleftharpoons}} \text{Ca·CaM·CaMBP}^*$$

In most experiments in the past, the species that was monitored was Ca·CaM·CaMBP* because the parameter measured was the ability of calmodulin, in the presence of increasing concentrations of calcium, to stimulate the activity of an enzyme. Clearly, there are conformational changes that can occur in both the calmodulin and calmodulin binding protein upon initial binding (formation of Ca·CaM·CaMBP). Additional contact points are required to get to the Ca·CaM·CaMBP* state, which is then capable of interacting with substrates and producing products. Much as the interaction of calmodulin with the calmodulin binding protein affects the interaction of calcium with calmodulin, it is not difficult to envision how the interaction of the activated calmodulin binding protein (Ca·CaM·CaMBP*) with its substrates and products might affect the interaction of calmodulin with the calmodulin binding protein. This alteration in the interaction would not necessarily be a physical dissociation, but may be a more relaxed structure with fewer or distinct sets of contact points. Thus, it is important to try to estimate the couplings among these various mechanistic states, some of which may exist very transiently.

One approach to simplifying this analysis would be a site-specific mutagenesis/protein engineering approach in which high-affinity calmodulin binding structures that lack catalytic activity are studied. The calmodulin binding structures could range in complexity from peptide analogs based on elucidated calmodulin binding sites to regions of calmodulin binding proteins produced from cDNAs. Recently, we have initiated such an approach to study the interaction of calmodulin and smooth muscle myosin light chain kinase (MLCK). Initially, a synthetic peptide analog of the calmodulin binding site was utilized (Lukas et al. 1986). More recently, synthetic nucleotide probes based on the amino acid sequence of the calmodulin binding site were used to isolate cDNA structures. The cDNA structures were characterized, and then expressed proteins were produced by an *E. coli* expression cloning system similar to that used for calmodulins. The use of expressed fragments of increasing complexity allows the elucidation of fine structural details (e.g., NMR, spectroscopy and crystallography) on the same structures tested for protein binding and calcium binding activity. The availability of the cDNAs also allow mutagenesis studies to be done on the calmodulin binding site (receptor) to complement the analysis of mutated calmodulins (ligand).

Clearly, the next stage is to initiate analysis of the interaction between substrates/inhibitors and the enzymes. In this regard, we have extended the studies of the MLCK system by the synthesis of peptide substrates and inhibitors that have micromolar K_m values. Analogs of the substrates and inhibitors, especially those with reporter groups (e.g., spin labels for ESR spectrometry), potentially can be very useful in attempts to monitor changes in the environment of the active site. Thus, a variety of engineered calmodulins (some with reporter groups such as tryptophan), engineered functional fragments of MLCK (also containing spectroscopic probes), and synthetic analogs that are competitive inhibitors or substrates potentially can provide a more simplified supramolecular complex amenable to experimental analysis and manipulation.

2.4 Activation of Calmodulin Binding Enzymes

As indicated in the previous sections, the investigation of how calmodulin brings about the activation of an enzyme is a complex problem that has several mechanistic steps. Based upon the continuing studies of calcium binding by calmodulin and how it is coupled to calmodulin binding and activation of its target proteins, there are logical starting points for more detailed investigations of the activation steps.

Initial studies (Cox et al. 1982; Huang et al. 1981; Malnoë et al. 1982) using the approaches discussed previously indicated that the active complex is Ca_3·CaM·CaMBP* or Ca_4·CaM·CaMBP* for phosphodiesterase, adenylate cyclase, MLCK and calcium ATPase. However, in these studies: (1) the enzyme concentration is much lower than the calmodulin concentration; (2) nothing is known about the calcium binding properties of the enzyme-calmodulin complex or calmodulin-enzyme-substrate complex; and (3) there is a population of calmodulin molecules, the average state of which is assumed to be three or four calciums bound. Therefore, it is difficult to know what is being monitored, much less its dependence on a given calcium-calmodulin complex. From the results gathered up to now, we cannot conclude that all these calmodulin-dependent enzymes interact with the same calmodulin conformation. On the contrary, comparative structure-function analyses, activity analysis of calmodulin fragments and chemically modified calmodulins, studies of the interaction of calmodulin with calmodulin binding peptides, and, more recently, site-specific mutagenesis and protein engineering studies indicate that:

1. Each enzyme, or group of enzymes, seems to possess its own set of requirements for interaction with, and activation by, calmodulin.

2. The interaction of calmodulin with an enzyme modifies the enzyme conformation and, also, the calmodulin conformation.

These observations raise questions about the conclusions from the earlier studies and indicate the need for further analysis of mutant calmodulins in which the calcium binding ligands are selectively altered. The mutant protein study might simplify the experimental analysis by providing a homogeneous population of Ca·CaM complexes.

3 Site-Specific Mutagenesis and Protein Engineering of Calmodulin: an Approach to Dissecting the Molecular Mechanism of Calmodulin Action

To analyze the structure-activity relationships of calmodulin, we have to tackle a problem similar to that of Champollion, who decoded the Egyptian writing from the Rosetta Stone. For calmodulin, the top part of our Rosetta Stone is the primary structure, the mid part the tertiary structure and the bottom part the function of calmodulin, i.e., its properties of interaction and activation of different enzymes. To increase the level of information of the Rosetta Stone of calmodulin, we may study calmodulins from closely and distantly related phylogenetic species, or calmodulins expressed from their cDNAs (Van Eldik et al. 1982; Watterson et al. 1980; Schaefer et al. 1987; Marshak et al. 1984; Lukas et al. 1985b; Lukas et al. 1984). However, these studies are limited by the presence of multiple differences among the calmodulin structures being compared. Therefore, it has become necessary to produce specific mutations at predetermined sites in the protein in order to gain new insight into the relationships among calmodulin structure, folding and function.

An in-depth study of calmodulin's interaction with metal ions and its binding proteins will require a number of "mutant" proteins. This, in turn, requires an efficient method for site-specific mutagenesis and protein engineering. In the case of calmodulin, a synthetic gene designed for efficient, cassette-mediated mutagenesis has been the starting material (Roberts et al. 1985). A similar approach has been initiated with another calcium binding protein, the intestinal calcium binding protein (Brodin et al. 1986). The design of the calmodulin gene was based on several considerations:

1. The gene was designed to code for calmodulin based on sequence analysis of naturally occurring calmodulin.
2. The gene was designed to incorporate multiple, unique restriction sites that are regularly spaced within the coding sequence and not found in commonly used recombinant DNA vectors. This allows for mutagenesis while the gene is resident in the vector and increases efficiency by forcing the correct orientation of the cassette. Using the cassette mutagenesis approach, it is possible to engineer and characterize a mutant gene in 2 to 4 weeks.
3. The coding sequence of the synthetic gene is flanked by unique EcoRI and BamH1 sites at the 5′ and 3′ termini for facile molecular cloning. These are popular insertion sites engineered into many recombinant DNA vectors.

The initial expressed protein, termed VU-1 calmodulin, has an amino acid sequence that is a hybrid of vertebrate and plant calmodulins (Roberts et al. 1985). In addition, the bacterially expressed protein lacks two of the post-translational modifications common to vertebrate, plant and some other calmodulins (acetylation of the amino terminus and trimethylation of lysine-115). Finally, VU-1 calmodulin activates all calmodulin-stimulated enzymes that have been tested. VU-1 can now be used as a standard of comparison in our studies of calmodulin structure and function. This allows a direct comparison of proteins that differ only in the specific amino acid under study.

3.1 Strategies for the Production of Mutant Calmodulins

The ability to produce site-specific alterations in calmodulin allows us to address two interrelated ligand-receptors problems:

1. The basic features of calcium (= ligand) interaction with calmodulin (= receptor).
2. The general and specific features of calmodulin (= ligand) interactions with its physiological binding proteins (= receptors).

To tackle the first problem, we are using mutant proteins to characterize the properties of the amino acids involved in calcium ligation and the amino acids (or patterns of amino acids) involved in the coupling between the different calcium sites. Modification of the calcium binding loop should provide insight into both ligand-receptor problems. Relatedly, modification of helical regions of calmodulin and connecting peptides among calcium binding sites should provide insight into the mechanisms of coupling between the calcium sites. Along these lines, at least three kinds of mutant proteins are required:

1. Mutants with reporter groups, such as tryptophan, located in or near a given calcium site in order to follow the modification of this specific site and its local environment. Introduction of such residues must perturb as little as possible the structure and the properties of calmodulin.
2. Mutants that lack, or have modified properties in, a single calcium binding site in order to check the coupling between the occupancy of this site and the others.
3. Mutants that are perturbed in their helical parts and/or their connecting peptide in order to try to modify the coupling between the different sites.

The availability of primary structures from other calcium binding proteins allows the detection of patterns of amino acids that are needed in order for a peptide to bind calcium and magnesium (Haiech and Sallantin 1985). Using logical rules to characterize these patterns, it is possible to optimize the experimental strategy for the introduction of specific amino acids to fulfill the requirements previously mentioned. In addition, the characterization of mutant calmodulins from mutant organisms, such as the *Paramecium* pantophobiac A mutant (Schaefer et al. 1987), can also provide logical starting points. After production of the mutant proteins, functional and structural analyses (e.g., X-ray crystallography, NMR spectrometry and spectroscopy) of the various mutant proteins should provide information about how the modifications alter the calcium binding properties, the calcium-induced conformational changes, the tertiary structure and the calcium-dependent activator activities.

As far as the interaction and the activation of enzymes are concerned, the problem is much more complex. In this case, we have to analyze several calmodulin-enzyme systems to be able to draw general rules. Our initial studies have included phosphodiesterase, adenylate cyclase, MLCK, NAD kinase, calcineurin and protein kinase II.

The study of the interaction of calmodulin with these enzymes, and how this results in activation, has at least two general requirements:

1. Sufficient amounts of homogeneous preparations of enzymes or their calmodulin binding domains must be available. For example, with MLCK the calmodulin binding domains of two isoforms of the enzyme have been synthesized (Edelman et al. 1985; Lukas et al. 1986) and functional fragments expressed from cDNAs are available.
2. Appropriately "tagged" derivatives of calmodulin, calmodulin binding proteins (the enzymes) and substrates or inhibitors of the enzymes must be available. Reporter groups, such as tryptophan, can be introduced into calcium binding sites as well as potential protein-protein interaction domains to facilitate monitoring of changes in environment. A similar approach is being used with analogs of calmodulin binding sites produced by peptide synthesis (Lukas et al. 1986) or expression cloning of cDNAs (Zimmer and Watterson, unpublished observations). To facilitate competition binding studies in which actual dissociation is monitored, we have metabolically labeled VU-1 calmodulin with ^{35}sulfate (Haiech and Asselin, unpublished observation). To facilitate monitoring of the active site environment, we have made recently a series of spin labeled analogs of peptide inhibitors (Lukas and Watterson, unpublished observations). With the availability of these various "tagged" molecules, it should be feasible to do studies on a series of analogs of the various ligands (calmodulin, inhibitor/substrate) and receptors (calmodulin, enzyme) under the same set of experimental conditions. These types of results combined with those from two-dimensional NMR and crystallography, should provide new insight into the molecular basis of binding and activation.

4 Conclusion

Site-directed mutagenesis and protein engineering provide new strategies for studies aimed at understanding the molecular mechanisms of calcium-calmodulin-protein interaction and gaining insight into the physiological function of calmodulin. This approach, therefore, has the potential of helping to elucidate a fundamental mechanism for eukaryotic cells: the translation of a quantitative calcium signal into a qualitative cellular response.

References

Brodin P, Grundstrom T, Hofmann T, Drakenberg T, Thulen E, Forsen S (1986) Expression of bovine intestinal calcium binding proteins from a synthetic gene in Escherichia coli and characterization of the product. Biochemistry 25:5371–5378

Cox JA, Comte M, Stein EA (1982) Activation of human erythrocyte Ca^{2+}-dependent Mg^{2+}-activated ATPase by calmodulin and calcium. Proc Natl Acad Sci USA 79:4265–4269

Cox JA, Comte M, Malnoë A, Burger D, Stein EA (1984) Mode of action of the regulatory calmodulin. Met Ions Biol Syst 17:215–293

Craig TA, Watterson DM, Prendergast FP, Haiech J, Roberts DM (1987) Site-specific mutagenesis of the alpha-helices of calmodulin: effects of altering a charge cluster in the helix that links the two halves of calmodulin. J Biol Chem 262:3278–3284

Crouch TH, Klee CB (1980) Positive cooperative binding of calcium to bovine brain calmodulin. Biochemistry 19:3692–3698

Edelman AM, Takio K, Blumenthal DK, Hansen RS, Walsh KA, Titani K, Krebs EG (1985) Characterization of the calmodulin-binding and catalytic domains in skeletal muscle myosin light chain kinase. J Biol Chem 260:11275–11285

Forsen S, Vogel HJ, Drakenberg T (1986) In: Cheung WY (ed) Calcium and cell function. Vol 6. Academic Press, NY, pp 113–157

Haiech J, Demaille JG (1981) Supramolecular organization of regulatory proteins into calcisomes: a model of the concerted regulation by calcium ions and cyclic adenosine 3′:5′-monophosphate in eukaryotic cells. In: Holzer (ed) Metabolic interconversion of enzyme. Springer, Berlin Heidelberg New York, pp 303–313 (Proc Life Sci)

Haiech J, Sallantin J (1985) Computer search of calcium binding sites in a gene data bank: use of learning techniques to build an expert system. Biochemie 5:555–560

Haiech J, Klee CB, Demaille JG (1981) Effects of cations on affinity of calmodulin for calcium: ordered binding of calcium ions allows the specific activation of calmodulin stimulated enzymes. Biochemistry 20:3890–3897

Huang CY, Chou V, Chock PB, Wang JH, Sharma RK (1981) Mechanism of activation of cyclic nucleotide phosphodiesterase: requirement of the binding of four Ca^{2+} to calmodulin for activation. Proc Natl Acad Sci USA 78:871–874

Kilhoffer MC, Haiech J, Demaille JG (1983) Ion binding to calmodulin. Mol Cell Biochem 51: 33–54

Klee CB, Crouch TH, Richman PG (1980) Calmodulin. Ann Rev Biochem 49:489–515

Klotz IM (1985) Ligand-receptor interactions: facts and fantasies. Q Rev Biophysics 18:227–259

Lukas TJ, Iverson DB, Schleicher M, Watterson DM (1984) Structural characterization of a higher plant calmodulin. Plant Physiol 75:788–795

Lukas TJ, Marshak DR, Watterson DM (1985a) Drug-protein interactions: isolation and characterization of covalent adducts of phenoxybenzamine and calmodulin. Biochemistry 26:151–157

Lukas TJ, Wiggins ME, Watterson DM (1985b) Amino acid sequence of a novel calmodulin from the unicellular alga Chlamydomonas. Plant Physiol 78:477–483

Lukas TJ, Burgess WH, Prendergast FG, Lau W, Watterson DM (1986) Calmodulin binding domains: characterization of a phosphorylation and calmodulin binding site from myosin light chain kinase. Biochemistry 25:1458–1464

Malnoë A, Cox SA, Stein EA (1982) Ca^{2+}-dependent regulation of calmodulin binding and adenylate cyclase activation in bovine cerebellar membranes. Biochim Biophys Acta 714:84–92

Marshak DR, Clarke M, Roberts DM, Watterson DM (1984) Structural and functional properties of calmodulin from the eukaryotic microorganism Dictyostelium discoideum. Biochemistry 23: 2891–2899

Milos M, Schaer JJ, Comte M, Cox JA (1986) Calcium-proton and calcium-magnesium antagonisms in calmodulin: microcalorimetric and potentiometric analyses. Biochemistry 25:6279–6287

Roberts DM, Crea R, Malecha M, Alvarado-Urbina G, Chiarello RH, Watterson DM (1985) Chemical synthesis and expression of a calmodulin gene designed for site-specific mutagenesis. Biochemistry 24:5090–5098

Schaefer WH, Lukas TJ, Blair IA, Schultz JE, Watterson DM (1987) Amino acid sequence of a novel calmodulin from Paramecium tetraurelia that contains dimethyllysine in the first domain. J Biol Chem 262:1025–1029

Van Eldik LJ, Zendegui JG, Marshak DR, Watterson DM (1982) Calcium-binding proteins and the molecular basis of calcium action. Int Rev Cytol 77:1–61

Walsh M, Stevens JC, Oikowa K, Kay CM (1978) Circular dichroism studies on Ca^{2+} dependent protein modulator oxidized with N-chlorosuccinimide. Biochemistry 17:3928–3930

Wang CL (1985) A note on Ca^{2+} binding to calmodulin. Biochem Biophys Res Com 130:426–430

Watterson DM, Sharief F, Vanaman TC (1980) The complete amino-acid sequence of the Ca^{2+}-dependent modulator protein (calmodulin) of bovine brain. J Biol Chem 255:962–975

The Role of Calcium in the Regulation of Plant Cellular Metabolism

D. MARMÉ[1]

1 Introduction

Living systems have established several mechanisms by which they transduce extracellular signals into an intracellular language which can be understood by the biochemical and biophysical machinery of the cell. One such mechanism was discovered in animal cells almost 20 years ago by Robinson et al. (1971) and is called the second messenger principle with cyclic 3′,5′-adenosine monophosphate (cAMP) as the intracellular second messenger. With the discovery of calmodulin, a cellular Ca^{2+}-binding protein, by Cheung (1980) it became apparent that Ca^{2+} fulfils all the criteria for being a second messenger in animals too. Recently Rasmussen (1981), in his monograph on Ca^{2+} and cAMP, has called them the "synarchic messengers", with Ca^{2+} being even more important than cAMP (Rasmussen 1981). Recently, a book has been published describing the manifold physiological reactions which are mediated and regulated by calcium (Marmé 1985).

There is convincing evidence now that cAMP also exists in higher plants (Van Onckelen et al. 1982) but there is no physiological evidence for a role as second messenger similar to that in animal cells. In particular, no cAMP-dependent protein kinase – the primary target molecule for cAMP in animal cells – has been reported so far. On the other hand, it has been reported that many physiological processes in plants are under the control of Ca^{2+} (e.g. cell elongation, cell division, protoplasmic streaming, enzyme secretion). However, the molecular mechanism of this Ca^{2+}-dependent regulation has not been convincingly elucidated in most of these cases. One of the reasons is the difficulty of measuring the free cytoplasmic Ca^{2+} concentration (Williamson 1981).

Over the last few years a new and equally important signal transducing system has been discovered in animal cells: the protein kinase C/phosphatidylinositol breakdown (Marmé and Matzenauer 1985). There are a few indications that a similar mechanism also operates in plants (Morré et al. 1984 and this paper).

In this article we will deal with the regulatory role of calcium in the cellular metabolism. We will describe the various mechanisms by which a plant cell controls the cytoplasmic free calcium concentration. The calcium, calmodulin-dependent enzymes and their control mechanisms are reviewed and the calcium-, calcium/calmodulin- and calcium/phospholipid-dependent protein phosphorylation will be discussed. A critical

1 University of Freiburg, Institute of Biology III, Schänzlestr. 1, D–7800 Freiburg, FRG

Ch. Gerday, R. Gilles, L. Bolis (Eds.)
Calcium and Calcium Binding Proteins

view on unsolved problems of plant physiology and how calcium and calcium-dependent mechanisms can contribute to their solution, will conclude this paper.

2 Regulation of Cellular Calcium Concentrations

The control and regulation of the free cytoplasmic Ca^{2+} concentration is a prerequisite for the second messenger concept for Ca^{2+}. It has been reported that the control of cytoplasmic Ca^{2+} in plant cells is achieved as in animal cells by ATP-dependent extrusion of Ca^{2+} out of the cell (Gross and Marmé 1978; Dieter and Marmé 1980d, 1983) and by accumulation of Ca^{2+} into cellular organelles like mitochondria (Hodges and Hanson 1965; Dieter and Marmé 1980c) and endoplasmic reticulum (Gross 1982). It has been reported recently that plants possess an additional Ca^{2+} transport system located in the tonoplast (Gross 1982).

We have been able to demonstrate that the Ca^{2+} uptake into a microsomal plasma membrane-enriched fraction from dark-grown plant tissues is stimulated by the addition of calmodulin (Dieter and Marmé 1980c,d). The membrane vesicles which accumulate Ca^{2+} upon hydrolysis of ATP in vitro are most probably inside-out vesicles derived from the plasma membrane (Dieter and Marmé 1980d), suggesting that, in situ, this Ca^{2+} transport system extrudes Ca^{2+} out of the cell. The calmodulin stimulation of the microsomal Ca^{2+} uptake is due to an increase in both the apparent maximum transport velocity and the affinity for Ca^{2+} (Dieter and Marmé 1983). The calmodulin-dependent Ca^{2+} transport ATPase has been partially purified by calmodulin-Sepharose affinity chromatography (see below).

We could show that mitochondria take up much more Ca^{2+} than do the microsomal vesicles (Dieter and Marmé 1980c, 1983). The determination of the kinetic constants K_m and V_{max} of the plant mitochondrial Ca^{2+} transport system reveals a V_{max} of 63 nmol/min/mg and a K_m of about 250 μM (Dieter and Marmé 1983). Comparison of the corresponding values of the microsomal Ca^{2+} transport system shows that the V_{max} of mitochondria is about 10–20 times higher, but that mitochondria have an affinity for Ca^{2+}, which is at most one-tenth that of microsomes. When the transport activity of mitochondria and microsomes are plotted as a function of the free Ca^{2+} concentration it becomes obvious that the mitochondrial Ca^{2+} pump operates only at much higher Ca^{2+} concentrations as compared to the microsomal Ca^{2+} pump. Therefore, one may assume that, as in animal cells, the low Ca^{2+} concentration in an unstimulated plant cell is maintained by the microsomal Ca^{2+} pump, which is most probably located in the plasma membrane.

The second messenger concept for Ca^{2+} in higher plants implies that the cytoplasmic free Ca^{2+} concentration can be changed by 'primary signals' like light or hormones. Any change of the transport properties of Ca^{2+} transport systems caused by these signals would result in a change of the free cytoplasmic Ca^{2+} concentration. Kubowitz et al. (1982) demonstrated that plant hormones like auxin, zeatin and kinetin alter the activity of the microsomal Ca^{2+} transport system isolated from soybean hypocotyl hooks. Recently, Olah et al. (1983) showed that the affinity of a

Ca^{2+}-ATPase towards Ca^{2+} and calmodulin increases when the plant seedlings were treated with the synthetic cytokinin benzylaminopurine.

We could show that the calmodulin stimulation of the microsomal Ca^{2+} uptake disappears when the intact corn seedlings are irradiated with far red light; the calmodulin-independent Ca^{2+} uptake, however, was not significantly changed (Dieter and Marmé 1981b). The determination of the kinetic constants V_{max} and K_m reveals that calmodulin does not increase the V_{max} or the affinity for Ca^{2+} when the microsomal vesicles are isolated from far red light-irradiated corn seedlings (Dieter and Marmé 1983). The mitochondrial Ca^{2+} uptake is also inhibited by far red light (Dieter and Marmé 1981b). Blue and ultra-violet light inhibit both the calmodulin-dependent and calmodulin-independent microsomal Ca^{2+} transport (E. Schnellbächer and D. Marmé, unpublished).

From these data it can be seen that the low Ca^{2+} concentration in the cytoplasm of a plant cell is maintained by a calmodulin-dependent Ca^{2+} transport ATPase located most probably in the plasma membrane. By changing Ca^{2+} fluxes through the membranes surrounding the cytoplasm, "primary signals" like light, hormones, etc., are able to alter $[Ca^{2+}]_{cyt}$ and consequently are able to change Ca^{2+}-dependent biochemical and physiological reactions.

Recently we were able to demonstrate that a radiolabelled calcium antagonist, ^{3}H-verapamil, which is known to block calcium channels in mammalian cells, also binds to plant membranes (Andrejauskas et al. 1985). Association of ^{3}H-verapamil to a membrane fraction was saturable and reversible. The apparent equilibrium dissociation constant was K_D = 102 nM and the maximal number of binding sites was B_{max} = 60 pmol/mg. Sucrose density fractionation of zucchini membrane preparations revealed that ^{3}H-verapamil binding sites are located primarily at the plasma membrane. The physiological function of these binding sites is not yet clear. However, it could well be – as in the case of mammalian cells – that they indicate the existance of specific calcium channels in plant cells.

3 Calcium and Calmodulin-Dependent Enzymes

Only a few enzymes have been discovered so far which are under the control of Ca^{2+} and calmodulin: NAD kinase (soluble in cytoplasm), NAD kinase (located in the outer mitochondrial membrane), NAD kinase (located in the chloroplast envelope), Ca^{2+}+ Mg^{2+}) ATPase and various protein kinases.

Soluble NAD kinase was found in a cytoplasmic fraction obtained from dark-grown zucchini (Dieter and Marmé 1980a). The enzyme has been partially purified by calmodulin-Sepharose affinity chromatography and could be stimulated about eightfold by calmodulin in a calcium-dependent manner. The stimulation could be achieved by either bovine brain or zucchini calmodulin at concentrations for half saturation of about 0.1 and 1.0 μM, respectively. At optimal calmodulin concentrations, half saturation occurs at a pCa of about 4 μM. The M_r of the NAD kinase from zucchini has been estimated by Sephadex G-100 chromatography to be about 50 kDa (Dieter and Marmé 1980b).

NAD kinase activity from dark-grown corn coleoptiles was shown to be almost totally dependent on Ca^{2+} and calmodulin. Nearly all of the enzyme activity was found in a particulate fraction (Dieter and Marmé 1984). Upon differential and density gradient centrifugation, the NAD kinase activity co-migrates with the mitochondrial cytochrome c oxidase, whereas marker activities for nuclei, etioplasts, endoplasmic reticulum, and microbodies could well be separated, indicating that the NAD kinase is associated with mitochondria. This NAD kinase, associated with intact mitochondria, can be activated by exogenously added Ca^{2+} and calmodulin. In order to investigate the submitochondrial localization of the NAD kinase, the organelles were ruptured by osmotic treatment and sonication and the submitochondrial fractions were separated by density gradient centrifugation. The NAD kinase activity exhibits the same density pattern as the antimycin A-insensitive, NADH-dependent, cytochrome c reductase, a marker enzyme of the outer mitochondrial membrane. Marker enzymes for the mitochondrial matrix and the inner mitochondrial membrane indicate that the Ca^{2+}, calmodulin-dependent NAD kinase from coleoptiles of dark-grown seedlings is located at the outer mitochondrial membrane.

In homogenates from light-grown pea seedlings, more than half of a Ca^{2+}, calmodulin-dependent activity and most of a Ca^{2+}, calmodulin-independent activity of the homogenate were associated with chloroplasts (Marmé and Dieter 1982). The Ca^{2+}, calmodulin-dependent activity could be detected by adding Ca^{2+} and calmodulin to the incubation medium containing intact chloroplasts. This activity could not be separated from the chloroplasts by successive washes or by phase partition in aqueous two polymer phase systems. After chloroplast fractionation, the Ca^{2+}, calmodulin-dependent NAD kinase activity was localized at the envelope, and a Ca^{2+}, calmodulin-independent activity was recovered from the stroma.

In all three cases the NAD kinase is able to sense changes of the calcium concentration in the cytoplasm, a prerequisite of the calcium messenger concept. Furthermore, the NAD kinase is of special interest because of its light dependence. We could show recently (Marmé and Dieter 1982) that in corn coleoptiles, when exposed to far red light, the NADP(H)/NAD(H) ratio increases. This reflects the activation of the NAD kinase located in the outer mitochondrial membrane by an increase of the free, cytoplasmic Ca^{2+} concentration which is due to the light inhibition of the Ca^{2+} transport activity (Dieter and Marmé 1981). The effect of far red light could be mimicked by incubating segments of corn coleoptiles at high or low Ca^{2+} concentrations in the presence of the divalent cationophore A23187 (Dieter et al. 1984). Similar data were obtained with segments from zucchini hypocotyls. These results suggest that the Ca^{2+}-dependent, regulatory mechanism is the same foı the soluble, cytoplasmic NAD kinase in zucchini and the outer mitochondrial membrane-bound NAD kinase from corn.

First evidence for a calmodulin-dependent Ca^{2+} transport ATPase came from Ca^{2+} transport experiments (see above). The calmodulin dependence of the $(Ca^{2+} + Mg^{2+})$-ATPase permits purification of the enzyme by calmodulin-Sepharose affinity chromatography (Dieter and Marmé 1981a). After solubilization and centrifugation of a microsomal fraction from coleoptiles of dark-grown corn seedlings, the supernatant was loaded on an affinity column in the presence of Ca^{2+}. The column was washed with buffer containing Ca^{2+} and NaCl to remove contaminating material. The bound proteins were eluted by chelating Ca^{2+} with EGTA. The wash fraction contains most of the ATPase activity but no calmodulin-dependent one. The eluate fraction has only a very small percentage of the total ATPase activity, which can be stimulated by calmodulin in the presence of Ca^{2+} by more than 130 percent as compared to only 8 percent in the supernatant. Comparison of the calmodulin-dependent ATPase activity and the calmodulin-dependent Ca^{2+} transport activity suggests that the partially purified ATPase is most probably identical with the calmodulin-dependent Ca^{2+} transport ATPase.

4 Calcium-Dependent Protein Phosphorylation

The control by Ca^{2+} of protein phosphorylation in animal cells is well documented. Recently, two short reports have been published indicating the regulatory effect of Ca^{2+} and calmodulin on plant protein phosphorylation (Hetherington and Trewavas 1982; Polya and Davies 1982). We have described the biochemical and regulatory properties of a protein kinase activity associated with a particulate fraction from zucchini hypocotyl hooks (Salimath and Marmé 1983). We could show Ca^{2+} and calmodulin-dependent protein phosphorylation in a membrane fraction from zucchini.

Because of its particular regulatory properties in plants, the influence of the concentration of free Ca^{2+} on the phosphorylation of the membrane-bound proteins is of special interest. Phosphorylation of one peptide (M_r 180,000) is inhibited at physiological Ca^{2+} concentrations between 0.1 μM and 1 mM. This may be due to the inhibition of a specific protein kinase or activation of a specific phosphoprotein phosphatase. The incorporation of ^{32}P into two other polypeptides of M_r 26,000 and M_r 36,000 is enhanced at the same Ca^{2+} concentration range, while peptides of M_r 45,000 and M_r 70,000 exhibit almost no change in phosphorylation between 0.1 μM and 1 mM free Ca^{2+}. Phosphorylation of six other proteins was enhanced by Ca^{2+}. In all these cases, a free Ca^{2+} concentration of about 0.3–3 μM was favourable for maximal phosphorylation. It is probable that the phosphorylation in the presence of Ca^{2+} alone is already due to the calmodulin contaminating the EDTA-washed membranes.

The effect of the Ca^{2+}-binding protein calmodulin on the phosphorylation of the membrane proteins was studied by varying the calmodulin concentration at a free Ca^{2+} concentration of 1 μM. The relative ^{32}P incorporation into at least five peptides was affected by calmodulin (Salimath and Marmé 1983). All bands of M_r 14,000, 26,000, 36,000, 45,000 and 100,000 were enhanced at calmodulin concentrations between 10 nM and 10 μM. We could show that specific calmodulin antagonists fluphenazine and R24571 inhibit the calmodulin-dependent stimulation of the phosphorylation of all five peptides at a drug concentration between 1 and 20 μM. When fractions obtained from density gradient centrifugation were assayed for the incorporation of ^{32}P, it was found that the protein substrates which are phosphorylated in a Ca^{2+}- and calmodulin-dependent fashion are located in the endoplasmic reticulum (M_r 14,000, 36,000 and 100,000) and mitochondria (M_r 45,000), and that one peptide

(M_r 26,000) is distributed between plasma membrane and endoplasmic reticulum, as identified by membrane and organelle markers (B.P. Salimath and D. Marmé, unpublished).

All attempts to demonstrate either cAMP-dependent of cGMP-dependent protein kinase have failed.

We conclude from these and other published data that protein phosphorylation in plants can be regulated by Ca^{2+} at physiological concentrations and that at least part of this regulation is mediated by calmodulin. We do not yet have any relevant information about the physiological significance of such a Ca^{2+}- and calmodulin-dependent protein phosphorylation.

Very recently we have separated protein kinase activities by DEAE cellulose ion exchange and by blue-Sepharose chromatography. A protein kinase could be identified which was activated by 0.3 μM Ca^{2+} and which was further stimulated by phosphatidylserine. The phospholipid alone had no effect (Schäfer et al. 1985). This enzyme appears to be similar to the protein kinase C which has been reported from animals to depend on phosphotidylserine and calcium. However, the plant enzyme needs further characterization to ensure this similarity.

5 Conclusion

This article has summarized information on the regulatory role of Ca^{2+} in plant physiology. The discoveries of plant calmodulin and Ca^{2+}-calmodulin-dependent enzymes were major contributions to our understanding of the underlying molecular mechanisms. The biochemical properties of cellular Ca^{2+}-transport mechanisms and their regulations are beginning to be explored. Detailed knowledge of the distribution and redistribution of cellular Ca^{2+} on extracellular signals is a prerequisite for the exact analysis of Ca^{2+}-dependent cellular processes.

There are certainly similarities between the second messenger mechanisms of animal and plant cells (e.g. both plant and animal calmodulin show a high degree of structural and functional homology). However, as can be extrapolated from the data presented here, the ways in which extracellular signals change cytoplasmic Ca^{2+} concentrations and also the equipment of animal and plant cells with Ca^{2+}-calmodulin-dependent enzymes seems to be very different. In addition to the regulatory Ca^{2+} system, a protein kinase activity, dependent on Ca^{2+} and phospholipids, has been demonstrated to occur in plants. This might indicate the existence of the protein kinase C/phosphatidylinositol metabolism type of regulatory system in the plant kingdom.

Future research should focus mainly on three topics: (a) the identification of external factors or signals that interfere with the cellular Ca^{2+}-transport mechanisms, (b) the search for additional Ca^{2+}-calmodulin-dependent enzymes and the elucidation of their regulatory potential in cellular metabolism and (c) the exploration of the phospholipid-dependent protein kinase system.

References

Andrejauskas E, Hertel R, Marmé D (1985) Specific binding of the calcium antagonist ^{3}H-verapamil to membrane fractions from plants. J Biol Chem 260:5411–5414

Cheung WY (1980) In: Cheung WY (ed) Calmodulin: an introduction in calcium and cell function. Vol 1. Academic Press, New York, pp 1–12

Dieter P, Marmé D (1980a) Partial purification of plant NAD kinase by calmodulin-Sepharose affinity chromatography. Cell Calcium 1:279–286

Dieter P, Marmé D (1980b) Calmodulin-activated plant microsomal Ca^{2+} uptake and purification of plant NAD kinase and other proteins by calmodulin-Sepharose affinity chromatography. Ann NY Acad Sci 356:371–373

Dieter P, Marmé D (1980c) Ca^{2+} transport in mitochondrial and microsomal fractions from higher plants. Planta 150:1–8

Dieter P, Marmé D (1980d) Calmodulin activation of plant microsomal Ca^{2+} uptake. Proc Natl Acad Sci USA 77:7311–7314

Dieter P, Marmé D (1981a) A calmodulin-dependent, microsomal ATPase from corn (Zea mays L.). FEBS Lett 125:245–248

Dieter P, Marmé D (1981b) Far-red light irradiation of intact corn seedlings affects mitochondrial and calmodulin-dependent microsomal Ca^{2+} transport. Biochem Biophys Res Commun 101: 749–755

Dieter P, Marmé D (1983) The effect of calmodulin and far-red light on the kinetic properties of the mitochondrial and microsomal calcium ion transport system from corn. Planta 159:277–281

Dieter P, Marmé D (1984) A Ca^{2+}, calmodulin-dependent NAD kinase from corn is located in the outer mitochondrial membrane. J Biol Chem 259:184–189

Dieter P, Salimath B, Marmé D (1984) The role of calcium and calmodulin in higher plants. In: Boudet A (ed) Annual proceedings of the phytochemical society of Europe. Vol 23. Oxford University Press, pp 213–229

Gross J (1982) Oxalate-enhanced active calcium uptake in membrane fractions from zucchini squash. In: Marmé D, Marré D, Hertel R (ed) Plasmalemma and tonoplast. Elsevier, Amsterdam, pp 369–379

Gross J, Marmé D (1978) ATP-dependent Ca^{2+} uptake into plant membrane vesicles. Proc Natl Acad Sci USA 75:1232–1236

Hetherington A, Trewavas A (1982) Calcium-dependent protein kinase in pea shoot membranes. FEBS Lett 145:67–71

Hodges TK, Hanson JB (1965) Calcium accumulation by maize mitochondria. Pl Physiol Lancaster 40:101–108

Kubowitz BP, Vanderhoef LN, Hanson JB (1982) ATP-dependent calcium transport in plasmalemma preparation from soybean hypocotyls. Pl Physiol Lancaster 69:187–191

Marmé D (1985) Calcium and cell physiology. Springer, Berlin Heidelberg New York Tokyo

Marmé D, Dieter P (1982) Calcium and calmodulin-dependent enzyme regulation in higher plants. In: Marmé D, Marré D, Hertel R (eds) Plasmalemma and tonoplast: their functions in the plant cell. Elsevier, Amsterdam, pp 111–118

Marmé D, Matzenauer S (1985) Protein kinase C and polyphosphoinositide metabolites: their role in cellular signal transduction. In: Marmé D (ed) Calcium and cell physiology. Springer, Berlin Heidelberg New York Tokyo, pp 377–386

Morré DJ, Gripshover B, Monroe A, Morré JT (1984) Phosphatidylinositol turnover in isolated soybean glycine-max membranes stimulated by the synthetic growth hormone 2 4-D. J Biol Chem 259:15346–15368

Olah Z, Berczi A, Erdei L (1983) Benzylaminopurine-induced coupling between calmodulin and Ca-ATP in wheat root microsomal membranes. FEBS Lett 154:395–399

Polya GM, Davies JR (1982) Resolution of Ca^{2+}-calmodulin-activated protein kinase from wheat germ. FEBS Lett 150:167–171

Rasmussen H (ed) (1981) Calcium and cAMP as synarchic messengers. Wiley & Sons, New York
Robinson GA, Butcher RW, Sutherland EW (1971) Cyclic AMP. Academic Press, New York, pp 1–47
Salimath BP, Marmé D (1983) Protein phosphorylation and its regulation by calcium and calmodulin in membrane fractions from zucchini hypocotyls. Planta 158:560–568
Schäfer A, Bygrave F, Matzenauer S, Marmé D (1985) Identification of a calcium and phospholipid-dependent protein kinase in plant tissue. FEBS Lett (submitted)
Simon P, Bonzon M, Grepping H, Marmé D (1984) Subchloroplastic localization of NAD kinase activity: evidence for a Ca^{2+}, calmodulin-dependent activity at the envelope and for a Ca^{2+}, calmodulin-dependent activity in the stroma of pea chloroplasts. FEBS Lett 167:332–338
Van Onckelen HA, Dupont M, de Greef JA (1982) High-performance liquid chromatographic identification and quantitation of cyclic adenosine 3′,5′-monophosphate in higher (Phaseolus vulgaris L.) and lower (Chlorella sp.) plants. Physiologia Pl 55:93–97
Williamson RE (1981) Free Ca^{2+} concentration in the cytoplasm: a regulator of plant cell function. What's New Pl Physiol 12:45–48

Calcium in Different Biological Systems

Calcium Homeostasis in Exocrine Secretory Cells

R. M. CASE, T.-A. ANSAH, S. DHO, A. MIZINIAK and L. WILSON[1]

1 Introduction

There are two major types of secretory process. One involves the discharge of macromolecules from the cell by exocytosis (e.g. digestive enzymes, transmitters and mucus). The other, which may be referred to as electrolyte secretion, involves the net unidirectional transport of ions across an epithelial membrane, accompanied by water, usually in isoosmotic proportions. Both secretory processes occur in exocrine glands which are therefore useful tissues in which to study stimulus-secretion coupling.

Ca^{2+} can play an important role in controlling both types of secretory process. To help define these roles, we have studied Ca^{2+}-homeostasis in dispersed acini prepared from the pancreas and the mandibular salivary gland. On the one hand, pancreatic acini secrete mainly digestive enzymes, and a little fluid, in response to stimulation with cholinergic agonists and the peptide hormone cholecystokinin-pancreozymin (CCK). On the other hand, cholinergic agonists evoke a copious electrolyte secretion, but only a little protein secretion from mandibular gland acini.

2 Methodology

Acini were prepared from the pancreas of guinea-pigs or rats (Ansah et al. 1986b) and mandibular glands of rabbit (Ansah et al. 1987) by enzymatic digestion and mechanical dispersion of the respective tissues. Some acini from each preparation were used for secretory studies and some for monitoring intracellular Ca^{2+} concentration ($[Ca^{2+}]_i$).

It is relatively easy to measure pancreatic protein secretion by assaying the incubation medium for its content of amylase. The content of amylase in the acini at the end of an experiment was also determined, allowing amylase release to be expressed as a percentage of the total present (Ansah et al. 1986b). It is impossible to assay electrolyte secretion in this way. However, since cholinergic stimulation of salivary glands leads to a large efflux of K^+ from the cell, K^+-loss was used as an index of agonist-induced cell activity (Ansah et al. 1987).

1 Department of Physiological Sciences, Stopford Building, University of Manchester, Manchester M13 9PT, U.K.

Ch. Gerday, R. Gilles, L Bolis (Eds.)
Calcium and Calcium Binding Proteins

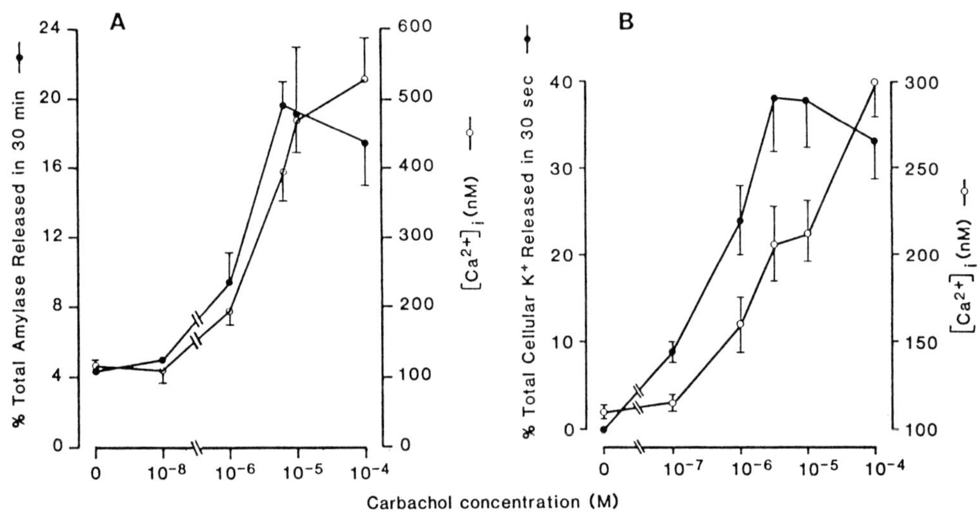

Fig. 1A,B. Relationship between $[Ca^{2+}]_i$ and secretory response. Secretory response (•) and peak fluorescence (○) achieved in the presence of carbachol were determined in parallel experiments in acini prepared from guinea pig pancreas (**A**) and rabbit mandibular gland (**B**). Data shown represent the mean ± S.E.M. of 5–8 separate experiments

$[Ca^{2+}]_i$ was monitored by quin-2 fluorescence as described previously (Ansah et al. 1986b, 1987). Fluorescence measurements were made from samples continuously stirred and maintained at 37°C. The fluorescence signal was calibrated at the end of each measurement as described by Tsien et al. (1982): the peak fluorescence following a given stimulus was used to define the response to that stimulus.

3 Results and Discussion

3.1 Secretion and $[Ca^{2+}]_i$

The resting $[Ca^{2+}]_i$ was the same in guinea pig pancreatic acini (116 ± 8 nM, n = 10), rat pancreatic acini (150 ± 28, n = 9) and rabbit mandibular gland acini (109 ± 4 nM, n = 18). Carbachol caused $[Ca^{2+}]_i$ to rise in all three preparations. At the concentration of carbachol which evoked a maximum secretory response, the rise in $[Ca^{2+}]_i$ was greater in guinea pig pancreas (430 ± 79 nM, n = 5) and least in rabbit mandibular acini (205 ± 10 nM, n = 4). These values may not represent absolute values as some buffering of Ca^{2+} by quin-2 probably occurs. However, any such buffering did not influence amylase secretion by guinea pig pancreatic acini. In spite of its buffering properties, quin-2 provides a relatively simple technique for directly monitoring cytosolic Ca^{2+}. Our studies and those of others using similar techniques (Merritt and Rubin 1985; Ochs et al. 1985) have demonstrated a direct relationship between increases in $[Ca^{2+}]_i$ and secretion (Fig. 1). Although the kinetics of the relationship are yet to be worked out, the results are in agreement with earlier studies in which $[Ca^{2+}]_i$ has been measured with Ca^{2+}-sensitive microelectrode (O'Doherty and Stark 1982) or the photoprotein aequorin (Dormer 1983).

In both pancreatic and mandibular acini, supramaximal doses of carbachol caused a further rise in $[Ca^{2+}]_i$, while the secretory response tended to decline (Fig. 1). This dissociation between $[Ca^{2+}]_i$ and both secretory responses has been observed previously (Ansah et al. 1986b), but no sufficient explanation for it can be offered.

The agonist-induced rise in $[Ca^{2+}]_i$ could be mediated by Ca^{2+} influx and/or the release of Ca^{2+} from intracellular stores. To address this issue, $[Ca^{2+}]_i$ was determined in mandibular acini in the presence (2 mM) and nominal absence (35 μM) of Ca^{2+} in the incubation medium. The absence of extracellular Ca^{2+} had no effect on resting $[Ca^{2+}]_i$ but, following stimulation with carbachol at the concentration which produces maximum K^+-release (10^{-5} M), $[Ca^{2+}]_i$ increased only to 134 ± 14 nM (n = 4). In pancreatic acini, by comparison, removal of extracellular Ca^{2+} has relatively little effect on at least the initial response to carbachol (Ochs et al. 1985).

These observations fit well with the effects of extracellular calcium deprivation on secretion by intact tissues. Fluid secretion from the isolated, perfused rabbit mandibular gland is very sensitive to changes in extracellular Ca^{2+} (Hunter et al. 1983) while enzyme secretion from the perfused cat pancreas (Hunter et al. 1983) and the incubated uncinate portion of the rat pancreas (Case and Clausen 1973; Argent et al. 1982) is not.

3.2 Intracellular Pool of Ca^{2+}

The early observations on rat pancreas just referred to (Case and Clausen 1973) suggested that stimuli which evoke pancreatic enzyme secretion do so by releasing Ca^{2+} from intracellular stores. Recent data reviewed by Schulz, this Volume, confirm this to be so. To obtain some information about this store(s), the response of pancreatic acini to sequential addition of carbachol and CCK-octapeptide (CCK-8) was determined (Fig. 2). As the dose of the first stimulus (be it CCK-8 or carbachol) was increased, and hence also its effect on $[Ca^{2+}]_i$ was increased, the response to the second stimulus was reduced. However, when the carbachol-stimulated rise in $[Ca^{2+}]_i$ was attenuated with atropine, the depleted Ca^{2+} pool was replenished allowing subsequent stimulation with CCK-8 to elevate $[Ca^{2+}]_i$ once again (Fig. 2). These findings indicate that the releasable pool of Ca^{2+} is finite and that the same pool is mobilised by both carbachol and CCK-8.

The nature of the pool remains uncertain, but data from Schulz's laboratory suggest that Ca^{2+} is mobilized from the endoplasmic reticulum. In other words, the endoplasmic reticulum (or perhaps a specialised part of it) may play a role in the pancreas (and other tissues) analogous to the role played by sarcoplasmic reticulum in muscle. The contractility of skeletal muscle is strongly dependent upon the thyroid status of an animal. Recent observations suggest that thyroid hormones increase the availability of Ca^{2+} in the cytoplasm of muscle cells by augmenting the mobilisation of calcium from the sarcoplasmic reticulum, perhaps by stimulating the proliferation of the sarcoplasmic reticulum (van Hardeveld and Clausen 1986).

We have therefore studied the effect of thyroid status on $[Ca^{2+}]_i$ and amylase secretion in rat pancreas (Ansah et al. 1986c). Hyperthyroidism was induced by daily subcutaneous injections of triiodothyronine (20 μg/100 g) for 8 days. Resting $[Ca^{2+}]_i$ was

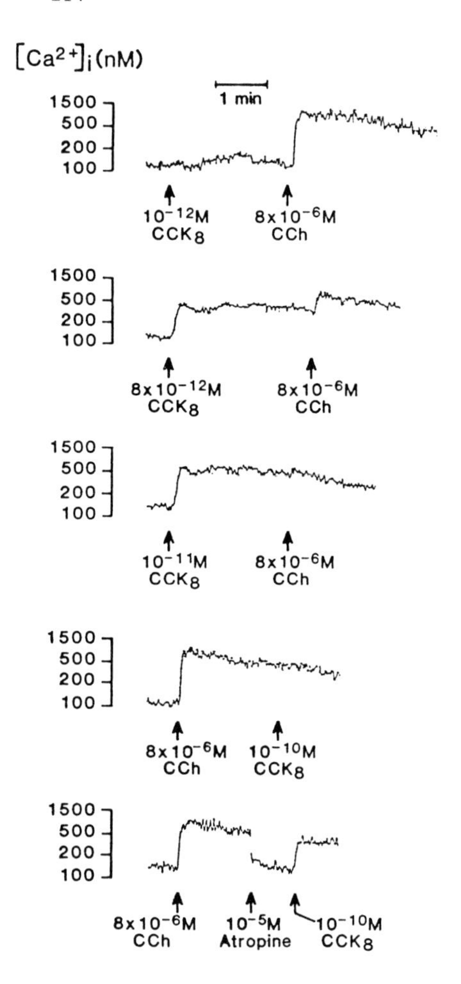

Fig. 2. Changes in $[Ca^{2+}]_i$ in response to sequential addition of CCK-8 and carbachol. Guinea pig pancreatic acini loaded with quin-2 were suspended in 2 ml of incubation medium at 37°C in the spectrofluorimeter. Secretagogues or atropine were added as shown. $[Ca^{2+}]_i$ was estimated from fluorescence measurements as described by Tsien et al. (1982). Each recording is of a single experiment representative of at least 5 others

elevated in pancreatic acini prepared from hyperthyroid animals (194 ± 21 nM, n = 23 versus 150 ± 28 nM, n = 9). In the presence of an optimal concentration of carbachol (5×10^{-7} M), $[Ca^{2+}]_i$ increased 4.6-fold in hyperthyroid animals compared with 2.6-fold in euthyroid animals (Fig. 3). However, thyroid status had no major effect on amylase secretion though, at a concentration of 10^{-7} M carbachol, secretion was higher in hyperthyroid animals. Although this small effect seems surprising, in the light of recent findings (Schick et al. 1984) it may not be. It is now evident that each pancreatic secretagogue produces its own pattern of anticoordinate changes in the synthetic rate of individual secretory proteins. In rats, caerulein (a cholecystokinin analogue) enhances synthesis of the proteases (i.e. trypsinogen and chymotrypsinogen) but decreases the synthesis of amylases (Schick et al. 1984). It is therefore conceivable that thyroid hormones could accelerate the release of proteins from zymogen granules but decrease the synthesis of amylase so that no net change in secretion (of amylase) would be observed. This needs to be determined.

The data on $[Ca^{2+}]_i$ suggest that the effect of thyroid hormones on cellular Ca^{2+} homeostasis is not restricted to muscle. They support the hypothesis that thyroid

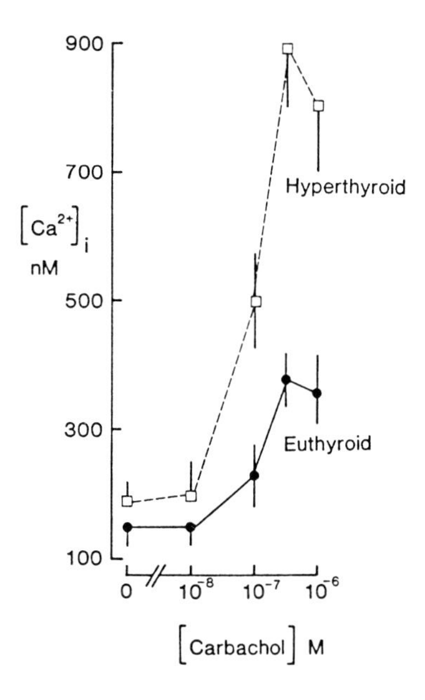

Fig. 3. Effect of carbachol on $[Ca^{2+}]_i$ in rat pancreatic acini. Changes in $[Ca^{2+}]_i$ in response to carbachol were determined in quin-2-loaded acini prepared from hyperthyroid (□) and euthyroid (•) rats. Data shown represent the mean ± S.E.M. of at least five experiments in each group

hormones have a general stimulating effect on intracellular Ca^{2+} mobilization (Storm and van Hardeveld 1986) perhaps through a trophic effect on intracellular Ca^{2+} stores.

3.3 Interactions with Phorbol Esters

As described by Cockcroft (1987), a key step in the action of calcium-mobilizing agonists involves hydrolysis by phospholipase C of phosphatidylinositol 4,5-bisphosphate to inositol 1,4,5-trisphosphate, which releases Ca^{2+} from intracellular stores, and 1,2-diacylglycerol, which activates protein kinase C. Tumour-promoting phorbol esters also activate protein kinase C and are generally accepted as stable substitutes for diacylglycerol (Ashendal 1985).

Synergism between the action of calcium ionophores (A23187 and ionomycin) and phorbol esters provides evidence for the involvement of these two separate pathways in evoking amylase secretion from the pancreas (Fig. 4; Dho et al. 1986; Ansah et al. 1986a) and other physiological responses. However, the nature of the interactions between these two signalling systems is not fully understood. We have therefore studied the effect of the phorbol ester 12-0-tetradecanoyl-phorbol 13 acetate (TPA) on $[Ca^{2+}]_i$ in guinea pig acini.

TPA alone caused pancreatic amylase secretion (Figs. 4, 5). The effect of TPA is particularly marked in the guinea pig. As with carbachol and CCK-8, supramaximal concentrations of TPA were inhibitory (Fig. 5). TPA had no effect on resting $[Ca^{2+}]_i$. However, when acini were pre-incubated with TPA for 5 min, the rise in $[Ca^{2+}]_i$ caused by carbachol was inhibited. The degree of inhibition depended upon the concentration of TPA (Fig. 5) and the dose of carbachol (Fig. 6). This inhibitory effect of TPA was not immediate: inhibition reached 50% within 1 min and was almost complete within 3 min.

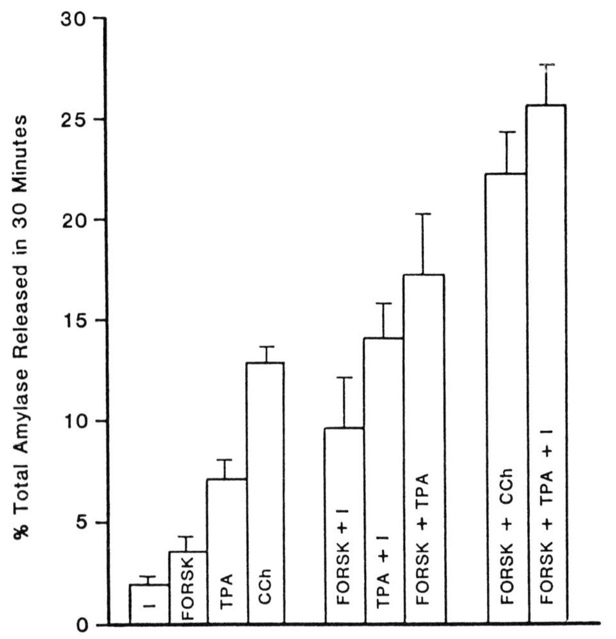

Fig. 4. Interactions between ionomycin (I; 10^{-6} M), forskolin (forsk; 10^{-5} M), 12-0-tetradecanoyl-phorbol-13-acetate (TPA; 5×10^{-7} M) and carbachol (CCh; 8×10^{-6} M) on amylase release from guinea pig pancreatic acini. Acini were incubated in the presence of one or more agonists for 30 min. Results represent the mean ± S.E.M. of at least seven experiments

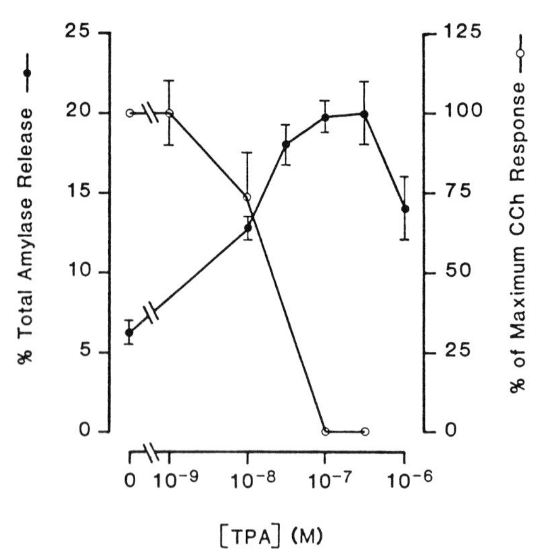

Fig. 5. Dose-response relationship of the effect of TPA on amylase secretion and carbachol-induced rise in $[Ca^{2+}]_i$. Dispersed guinea pig pancreatic acini were incubated at 37°C in the presence of varying concentrations of TPA and amylase secretion (•) determined over a 30-min period. In parallel experiments acini loaded with quin-2 were preincubated for 5 min with varying concentrations of TPA in the spectrofluorimeter. Changes in $[Ca^{2+}]_i$ in response to 2×10^{-6} M carbachol were determined (○). Values are expressed as percentage of maximum response in the absence of TPA and in the presence of DMSO (0.1%). Results are the means ± S.E.M. of 4–6 separate experiments

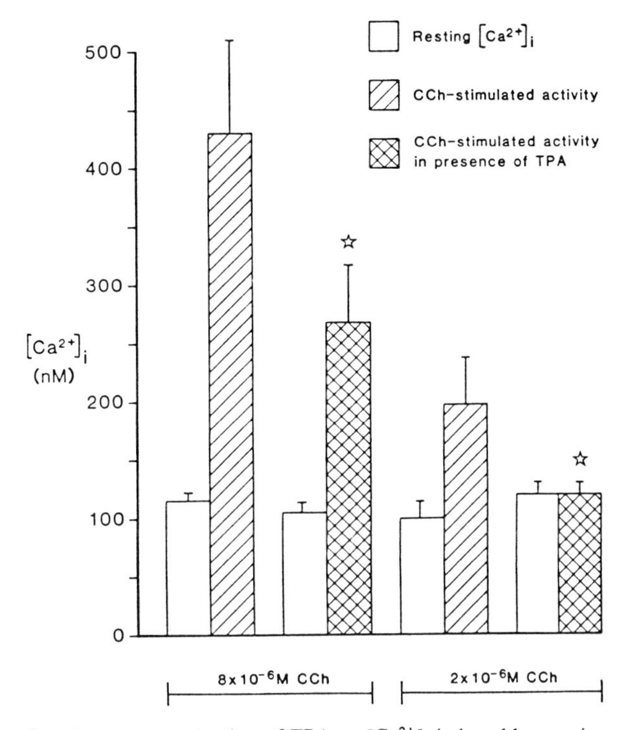

Fig. 6. Comparison of the effects of optimum concentration of TPA on $[Ca^{2+}]_i$ induced by varying concentrations of carbachol. Quin-2-loaded acini were preincubated in the spectrofluorimeter in the presence of TPA (5×10^{-7} M in 0.1% DMSO) or in 0.1% DMSO alone. Carbachol (8×10^{-6} M) was added and changes in $[Ca^{2+}]_i$ monitored. Data represent means ± S.E.M. of 8–9 separate determinations. *Asterisk* indicates that the value is significantly less ($p > 0.05$) than that for the control value by Student's unpaired t test

The mechanism of interaction remains uncertain. TPA did not inhibit the rise in $[Ca^{2+}]_i$ brought about by ionomycin, which argues against a non-specific interaction between TPA and Ca^{2+}. The receptor is a possible site of action but studies in PC12 pheochromocytoma cells show TPA to inhibit the carbachol-stimulated rise in $[Ca^{2+}]_i$ without affecting the number or affinity of receptor sites (Vincentini et al. 1985).

There is some evidence that TPA pretreatment inhibits inositol trisphosphate formation (Rittenhouse and Sasson 1985; Orellana et al. 1985). If TPA acts at the level of phospholipase C it could presumably act either directly by uncoupling the receptor-signal transduction mechanism, or by the equivalent of end-product inhibition (i.e. due to its structural similarity to diacylglycerol).

It has been suggested that the signal transduction mechanisms of the Ca^{2+}-mobilizing system is similar to that of the adenylate cyclase system in that a guanine nucleotide-binding protein (G protein) has been shown to regulate phospholipase C activity (Cockcroft 1987). In the rat and guinea-pig pancreas, amylase secretion can be evoked by activation of Ca^{2+}-mobilizing and adenylate cyclase-linked receptors. Therefore the possibility of an interaction between forskolin (a potent activator of adenylate cyclase) and the Ca^{2+}-signalling system was investigated in the guinea-pig. When the activities obtained by the combination of 10^{-5} M forskolin with maximally effective doses of ionomycin, TPA and carbachol were compared, varying degrees of potentiation were observed (Fig. 4). In the presence of TPA, forskolin evoked a 140 percent increase in amylase secretion. However the potentiating effect of forskolin on either carbachol or TPA plus ionomycin were similar (71 percent and 81 percent). These results suggest that the site of interaction between forskolin and carbachol exists distal to the accumulation of second messengers.

In conclusion Ca^{2+} plays a critical role in electrolyte and protein secretion in exocrine secretory cells. Both extracellular and intracellular Ca^{2+} are important in secretion. Second messenger interactions also contribute to the overall secretory response.

References

Ansah T-A, Dho S, Case RM (1986a) Phorbol ester (TPA), intracellular Ca^{2+} concentration and amylase secretion in pancreatic acini. Biomed Res 7 Suppl 2:109–112

Ansah T-A, Dho S, Case RM (1986b) Calcium concentration and amylase secretion in guinea pig pancreatic acini: interactions between carbachol, cholecystokinin octapeptide and the phorbol ester, 12-0-tetradecanoylphorbol 13-acetate. Biochim Biophys Acta 889:326–333

Ansah T-A, Dho S, Miziniak A, Case RM (1986c) Effects of thyroid hormone on cytosolic free calcium and amylase secretion in rat pancreatic acini. Digestion 35:5

Ansah T-A, Wilson L, Case RM (1987) Cytosolic free calcium in rabbit mandibular salivary gland cells following cholinergic stimulation. In: Davison JS, Shaffer EA (eds) Gastrointestinal and hepatic secretion. Calgary University Press, Calgary (in press)

Argent BE, Case RM, Hirst FC (1982) The effect of extracellular calcium deprivation on amylase secretion and ^{45}Ca efflux from rat pancreas. J Physiol 323:339–352

Ashendal CL (1985) The phorbol ester receptor: a phospholipid regulated protein kinase. Biochim Biophys Acta 818:219–242

Case RM, Clausen T (1973) The relationship between calcium exchange and enzyme secretion in the isolated rat pancreas. J Physiol 235:75–102

Cockcroft S (1987) Activation of polyphosphoinositide phosphodiesterase by hormones and guanine nucleotides: role of guanine nucleotide regulatory protein, Gp, in signal transduction. (This symposium)

Dho S, Ansah T-A, Case RM (1986) Effect of interactions between forskolin, phorbol ester and ionomycin on amylase secretion in guinea-pig acini. Biochem Soc Trans 14:1030

Dormer RL (1983) Direct demonstration of increases in cytosolic free Ca^{2+} during stimulation of pancreatic enzyme secretion. Biosci Rep 3:233–240

Hunter M, Smith PA, Case RM (1983) The dependence of fluid secretion by mandibular salivary gland and pancreas on extracellular calcium. Cell Calcium 4:307–317

Merritt JE, Rubin RP (1985) Pancreatic amylase secretion and cytoplasmic free calcium. Effects of ionomycin, phorbol dibutyrate and diacylglycerols alone and in combination. Biochem J 230:151–159

Ochs DL, Korenbrot JI, Williams JA (1985) Relation between free cytosolic calcium and amylase release by pancreatic acini. Am J Physiol 249:G389–G398

O'Doherty J, Stark RJ (1982) Stimulation of pancreatic acinar secretion: increases in cytosolic calcium and sodium. Am J Physiol 242:G513–G521

Orellana SA, Solski PA, Brown JH (1985) Phorbol ester inhibits phosphoinositide hydrolysis and calcium mobilization in cultured astrocytoma cells. J Biol Chem 260:5236–5239

Rittenhouse SE, Sasson JP (1985) Mass changes in myoinositol trisphosphate in human platelets stimulated by thrombin. J Biol Chem 260:8657–8660

Schick J, Kern H, Scheele G (1984) Hormonal stimulation in the exocrine pancreas results in coordinate and anticoordinate regulation of protein synthesis. J Cell Biol 99:1569–1574

Storm H, van Hardeveld C (1986) Effect of hypothyroidism on the cytosolic free Ca^{2+} concentration in rat hepatocytes during rest and following stimulation by noradrenaline or vasopressin. Biochim Biophys Acta 885:206–215

Tsien RY, Pozzan T, Rink TJ (1982) Calcium homeostasis in intact lymphocytes: cytoplasmic free calcium monitored by a new intracellularly trapped fluorescent indicator. J Cell Biol 94:325–334

Van Hardeveld C, Clausen T (1986) Ca^{2+} and thyroid hormone action. In: Bader H, Gietzen K, Rosenthal R, Rüdel R, Wolf HU (eds) Intracellular calcium regulation. Manchester University Press, Manchester, pp 355–365

Vencentini LM, Di Virgilio F, Ambrosini A, Pozzan T, Meldolesi J (1985) Tumour promotor phorbol 12-myristate, 13-acetate inhibits phosphoinositide hydrolysis and cytosolic Ca^{2+} rise induced by the activation of muscarinic receptors in PC12 cells. Biochem Biophys Res Commun 127:310–317

The Regulation of the Ca-Pumping Activity of Cardiac Sarcoplasmic Reticulum by Calmodulin

M. Chiesi and J. Gasser[1]

1 Introduction

The rapid regulation of the cytosolic free calcium concentration in striated muscles is under the control of the sarcoplasmic reticulum (SR) activity. The major protein constituent of the SR membrane is a calcium-dependent ATPase of molecular mass 110,000, whose activity is coupled with the vectorial translocation of calcium ions from the cytosol into the compartment enclosed by the membranes. This transport activity is responsible for the relaxation of the myofibrils in striated muscles. During the process of excitation-contraction coupling, calcium ions, stored against a large concentration gradient inside the SR compartment, are readily mobilized and flow back to the cytosol, thus inducing contraction. In this process, electrical stimuli travelling along the T-tubular system during the action potential, and/or intracellular chemical messengers (such as inositoltriphosphate or calcium), released in the vicinity of the terminal cisterne of the SR network, are assumed to induce the transient opening of putative calcium channels located on the SR membrane. Regulatory mechanisms of the calcium fluxes across the SR membrane play a major role in the regulation and modulation of the mechanical activity of muscle cells. cAMP is a well-known modulator of myocardial contractility. The effect of this intracellular messenger is achieved, at least in part, by augmenting calcium transport into the SR. A stimulation of the calcium pump is expected to increase both the rate of relaxation of the muscle cells and the amount of calcium ions stored in the SR compartment. More calcium ions can be released to the myofibrils upon excitation, thus increasing the force of contraction. Also calmodulin was shown to stimulate the rate of calcium uptake by SR membranes isolated from cardiac tissue (Katz and Remtulla 1978). The molecular mechanism of this stimulation is still a controversial issue, possibly because calmodulin has more than one site of action on the SR membranes. Calmodulin is known to interact directly with several target proteins and to be able to stimulate in this way enzymatic activity. The possibility that the SR calcium pump is stimulated by calmodulin via direct interaction has been taken into consideration (Louis and Jarvis 1982; Mas Oliva et al. 1983). The experimental support obtained, however, is inconclusive at best and direct binding has never been demonstrated. On the other hand, the involvement of calmodulin via indirect mechanisms leading to the phosphorylation of phospholamban (PLB) is widely accepted.

---

1 Department of Research, Pharmaceutical Division, CIBA-GEIGY Ltd., Basel, and Institute of Biochemistry, Swiss Federal Institute of Technology, 8057 Zürich, Switzerland

Ch. Gerday, R. Gilles, L. Bolis (Eds.)
Calcium and Calcium Binding Proteins

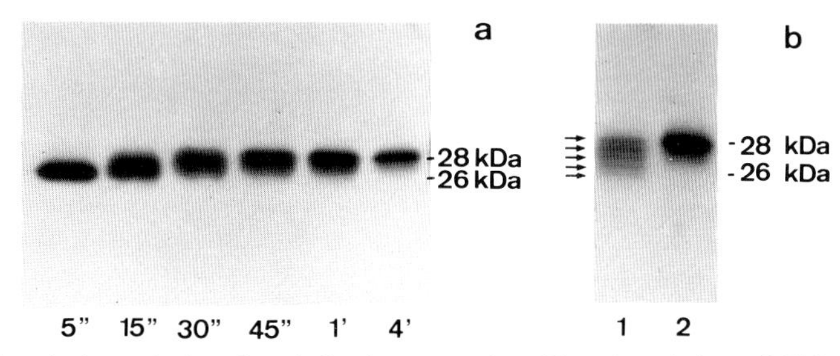

Fig. 1a,b. Multiple phosphorylation of the phospholamban complex. Phosphorylation of PLB induces a shift in the apparent molecular weight when analyzed by gel electrophoresis. **a** Effect of cAMP-dependent phosphorylation as a function of time on the mobility of the PLB complex on 15% SDS-polyacrylamide gels (Laemmli). PLB was pulse labelled with (γ-^{32}P)ATP and then chased with excess cold ATP in the presence of the cAMP-dependent kinase. After various intervals of time, aliquots were withdrawn and analyzed by electrophoresis. **a** Shows that the apparent molecular weight of PLB shifts from 26 kDa to 28 kDa. This indicates that the cAMP-dependent kinase is able to catalyze the incorporation of several phosphate groups into the same radioactive labelled PLB-complex. On particular gels (long polymerization time, low current during run) it is possible to see that the PLB complex migrates to maximally five distinct positions when phosphorylated with either the calmodulin-dependent or the cAMP-dependent kinase. These positions correspond to the various PLB populations containing from 1 up to 5 phosphate groups/complex. As an example, **b**, line *1*, shows PLB phosphorylated by the cAMP-dependent kinase (4 min phosphorylation in the presence of 120 units kinase/ml). Under these conditions all possible phosphorylation levels are apparent. When PLB is maximally phosphorylated by the cAMP-dependent kinase most of it migrates to the position corresponding to 5 phosphate groups incorporated (see **b**, line *2*)

2 Calmodulin and Phospholamban

Phospholamban is an intrinsic protein component of cardiac SR membranes where it is found in a ratio of about 1:1 with the calcium pump (Kirchberger and Tada 1976; Tada et al. 1983). The protein is completely absent in fast skeletal muscles but has been observed in slow twitch muscles also (Suzuki et al. 1986; Jorgensen and Jones 1986). PLB serves as a substrate to several protein kinases, such as the calmodulin-dependent kinase (Le Peuch et al. 1979; Bilezikjian et al. 1981), the cAMP-dependent kinase (Wray and Gray 1977; Tada et al. 1975), and the kinase-C (Limas 1980; Movsesian et al. 1984). Detailed information on the molecular properties of PLB (Fujii et al. 1986; Simmerman et al. 1986) and its complete cDNA-derived amino acid sequence (Fujii et al., submitted) have recently become available. The molecule is composed of 5 identical subunits, each consisting of 52 amino acid residues and having a molecular mass of 6080 Da. A highly hydrophobic region at the C-terminus (which is probably embedded in the SR membrane) confers to PLB its typical lipophylic character. Starting from the amino terminus, on the other hand, there is a hydrophylic domain about 30 amino acids long, which probably protudes from the SR membrane. Two adjacent serine and threonine residues in the latter domain are selectively phosphorylated by a cAMP-dependent and a calmodulin-dependent kinase, respectively. Phosphorylation of PLB induces a change in its mobility on SDS-gels after electrophoretic separation. This property has been exploited to show that it is possible, by using either the calmodulin or the cAMP-dependent kinase, to phosphorylate simultaneously all 5 subunits of PLB

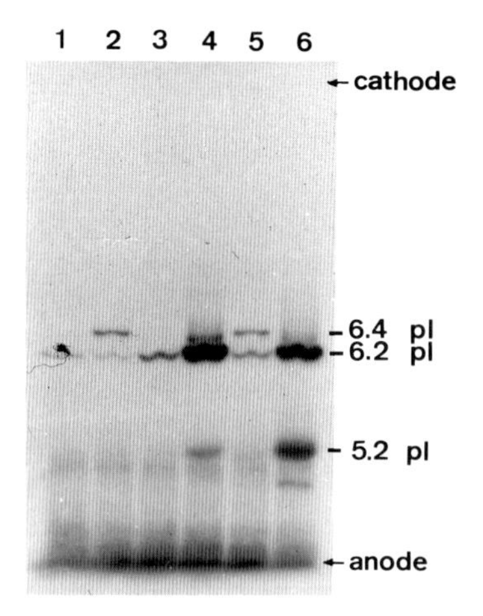

Fig. 2. Isoelectric focusing analysis of phospholamban phosphorylation. An isoelectric focusing system that allows the analysis of the single subunits of PLB, depending on their pI value, has been developed (see Gasser et al. 1987). Basically, the system consisted of flat bed agarose gels containing ampholines and neutral detergents. PLB subunits were phosphorylated as described in the legend to Table 1 in the presence of calmodulin and/or cAMP-dependent kinase, as indicated: *line 1* no additon; *line 2* calmodulin (3 μM); *line 3* kinase-A (60 units/ml); *line 4* high kinase-A (600 units/ml); *line 5* calmodulin and kinase-A; *line 6* calmodulin and high kinase-A

(see Fig. 1). In the amino terminus of PLB the presence of several positively charged aminoacids (3 arginines and 1 lysine pro subunit) confers to the molecule a rather alkaline character (Fujii et al. 1986). Indeed, a pI value of about 10 has been reported for PLB (Jones et al. 1985). A strong acidic shift, however, is observed when PLB becomes phosphorylated. The pI value shifts down to 6.4 or 6.2 after phosphorylation of PLB subunits with the calmodulin-dependent or the cAMP-dependent kinase, respectively (Fig. 2). Double phosphorylation by the two kinases of the same PLB subunit population induces a further acidic shift to a value of 5.2, as shown in Fig. 2, line 6.

3 Phospholamban and the Calcium Pumping ATPase

The levels of phosphorylation of PLB and of stimulation of the calcium ATPase correlate closely, suggesting that PLB might indeed mediate the effect of the various kinases on calcium transport in cardiac SR. Even though a functional association between PLB and ATPase had been proposed long ago, only recently has it been possible to demonstrate with immunological techniques that PLB is directly involved in the modulation process. The calcium pumping activity of cardiac SR can be stimulated by PLB-specific monoclonal antibodies (Suzuki and Wang 1986). The regulatory effect induced by the protein kinases can be mimicked by the antibodies, thus showing that PLB and ATPase are indeed interacting. PLB modulates the turnover rate of the calcium-dependent ATPase of cardiac SR, in particular when calcium ions are present in the submicromolar range (Bilezikjian et al. 1981; Tada et al. 1979). The major effect is a shift of the apparent Km(Ca) from 1–2 μM down to 0.5–0.7 μM. The calcium uptake rate in the presence of 0.7 μM free calcium is increased four- to fivefold, when PLB is maximally phosphorylated (see Fig. 3). Whether the amount of activated kinases in vivo is adequate to produce the phosphorylation levels necessary to induce such an activation is impossible to decide. However, this observation clearly shows the

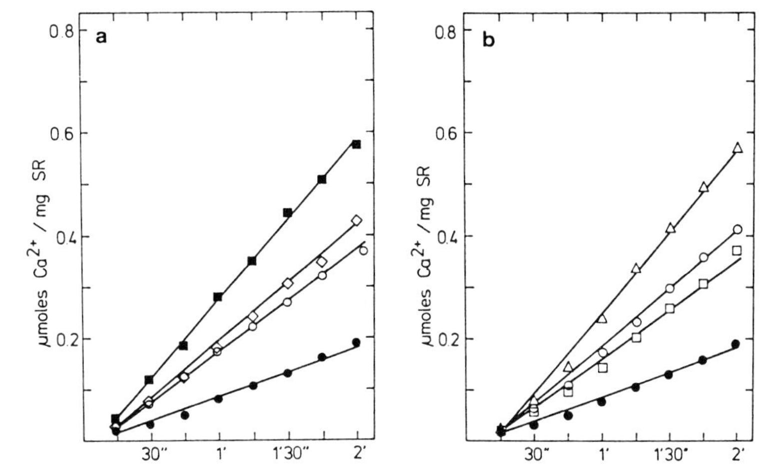

Fig. 3. Stimulation of the calcium pumping activity of cardiac SR by phosphorylation. SR vesicles were phosphorylated in the standard medium (see legend to Table 1) in the presence of the following additions: **a** none (*closed circles*), 3 μM calmodulin (*open circles*), 60 units/ml protein kinase-A (*open rectangles*), and 3 μM calmodulin and 60 units/ml kinase-A (*closed rectangles*). Stimulation of Ca-uptake by the two kinases is additive. **b** None (*closed circles*), 60 units/ml (*open rectangles*), 200 units/ml (*open circles*), or 600 units/ml protein kinase-A (*open triangles*). Calcium uptake was then assayed in the presence of ^{45}Ca and 3 mM oxalate with the Millipore filtration method. The free Ca concentration in the uptake medium was 0.7 μM. The phosphorylation level of PLB under the various conditions can be viewed in Fig. 2

physiological significance of phosphorylation processes in the regulation of the intracellular calcium concentration in the heart. The molecular details on how PLB affects the ATPase activity are not known. In principle, PLB could also be viewed as an inhibitor of the calcium pump (Katz 1980). Its phosphorylation would remove the inhibition, as suggested by recent experiments on reconstitution (Inoui et al. 1986) and proteolysis (Kirchberger et al. 1986). It is likely that the amphiphilic characteristics of PLB might be modified by the phosphorylation process, thus changing the hydrophobic microenvironment of the ATPase. In addition, the density of negative charges in the vicinity of the calcium binding sites of the ATPase is expected to increase dramatically after inserting several phosphate groups into the nearby PLB molecules. Due to a screening effect the local concentration of calcium ions is expected to increase, thus explaining the apparent shift in the K_m(Ca).

4 Concerted Action of cAMP and Calmodulin

The action of the calmodulin-dependent and the cAMP-dependent kinase on SR membranes is usually independent and additive (see for instance Fig. 3a), indicating that either (1) PLB contains different sites for the cAMP-dependent and the calmodulin-dependent phosphorylations, or (2) different populations of PLB are specifically under control of one kinase or the other. The first possibility is supported by the recent

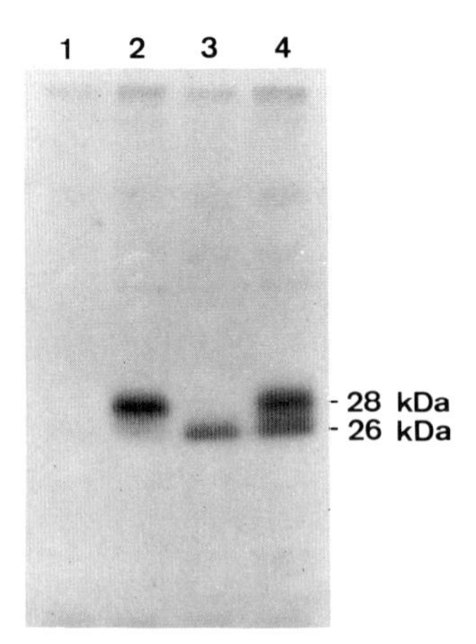

Fig. 4. Functionally distinct phospholamban populations. Cardiac SR membranes were phosphorylated as described in the legend to Table 1. Line *1* control; line *2* 3 μM calmodulin; line *3* 60 units/ml protein kinase-A; line *4* calmodulin and protein kinase-A. Radioactive labelled proteins were than separated by SDS-gel electrophoresis and visualized by autoradiography. The apparent mobility of the PLB complexes phosphorylated in the presence of calmodulin or cAMP is different under these experimental conditions. When both, cAMP and calmodulin-dependent phosphorylations were carried out simultaneously, it became apparent that a population of PLB was phosphorylated by the exogenous kinase-A only, while another portion of PLB was phosphorylated specifically by the endogenous calmodulin-dependent kinase. The two kinases could not act on the same PLB complex

investigations at the molecular level, which have clearly demonstrated that each PLB subunit contains two distinct phosphorylation sites for the calmodulin-dependent and the cAMP-dependent kinase, respectively. On the other hand, the analysis of the phosphorylation pattern of SR membranes does not show the presence of PLB molecules simultaneously phosphorylated by the two kinases (see Fig. 4). Such experiments rather support the concept of two functionally distinct PLB populations, each of them specifically and exclusively phosphorylated by one kinase or the other. The most likely explanation for this apparent contradiction is a non-homogeneous distribution of the kinases in the SR membranes and steric hinderance phenomena. Indeed, the subfractionation of the SR membranes into light and heavy fractions (deriving mainly from the longitudinal system and the cisternal compartments, respectively) has shown that the endogenous calmodulin-dependent phosphorylation is restricted mainly to the cisternal elements (see Fig. 5). The calmodulin-dependent kinase involved is membrane-bound and catalyzes the phosphorylation of the specific population of PLB with which it interacts only. On the other hand, the cAMP-dependent kinase is soluble and is exogenously added to the membranes. In principle, it should reach all available substrate molecules. The phosphorylation sites located on the PLB molecules interacting with the endogenous calmodulin-dependent kinase, however, seem to be protected. They are not phosphorylated by the exogenous cAMP-dependent kinase. Thus, in the presence of both, calmodulin and the cAMP-dependent kinase, the phosphorylation of PLB is usually additive (Table 1), but different PLB populations are involved (Fig. 4). However, when the cAMP-dependent kinase is added in very high concentra-

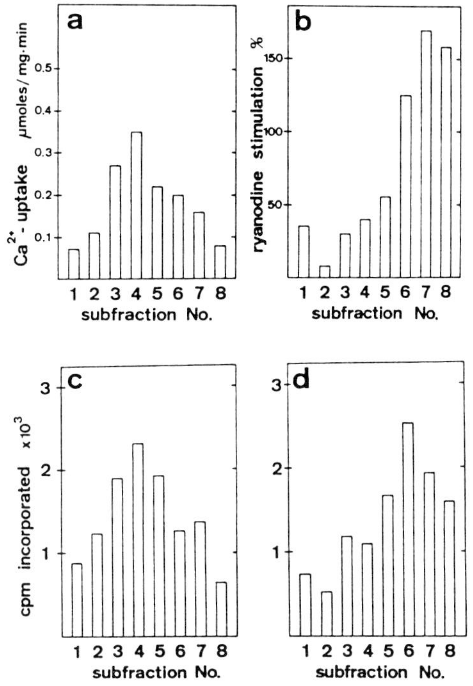

Fig. 5a–d. Distribution of calmodulin and cAMP-dependent phosphorylation of phospholamban in the various subfractions of cardiac SR membranes. Cardiac SR membranes were subfractionated on a sucrose-dextran gradient (ranging from 5–7% Dextran T-10 and 10–26% sucrose). Various fractions were separated and analyzed. The activity of Ca-uptake in the absence of ryanodine is a characteristic of SR vesicles originating from the longitudinal system (**a**). Ryanodine-dependent stimulation of the Ca-transport ability, on the other hand, is a marker of SR membranes deriving from the cisternal compartments (**b**). The figure shows that the cAMP-dependent (**c**) and the calmodulin-dependent (**d**) phosphorylation of PLB is differently distributed in the SR subfractions: the former phosphorylation is occurring predominantly in the cisternal compartments while the latter phosphorylation is found in particular in the longitudinal system

Table 1. Phosphorylation of phosphalamban by the concerted action of cAMP-dependent kinase and calmodulin. The phosphorylation medium contained 100 mM NaCl, 5 mM $MgCl_2$, 20 mM Hepes, pH 7.4, 1 mM EGTA, and 0.9 mM $CaCl_2$. SR protein (0.5 mg/ml) was preincubated in the phosphorylation medium for 5 min in the presence, when required, of 3 μM calmodulin (CAL), or 60 units/ml (kin-A), or 600 units/ml (high kin-A) of cAMP-dependent kinase. The reaction was started by adding 0.3 mM (γ-^{32}P)ATP. After 4 min phosphorylation, the proteins were separated by SDS-gel electrophoresis, the region corresponding to PLB was cut from the gel and radioactivity incorporated was counted in a scintillation cocktail. The results represent relative units normalized to the amount of radioactivity incorporated in the presence of high concentrations of cAMP-dependent kinase (kinase-A)

CAL	Kin-A	CAL + kin-A	High kin-A	CAL + high kin-A
35 ± 5	34 ± 4	65 ± 9	100 ± 8	174 ± 5

tions together with calmodulin, the level of phosphorylation obtained is more than additive, as if a new set of phosphorylation sites of PLB becomes available (Table 1). Indeed, the PLB population which is interacting with the endogenous calmodulin-dependent kinase becomes now available also to the exogenous cAMP-dependent kinase, provided that it is first phosphorylated by the calmodulin-dependent kinase. Under these conditions (i.e in the presence of both calmodulin and very high concentrations of cAMP-dependent kinase) the PLB subunits having a pl of 6.4 (i.e phosphorylated by the calmodulin-dependent kinase) completely desappear and a new double phosphorylated species with pI 5.2 now becomes apparent (see Fig. 2). These observations might explain previous reports (LePeuch et al. 1979) in which the stimulatory effect of the cAMP-dependent kinase on the transport activity of particular SR preparations (probably heavy SR) required the presence of calmodulin.

According to our experience, this synergistic action between calmodulin and cAMP-dependent kinase is observable only at very high concentrations of the cAMP-dependent kinase (500 or more units/ml) and we feel that it has no physiological significance. On the other hand, our investigations indicate that cardiac cells might be utilizing different kinases to regulate the calcium pumping activity of the SR system; the cisternal compartments being sensitive to calmodulin, while the longitudinal system responds mainly to cAMP.

In conclusion, calmodulin in heart muscle stimulates via a phosphorylation mechanism the calcium pumping activity of the SR. Calmodulin could mediate a negative feed back action of calcium ions in the cytosol by favoring their accumulation into the SR compartments. Conceivably, the action of calmodulin is effective only in the long term regulation of striated muscles. The relatively slow kinetics of phosphorylation-dephosphorylation reactions preclude an involvement of calmodulin in the beat-to-beat regulation of cardiac activity.

References

Bilezikjian LM, Kranias EG, Potter JD, Schwarz A (1981) Studies on phosphorylation of canine cardiac sarcoplasmic reticulum by calmodulin-dependent protein kinase. Circ Res 49:1356–1362

Fujii J, Kadoma M, Tada M, Toda H, Sakiyama F (1986) Characterization of structural unit of phospholamban by amino acid sequencing and electrophoretic analysis. Biochem Biophys Res Comm 138:1044–1050

Fujii J, Ueno A, Kitano K, Tanaka M, Kadoma M (1987) Complete complementary DNA-derived amino acid sequence of canine cardiac phospholamban. J Clin Inv 79:301–304

Gasser J, Chiesi M, Carafoli E (1986) Concerted phosphorylation of the 26-kilodalton phospholamban oligomer and of the low molecular weight phospholamban subunits. Biochemistry 25: 7615–7623

Inoui M, Chamberlain BK, Saito A, Fleischer S (1986) The nature of the modulation of Ca transport as studied by reconstitution of cardiac sarcoplasmic reticulum. J Biol Chem 261:1794–1800

Jones LR, Simmermann HK, Wilson WW, Gurd FR, Wegener AD (1985) Purification and characterization of phospholamban from canine cardiac sarcoplasmic reticulum. J Biol Chem 260: 7721–7730

Jorgensen AO, Jones LR (1986) Localization of phospholamban in slow but not in fast canine skeletal muscle fibres. J Biol Chem 261:3775–3781

Katz AM (1980) Relaxing effects of catecholamines in the heart. Trends Pharmacol Sci 1:434–436

Katz S, Remtulla MA (1978) Phosphodiesterase protein activator stimulates calcium transport in cardiac microsomal preparations enriched in sarcoplasmic reticulum. Biochem Biophys Res Comm 83:1373–1379

Kirchberger M, Tada M (1976) Effects of cAMP-dependent protein kinase on sarcoplasmic reticulum isolated from canine cardiac and slow and fast contracting skeletal muscles. J Biol Chem 251:725–729

Kirchberger M, Borchman D, Kasinathan C (1986) Proteolytic activation of the canine cardiac sarcoplasmic reticulum calcium pump. Biochemistry 25:5484–5492

LePeuch CJ, Haiech J, Demaille JG (1979) Concerted regulation of cardiac sarcoplasmic reticulum calcium transport by cAMP-dependent and calcium-calmodulin-dependent phosphorylations. Biochemistry 18:5150–5157

Limas CJ (1980) Phosphorylation of cardiac sarcoplasmic reticulum by a calcium-activated phospholipid-dependent protein kinase. Biochem Biophys Res Comm 96:1378–1383

Louis CF, Jarvis B (1982) Affinity labelling of calmodulin-binding components in canine cardiac sarcoplasmic reticulum. J Biol Chem 257:15187–15191

Mas Oliva J, DeMeis L, Inesi G (1983) Calmodulin stimulates both ATP-hydrolysis and synthesis catalyzed by a cardiac calcium ion dependent ATPase. Biochemistry 22:5822–5825

Movsesian MA, Nishikawa M, Adelstein RS (1984) Phosphorylation of phospholamban by calcium-activated, phospholipid-dependent protein kinase. J Biol Chem 259:8028–8032

Simmerman HK, Collins JH, Theibert JL, Wegener AD, Jones LR (1986) Sequence analysis of phospholamban. J Biol Chem 261:1333–1341

Suzuki T, Lui P, Wang JH (1986) The use of monoclonal antibodies for the species and tissues distribution of phospholamban. Cell Calcium 7:41–47

Suzuki T, Wang JH (1986) Stimulation of bovine cardiac SR Ca pump and blocking of phospholamban phosphorylation by a phospholamban monoclonal antibody. J Biol Chem 261: 7018–7023

Tada M, Inui M, Yamada M, Kadoma T, Abe H, Kakiuchi S (1983) Effects of phospholamban phosphorylation catalyzed by cAMP- and calmodulin-dependent protein kinases on calcium transport ATPase of cardiac sarcoplasmic reticulum. J Molec Cell Cardiol 15:335–346

Tada M, Kirchberger MA, Katz AM (1975) Phosphorylation of a 22'000 dalton component of the cardiac sarcoplasmic reticulum by cAMP-dependent protein kinase. J Biol Chem 250:2640–2647

Tada M, Yamada M, Ohmori F, Kuzuya T, Inui M, Abe H (1979) Transient state kinetic studies of Ca-dependent ATPase and calcium transport by cardiac sarcoplasmic reticulum. J Biol Chem 255:1985–1992

Wray HL, Gray RR (1977) cAMP stimulation of membrane phosphorylation and Ca-activated ATPase in cardiac sarcoplasmic reticulum. Biochim Biophys Acta 461:441–459

Calcium Binding and Structural Characteristics of Muscle Calsequestrins

S. SALVATORI, M. BRAVIN and A. MARGRETH[1]

1 Introduction

Skeletal muscle sarcoplasmic reticulum (SR) is a network of tubules surrounding myofibrils, which, by regulating the intracellular calcium concentration controls the contraction-relaxation cycle. Efflux of calcium from SR causes the rise of calcium concentration in the myofibers inducing muscle contraction, through a complex series of events (Perry 1986). Muscle relaxation occurs when the calcium concentration is reduced by the Ca-pump, which transports calcium from the cytosol into the SR lumen. The stoichiometry of the transport is two calcium ions per one molecule of ATP hydrolysed.

A great deal of evidence has been accumulated in the last 15 years, concerning a functional subspecialization of the SR even though the Ca-ATPase protein is present in the longitudinal tubules and the terminal cisternae (TC). Meissner (1975) showed that SR vesicles derived from TC are filled with a fibrous matrix. This electron-dense material is largely constituted of calsequestrin (Jorgensen et al. 1979), an extrinsic calcium-binding protein first described and biochemically characterized by MacLennan and Wong (1971) in rabbit skeletal muscle.

1.1 Calsequestrin

Calsequestrin (CS) can be isolated from SR membranes with mild detergent treatments (MacLennan 1974) or with incubation at alkaline pH, in the presence of EDTA (Duggan and Martonosi 1970). Further purification may be accomplished by chromatographic procedures (MacLennan et al. 1972; Cala and Jones 1983) or by precipitation in the presence of millimolar Ca (Ikemoto et al. 1971; Maurer et al. 1985). The properties of purified calsequestrin have been investigated extensively, in view of the possible role of this protein in the excitation-contraction coupling (MacLennan et al. 1983). Maurer et al. (1985) demonstrated that calsequestrin exhibits two different states of aggregation, depending on the presence of calcium; Saito et al. (1984) showed that calsequestrin is apparently tightly bound to the junctional membrane of terminal cisternae; Somlyo et al. (1985) demonstrated that during tetanus the calcium, previously

1 C.N.R. Centro per lo Studio della Biologia e Fisiopatologia muscolare, Istituto di Patologia generale, Università di Padova, 35100-Padova, Italy

Ch. Gerday, R. Gilles, L. Bolis (Eds.)
Calcium and Calcium Binding Proteins

bound inside the terminal cisternae, is almost completely released. All this suggests that the role of calsequestrin inside the terminal cisternae could be to concentrate calcium in a specialized area involved in calcium release.

The main structural and biochemical features of rabbit calsequestrin are the following: (a) CS is able to bind 43 moles of Ca per mole, with a dissociation constant of about 10 μM and 1 mM, at low and high ionic strength, respectively; (b) amino acid composition shows that CS is a very acidic protein, because of the high percentage of aspartate and glutamate (see below); (c) in different electrophoretic systems, CS mobility varies depending on the pH used (Michalak et al. 1980); (d) in the presence of the cationic dye Stains All, CS exhibits a metachromatic staining, both in electrophoresis slab gel and in aqueous solution (Campbell et al. 1983); (e) CS precipitates as Ca-calsequestrin complex when the Ca concentration is higher than 1 mM.

We have recently reported (Damiani et al. 1986) that calsequestrins purified from frog, chicken and mammals skeletal muscle differ in many respects, i.e. maximum calcium-binding capacity, electrophoretic mobility at alkaline pH and metachromatic staining with Stains All. In the present report we present some biochemical characteristics of TC and calsequestrin from these and other species. For this study we have used the preparative procedure recently published by Saito et al. (1984) which is very useful to obtain isolated TC. These fractions are characterized by the high calsequestrin content, which is twice that of Ca-ATPase, and by the fact that calsequestrin appears to be bound intraluminally to the membrane junctional area, where the outer feet projections are attached.

2 Results and Discussion

Figure 1A shows the electrophoretic protein pattern of rabbit and frog TC (lanes d and e) in Laemmli's gel system run at alkaline pH. In addition to the reported differences in molecular weight (Damiani et al. 1986), a different calsequestrin to Ca-ATPase ratio may be calculated from densitometric traces, being 0.7 and 0.4 for frog and rabbit TC, respectively. Two-dimensional gel electrophoresis run according to Michalak et al. (1980), demonstrated that calsequestrin band was not contaminated by other proteins (not shown).

In order to compare calcium-binding capacities of frog and rabbit calsequestrins, the Ca-binding overlay technique (Maruyama et al. 1984), was used. In agreement with equilibrium dialysis experiments (MacLennan and Wong 1971; Damiani et al. 1986), the results indicate that frog and rabbit calsequestrins, have a high Ca-binding capacity (Fig. 1B, lanes d' and e').

Figure 1A also shows SDS gel electrophoresis of TC fractions, isolated from lizard, pigeon and chicken muscles, run at alkaline pH (Laemmli 1970) and stained with Coomassie Blue. Under our conditions, calsequestrins of different species have different electrophoretic mobilities. Figure 1B also shows that chicken and pigeon calsequestrins, that bind very little calcium in equilibrium dyalysis experiments, bind an amount of calcium comparable to that of rabbit calsequestrin under these experimental conditions.

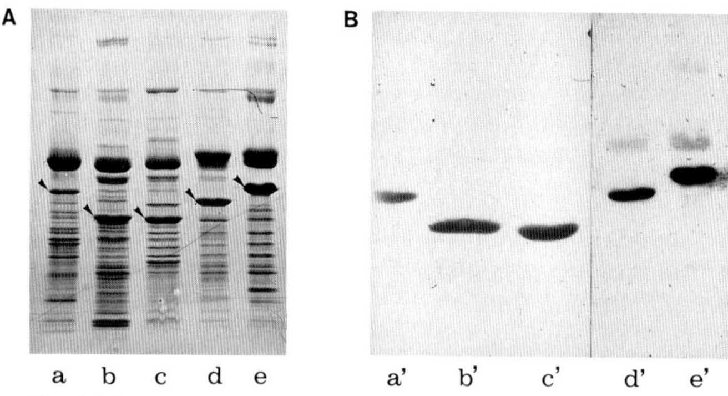

Fig. 1A,B. Identification of calcium binding proteins in terminal cisternae preparations from skeletal muscle of several species. **A** Coomassie Blue staining of electrophoresis of terminal cisternae from lizard (*a*), pigeon (*b*), chicken (*c*), rabbit (*d*) and frog (*e*), run at alkaline pH in a 5–15% polyacrylamide-gel linear gradient, according to Laemmli's system (1970). 100 μg of samples were loaded per well. Calsequestrin bands are indicated by ***arrowheads.*** **B** Autoradiography of terminal cisternae from lizard (*a'*), pigeon (*b'*), chicken (*c'*), rabbit (*d'*) and frog (*e'*) after electroblotting and Ca binding, according to Maruyama et al. (1984). Electrophoresis conditions were as in **A**. The incubation medium was as follows: 5 mM imidazole, pH 7.4, 60 mM KCl, 6 mM $MgSO_4$, 10 μM $CaCl_2$, 1 mCi/l ^{45}Ca; the time of incubation was 20 min and the washing solution was 30% ethanol

The already mentioned capacity of rabbit calsequestrin to bind 43 moles of Ca per mole of protein is based on a molecular weight of 45 kDa, measured in the electrophoretic system of Weber and Osborn (1969), i.e. at neutral pH. Despite the great difference in terms of relative mobility at alkaline pH (Fig. 1A), all calsequestrins examined exhibit a similar apparent molecular weight when electrophoresed at neutral pH (Fig. 2). This pH-dependent mobility of calsequestrin is largely due to its well-known ellipticity, which increases at alkaline pH (Cozens and Reithmeier 1984). Therefore, when electrophoresed at neutral pH (first dimension) and at alkaline pH (second dimension), according to Michalak et al. (1980), the spot of calsequestrin falls off the diagonal line. By examining calsequestrins from different species, it is evident that the distance of the calsequestrin spot from the diagonal line varies from species

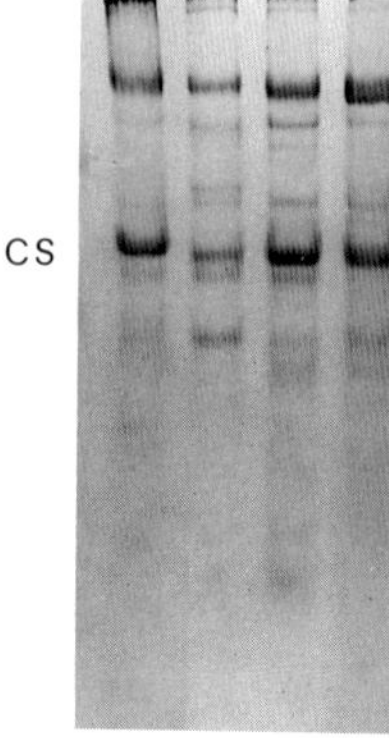

Fig. 2. Electrophoresis of SR terminal cisternae, at neutral pH. Coomassie Blue staining of 10% polyacrylamide gel electrophoresis of terminal cisternae of frog (*a*), chicken (*b*), pigeon (*c*) and rabbit (*d*), run at neutral pH, according to Weber and Osborn system (1969). 30 μg of sample were loaded per well

Table 1. Relationship between electrophoretic properties and amino acid composition of calsequestrins

Calsequestrin source	M_r (kDa) (W.&O.)	(L.)	L./W.&O.	GLU/ASP	Reference
Rabbit	44	65	1.5	0.9	Campbell et al. (1983)
Pig	48	64	1.3	0.9	White et al. (1983)
Dog (heart)	44	55	1.3	1.0	Campbell et al. (1983)
Chicken	46	55	1.2	1.0	Damiani et al. (1986)
Frog	47	73	1.6	0.7	Damiani et al. (1986)
Pigeon	45	53	1.2	–	Present work

to species. If this deviation is assumed as an index of the molecular asymmetry, as shown in Table 1, frog calsequestrin seems to have the most asymmetric structure among those so far investigated.

The electrophoretic behaviour is not accounted for by differences in the amino acid compositions. This seems to be quite homogeneous for all calsequestrins listed in Table 2. As expected for a Ca-binding protein, glutamate and aspartate are the most abundant amino acids, even though with different relative ratios as compared to

Table 2. Amino acid composition of calsequestrins

Amino acid	% of total residues Rabbit[a]	Pig[b]	Cardiac[c]	Chicken[d]	Frog[d]
Lysine	7	9	8	8	6
Histidine	2	2	3	2	2
Arginine	2	2	3	4	1
Aspartic acid	19	18	16	16	20
Threonine	3	3	4		3
Serine	4	5	5	6	3
Glutamic acid	18	16	16	16	13
Proline	5	3	4	5	3
Glycine	4	6	7	8	15
Alanine	7	7	6	7	5
Cysteic acid	1			1	3
Valine	7	5	7	7	6
Methionine	2	2	1	1	
Isoleucine	5	5	5	5	4
Leucine	9	9	8	10	7
Tyrosine	2	2	3		3
Phenylalanine	6	6	5	5	4
Asp + Glu	37	34	32	32	33
Lys + Arg	9	11	11	12	7
Acidic – basic	28	23	21	20	26

[a] MacLennan and Wong (1971); [b] White et al. (1983); [c] Campbell et al. (1983); [d] Damiani et al. (1986).

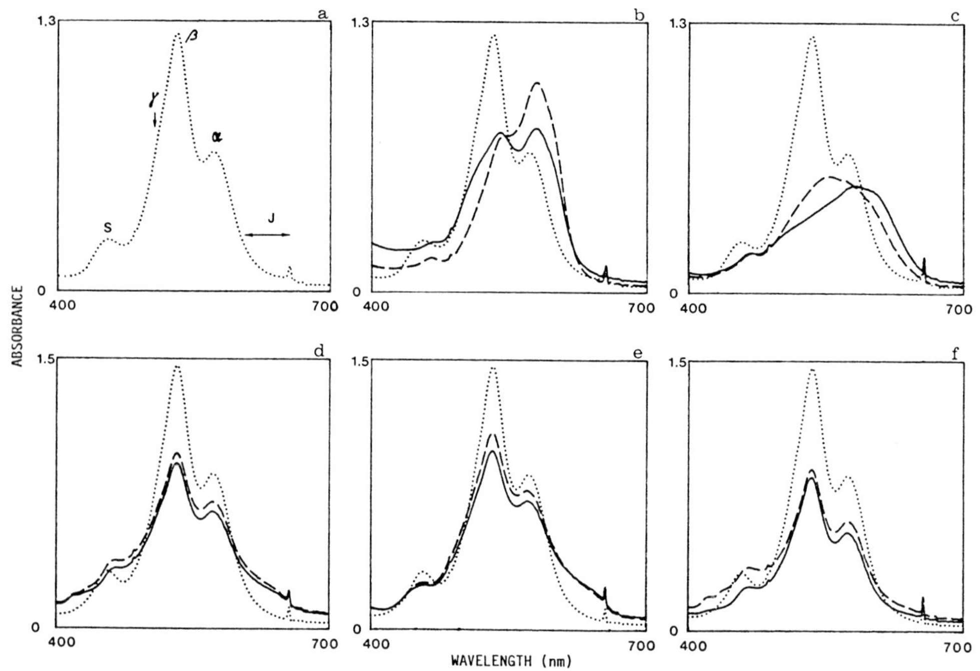

Fig. 3a–f. Calcium-induced spectral shifts of CS-Stains All complexes in aqueous solution. The spectra were obtained essentially as described by Caday and Steiner (1985), in 2 mM MOPS, 25 μM Stains All (Sigma Chemical Co., St. Louis, USA), 30% ethylene glycol. Purified calsequestrins were added to obtain a dye-to-protein molar ratio of about 50. Samples were allowed to equilibrate in the dark at 25°C for 10 min before measurements. When present calcium was 1 mM. **a** 25 μM Stains All spectrum with indication of alfa, beta, gamma, S and J states; **b** rabbit CS; **c** frog CS; **d** chicken CS; **e** pigeon CS; **f** human CS. (- - -) spectra in the absence of calcium; (——) spectra in the presence of 1 mM calcium; (· · · ·) spectrum of the dye

parvalbumins and troponin C. Nevertheless, frog calsequestrin seems to be unique in that the ratio of glutamate to aspartate is about 1:2 rather than 1:1, as in the other species investigated (Damiani et al. 1986).

Another striking characteristic of calsequestrin is its ability to stain metachromatically with the cationic dye Stains All (Campbell et al. 1983). Depending on the dye to protein ratio, it is possible to obtain complexes of different colours, and hence of different spectra (Kay et al. 1964; Bean et al. 1965). Therefore, it is possible to follow the interaction between calsequestrin and Stains All in aqueous solution spectrophotometrically (Campbell et al. 1983), by monitoring the metachromatic deviation from the Beer law (Michaelis and Granick 1945). The different spectral band maxima seem to correspond to different aggregation states of the dye, at least for alpha, beta and gamma states. The J state, that corresponds to the blue staining, appears somehow to be a particular case, since the spacing between anionic sites is also important for the resulting colour (Bean et al. 1965). When calsequestrins are examined for their ability to interact with Stains All in aqueous solution, spectral shifts are observed in the case of rabbit (Campbell et al. 1983), frog and other calsequestrins (Damiani et al. 1986). It has been established that the presence of some cations may induce changes in the

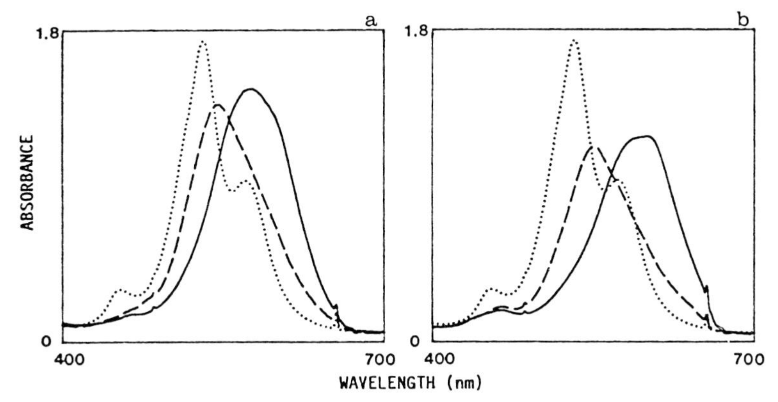

Fig. 4a,b. Calcium-induced spectral shifts of polyanions-Stains All complexes in aqueous solution. The spectral conditions were as indicated in Fig. 3. **a** Poly-aspartate (Sigma Chemical Co., St. Louis, USA); **b** poly-glutamate:tyrosine (9:1) (Miles Laboratories Inc., Kankakee, USA)

polyanions spectra (Kay et al. 1964). When experiments are carried out in the presence of 30% ethylene glycol as indicated by Caday and Steiner (1985), calcium-induced spectral shifts are observed in frog calsequestrin (Fig. 3c) but only partially in rabbit calsequestrin (Fig. 3b); the other species exhibited only a decrease of the maximal values, without shift (Fig. 3d, e, and f). Thus, frog calsequestrin again seems to be unique in that a beta state or a beta-J state is more easily formed, especially when calcium is present. This behaviour recalls that of some polyanions, namely poly-aspartate (Kay et al. 1964 and Fig. 4a) and hence the fact that frog calsequestrin is the richest in aspartate residues might be of significance. On the other hand, the content of aspartate residues can not be the only valid criterion since also copolymers of glutamate with other amino acids tends to give similar spectral shifts (Fig. 4b). This probably means that not only are the number of calcium binding sites important, but also the spacing between them. Moreover, taking into account that ions, and namely calcium, induces dramatic conformational changes in calsequestrin (Cozens and Reithmeier 1984), we can infer that little modifications in the primary structure could result in quite large modifications in calcium binding capacity. On the other hand, the SDS-gel electrophoresis densitometry of TC fractions isolated from different species (Fig. 1) shows that frog calsequestrin, which has the highest calcium binding capacity, is also present at higher concentration, at least in terms of calsequestrin to Ca-ATPase ratio. It is well known that all skeletal muscles so far investigated have terminal cisternae, more or less developed, and that these cisternae, isolated from different muscles and species, always contain calsequestrin, even though the existence of calsequestrin in lobster muscle is up to now controversial (Campbell et al. 1983; MacLennan et al. 1983). On the other hand, the calcium binding capacity and the amount of calsequestrin vary, depending on muscles and species.

Because of this, the question arises why the maximum calcium-capacity of calsequestrin and its amount within terminal cisternae vary from species to species. One possibility is that the binding capacity might be related to a different concentration gradient inside the terminal cisternae and to different amounts of calcium released.

Moreover, one must consider that the number of triads per sarcomere vary with species or muscles, being two in frog and four in mammals. Another interesting aspect is the difference in body temperature which is higher in birds than in mammals, while it is variable in heterotherms like frogs.

Calsequestrin seems, at least partially, to be bound in the junctional membrane, where calcium is released, but no direct explanations for this location are available at present. More investigations on the structure of calsequestrin, and particularly the knowledge of its primary structure, will shed light on its biochemical function.

Acknowledgements. We gratefully acknowledge the skilful technical assistance of Miss S. Furlan. This work was supported by institutional funds from Consiglio Nazionale delle Ricerche.

References

Bean RCM, Sheperd WC, Kay RE, Walwick ER (1965) Spectral changes in a cationic dye due to interactions with macromolecules. III. Stoichiometry and mechanism of the complexing reaction. J Phys Chem 69:4368–4379

Caday CG, Steiner RF (1985) The interaction of calmodulin with the carbocyanine dye (Stains-all). J Biol Chem 260:5985–5990

Cala SE, Jones LR (1983) Rapid purification of calsequestrin from cardiac and skeletal muscle sarcoplasmic reticulum vesicles by Ca-dependent elution from phenyl-sepharose. J Biol Chem 258:11932–11936

Campbell KP, MacLennan DH, Jorgensen AO (1983) Staining of the Ca-binding proteins, calsequestrin, calmodulin, troponin C, and S-100 with the cationic dye "Stains All". J Biol Chem 258:11267–11273

Cozens B, Reithmeier RAF (1984) Size and shape of rabbit skeletal muscle calsequestrin. J Biol Chem 259:6248–6252

Damiani E, Salvatori S, Zorzato F, Margreth A (1986) Characteristics of skeletal muscle calsequestrin: comparison of mammalian, amphibian and avian muscles. J Muscle Res Cell Mot 7: 435–445

Duggan PF, Martonosi A (1970) IX. The permeability of sarcoplasmic reticulum membranes. J Gen Physiol 56:147–157

Ikemoto N, Bhatnagar GM, Gergely J (1971) Fractionation of solubilized sarcoplasmic reticulum. Biochem Biophys Res Comm 44:1510–1517

Jorgensen AO, Kalnins V, MacLennan DH (1979) Localization of sarcoplasmic reticulum proteins in rat skeletal muscle by immunofluorescence. J Cell Biol 80:372–384

Kay RE, Walwick ER, Gifford CK (1964) Spectral changes in a cationic dye due to interaction with macromolecules. I. Behaviour of dye alone in solution and the effect of added macromolecules. J Phys Chem 68:1896–1906

Laemmli UK (1970) Cleavage of structural proteins during the assembly of the head of bacteriophage T4. Nature (Lond) 227:680–685

MacLennan DH (1974) Isolation of proteins of the sarcoplasmic reticulum. In: Fleischer S, Packer L (eds) Methods in enzimology. Vol 32. Academic Press, New York London, pp 291–302

MacLennan DH, Wong PTS (1971) Isolation of a calcium-sequestering protein from sarcoplasmic reticulum. Proc Natl Acad Sci US 68:1231–1235

MacLennan DH, Yip CC, Iles GH, Seeman P (1972) Isolation of sarcoplasmic reticulum proteins. Cold Spring Harbor Symp Quant Biol 37:469–478

MacLennan DH, Campbell KP, Reithmeier RAF (1983) Calsequestrin. In: Martonosi A (ed) Calcium and cell regulation. Vol 4. Academic Press, New York, pp 151–173

Maruyama K, Mikawa T, Ebashi S (1984) Detection of calcium binding proteins by ^{45}Ca autoradiography on nitrocellulose membrane after sodium dodecyl sulfate gel electrophoresis. J Biochem (Tokyo) 95:511–519

Maurer A, Tanaka M, Ozawa T, Fleischer S (1985) Purification and cristallization of the calcium binding protein of sarcoplasmic reticulum from skeletal muscle. Proc Natl Acad Sci USA 82:4036–4040

Meissner G (1975) Isolation and characterization of two types of sarcoplasmic reticulum vesicles. Biochim Biophys Acta 389:51–68

Michaelis L, Granick S (1945) Metachromasy of basic Dyestuffs. S Amer Chem Soc 67:1212–1219

Michalak M, Campbell KP, MacLennan DH (1980) Localization of the high affinity calcium binding protein and an intrinsic glycoprotein in sarcoplasmic reticulum membranes. J Biol Chem 255:1317–1326

Perry SV (1986) Activation of the contractile mechanism by calcium. In: Engel AG, Banker BQ (eds) Myology. Vol 1. McGraw-Hill, New York, pp 613–641

Saito A, Seiler S, Chu A, Fleischer S (1984) Preparation and morphology of sarcoplasmic reticulum terminal cisternae from rabbit skeletal muscle. J Cell Biol 99:875–885

Somlyo AV, McClellan G, Gonzalez-Serratos H, Somlyo AP (1985) Electron probe X-ray microanalysis of post-tetanic Ca and Mg movements across the sarcoplasmic reticulum *in situ*. J Biol Chem 260:6801–6807

Weber K, Osborn M (1969) The reliability of molecular weight determination by dodecyl sulphate polyacrylamide gel electrophoresis. J Biol Chem 244:4406–4412

White DM, Colwyn RT, Denborough MA (1983) A novel method for the isolation of calsequestrin from procine skeletal muscle sarcoplasmic reticulum. Biochim Biophys Acta 744:1–6

Studies on Voltage-Gated Calcium Channels by Means of Fluorescent $[Ca^{2+}]_i$ Indicators

J. Meldolesi[1], A. Malgaroli[1], A. Pandiella[1], A. Ambrosini[1], L. M. Vicentini[1], D. Milani[2], F. Di Virgilio[2] and T. Pozzan[2]

1 Introduction

Most excitable cells respond to depolarization with a rapid increase of their Ca^{2+} permeability, due to the opening of voltage-gated Ca^{2+} channels (Ca^{2+} VOCS) in the plasma membrane (Hagiwara and Byerly 1981; Reuter 1984). Opening of these channels causes the cytosolic concentration of Ca^{2+}, $[Ca^{2+}]_i$, to rise quickly, beginning in the cytosolic region(s) immediately adjacent to the plasma membrane. The $[Ca^{2+}]_i$ rise is opposed by processes that expell Ca^{2+} from the cell (active pumping by a Ca^{2+} ATPase and Na^+/Ca^{2+} exchange) or cause its segregation within intracellular, membrane-bounded campartments (a high affinity microsomal store as well as the mitochondria). Up to now, studies on Ca^{2+} VOCs were carried out primarily by electrophysiology (classical techniques as well as patch clamping) and by Ca^{2+} influx experiments. By the use of these techniques, a number of important achievements were obtained (Hagiwara and Byerly 1981; Reuter 1984), including the recent demonstration of the heterogeneous nature of Ca^{2+} VOCs. Following the simple nomenclature proposed by Nowicky et al. (1985), the various types of Ca^{2+} VOCs so far identified will be referred to here as the T (transient, or low-threshold) VOCs, probably involved in pacemaking; the L (long-lasting, or high-threshold), the classical Ca^{2+} VOCs that are affected by dehydropyridine drugs; and the N (neither transient, nor long-lasting) VOCs, present in neurons, that might be preferentially located in the presynaptic membrane (Nowicky et al. 1985). The aim of the present article is to demonstrate the applicability to Ca^{2+} VOC studies of the techniques making use of fluorescent, trappable Ca^{2+} indicators. The prototype of these indicators, quin-2, was introduced by Roger Y. Tsien and his collegues in 1982, and a newer congener, fura-2, that can be applied to single cell studies, was proposed in 1985 (Grynkiewicz et al. 1985). The data that can be learned by the use of these precious experimental tools appear complementary to those obtained by electrophysiology, and can therefore lead to a more profound appreciation of the functioning and regulation of Ca^{2+} VOCs.

1 Department of Pharmacology, CNR Center of Cytopharmacology and Scientific Institute S. Raffaele, University of Milano, 20129 Milano, Italy

2 Institute of General Pathology, CNR Center for the Physiology of Mitochondria, University of Padova, Italy

Ch. Gerday, R. Gilles, L. Bolis (Eds.)
Calcium and Calcium Binding Proteins

2 Activation and Inactivation of L-Type Ca^{2+} Channels in PC12 Cells

PC12 is a culture cell line, originally developed from a rat pheochromocytoma, that is able to release catecholamines and acetylcholine by exocytosis. The release process is under the concerted control of both $[Ca^{2+}]_i$ and protein phosphorylation, the latter due to both cAMP-dependent and C kinases (Pozzan et al. 1984; Rabe and McGee 1983). When PC12 cells are cultured without NGF ($PC12^-$ cells) they exhibit a chromaffin-like phenotype, with small, dense granules preferentially located beneath the plasma membrane, and are endowed with Ca^{2+} VOCs exclusively of the L type, inhibitable by both verapamil and dihydropyridines (such as nitrendipine; concentrations around 10 and 0.1 μM, respectively) (Meldolesi et al. 1984; Kogsamut and Miller 1986).

In the experiments reported in detail by DiVirgilio et al. (1987) $PC12^-$ cells were loaded with either quin-2 or fura-2 (the new $[Ca^{2+}]_i$ indicator that, because of its greater fluorescence, can be loaded within the cells at concentrations much lower than those of its more popular congener, Grynkiewicz et al. 1985) and then treated with a depolarizing (50 mM) concentration of K^+, followed after some time by either EGTA (to chelate extracellular Ca^{2+}) or a Ca^{2+} VOC blocker. In separate experiments, parallel batches of $PC12^-$ cells loaded with $[Ca^{2+}]_i$ indicators were treated first with EGTA and then with high K^+, before final readdition of Ca^{2+} to the incubation medium. Examples of the results obtained with fura-2 are given in Fig. 1. Depolarization by high K^+, applied in the Ca^{2+}-containing medium, caused $[Ca^{2+}]_i$ to rise almost instantaneously, but transiently, from the resting (0.1 μM) to very high (>1 μM) values. This initial $[Ca^{2+}]_i$ spike was followed by a plateau around 0.5 μM $[Ca^{2+}]_i$, that was maintained for prolonged (up to tens of min) periods of time. Both the spike and the plateau were due to the opening of Ca^{2+} VOCs as they were prevented, or rapidly dissipated, by Ca^{2+} channel blockers, applied before or after K^+, respectively. Thus, the plateau represents a new steady state, achieved sometime after depolarization, between influx and removal of Ca^{2+} from the cytosol. Under these conditions, the initial rate of influx should therefore be identical to the reciprocal of the initial rate of $[Ca^{2+}]_i$ decrease measured when the channels are blocked (or Ca^{2+} withdrawn from the incubation medium). On the other hand, in the cells incubated in Ca^{2+}-free medium, depolarization by high K^+ yielded no changes of $[Ca^{2+}]_i$. Even under these conditions, however, Ca^{2+} VOCs opened, as demonstrated by the $[Ca^{2+}]_i$ rise observed when excess Ca^{2+} was subsequently reintroduced into the medium. Notice, however, that this last rise occurred at rates much slower than those observed in the cells depolarized in the Ca^{2+}-containing medium, and led directly to a plateau, without any peak being visible (Fig. 1). The slowing down of the $[Ca^{2+}]_i$ rise observed when the cells were depolarized in the Ca^{2+}-free medium tended to increase with the length of the period between the K^+ and Ca^{2+} additions (i.e., the period of depolarization in Ca^{2+}-free medium), but the final plateau reached remained the same in all experimental conditions used. We conclude from these experiments that Ca^{2+} VOCs of $PC12^-$ cells (which are of the L type), after depolarization-induced activation, tend to inactivate within seconds both in Ca^{2+} containing and in Ca^{2+} free media (i.e., irrespective on whether $[Ca^{2+}]_i$ rises or not). Thus, they are affected by a time- and voltage-dependent

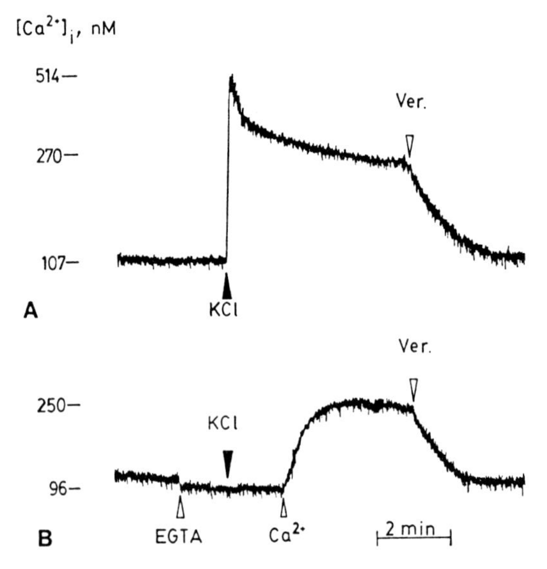

Fig. 1A,B. $[Ca^{2+}]_i$ rises induced by K^+ depolarization in PC12⁻ cells loaded with fura-2. Two parallel batches of cells were depolarized with KCl (50 mM). Cells shown in **A** were depolarized while bathed in a Ca^{2+} containing medium, those of **B** after extracellular Ca^{2+} had been chelated by excess EGTA; in the experiment of **B** Ca^{2+} was reintroduced to the medium 2 min after depolarization. Notice that the $[Ca^{2+}]_i$ transient in **A** is composed by an initial spike followed by a plateau, while that in **B** lacks the spike, and the plateau is reached very slowly. Verapamil (Ver., 20 μM) causes the dissipation of the plateau, demonstrating that the latter is due to the persistent opening of a fraction of Ca^{2+} VOCs

inactivation. Such an inactivation is however incomplete, because a few channels remain active during prolonged depolarization, and are responsible for the observed $[Ca^{2+}]_i$ plateau.

Based on previous results in various cellular systems, it had been repeatedly proposed that inactivation of Ca^{2+} VACs occurs as a direct consequence of the rised $[Ca^{2+}]$ at the cytosolic opening of the channels (see, for example, Hagiwara and Byerly 1981). In order to establish whether this kind of mechanism of inactivation exists in PC12 cells, measurements of initial rates of $[Ca^{2+}]_i$ rise and decline were carried out in cells depolarized in the Ca^{2+} free medium, treated with excess Ca^{2+} 2 min later, and with verapamil after two additional min. Under these conditions the initial rate of $[Ca^{2+}]_i$ increase is expected to be proportional to the activity of the Ca^{2+} VOCs at the moment Ca^{2+} is reintroduced into the medium, the initial rate of Ca^{2+} decline to the activity of the VOCs at the moment immediately preceding the pharmacological blockade of the channel. The results showed that the two rates were almost identical as long as the $[Ca^{2+}]_i$ rises remained below 0.35 μM. With higher $[Ca^{2+}]_i$ rises, however, the decline rate appeared appreciably slower than the increase rate, demonstrating the existence of an additional, $[Ca^{2+}]_i$-dependent mechanism of Ca^{2+} VOC inactivation, with threshold around 0.35 μM (for further details on these experiments see DiVirgilio et al. 1987).

In the past, the mechanism(s) of Ca^{2+} VOCs inactivation were open to a large degree of uncertainty (see for example Hagiwara and Byerly 1981). The two mechanisms we have discussed: voltage and $[Ca^{2+}]_i$-dependent mechanisms, were repeatedly proposed, based on evidence obtained by various techniques in different cell types. To our knowledge, this is the first time that the coexistence of the two mechanisms is demonstrated in one single cell type, and using one single technique. It might also be added that our findings make sense in terms of cell physiology. In fact, a $[Ca^{2+}]_i$-dependent inactivation can play an important role in preventing the rise of $[Ca^{2+}]_i$ to levels that could be deleterious for the cells, and even dangerous to its survival. On the other hand, a voltage-dependent inactivation appears also necessary, in order to assure in all cases the transiency of the Ca^{2+} permeability response of the plasmalemma.

A final note related to activation and inactivation of Ca^{2+} VOCs concerns the regulation of catecholamine release. When release is induced in $PC12^-$ cells by high concentrations of K^+, its rate appears biphasic: an initial burst followed by a persistent but slight stimulation (Meldolesi et al. 1984). These two phases correlate very nicely with the $[Ca^{2+}]_i$ changes now revealed by fura-2. Actually, by a closer analysis of this correlation it was established that (1) the secretion bursts are sustained by intracellular $[Ca^{2+}]$ gradients, with maxima beneath the plasma membrane, caused by the sudden opening of Ca^{2+} VOCs at depolarization. The local $[Ca^{2+}]_i$ maxima during the initial periods (of the order of a few hundred ms) of the response were estimated to be very high, certainly exceeding 10 and possibly approaching 100 μM; and (2) the $[Ca^{2+}]_i$ threshold for release, deduced by correlating $[Ca^{2+}]_i$ plateaus and slow-rate release stimulations in $PC12^-$ cells, was found to be around 0.5 μM. It is worth to emphasize that, consistent with the $[Ca^{2+}]_i$ data discussed above, only this second, slow phase of release was induced when excess Ca^{2+} was reintroduced to the medium bathing cells depolarized with K^+ in Ca^{2+}-free conditions.

2 Inhibition of Ca^{2+} VOCs by Activators of Protein Kinase C

Functioning of VOCs can be regulated by intracellular reactions. Classical studies in heart cells (Reuter 1984) already demonstrated the existence of a cAMP-dependent regulation of Ca^{2+} VOCs, that consists in changing the probability of their opening as a function of the concentration of the second messenger within the cytosol. This effect, which might be mediated by cAMP-dependent phosphorylation of a protein associated with the cytosolic surface of the Ca^{2+} VOC, has been demonstrated in both heart and skeletal muscle cells as well as in invertebrate neurons. In some other cells, however, cAMP has no effect on Ca^{2+} VOCs. cAMP-dependent protein kinases are not the only cellular protein phosphorylating enzymes. Recently, we have shown that another protein kinase, protein kinase C, is also involved in Ca^{2+} VOC regulation. The experiments were carried out in PC12, in another cell line (RINm5F, derived from a rat insulinoma) (DiVirgilio et al. 1986) and, more recently, in freshly dissociated bovine chromaffin cells (unpublished results), that were studied after loading with quin-2. In all these cell types we found that treatment with activators of protein

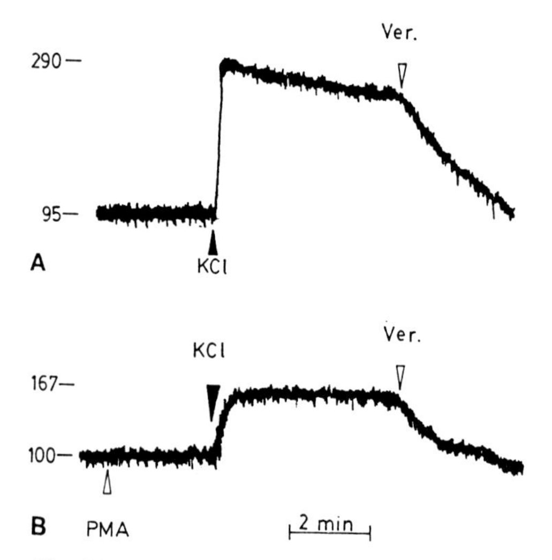

Fig. 2A,B. Inhibition by pretreatment with phorbol myristate acetate of the $[Ca^{2+}]_i$ rise triggered by depolarization with KCl in PC12 cells. Two parallel batches of cells were loaded with quin-2 and depolarized with KCl (50 mM) in Ca^{2+} containing medium without (**A**) or with (**B**) pretreatment with phorbol myristate acetate (PMA, 0.1 μM). Other symbols as in Fig. 1. Notice that, when using quin-2 (a $[Ca^{2+}]_i$ indicator with large Ca^{2+} buffering capacity), the initial spike of $[Ca^{2+}]_i$ rise by depolarization, visible when using fura-2, is blunted (cf. **A** of Figs. 1 and 2)

kinase C, such as phorbol-miristate-acetate (PMA), phorbol-dibutyrate or diacylglycerol, although without effect on the resting $[Ca^{2+}]_i$, causes a marked inhibition of the $[Ca^{2+}]_i$ rise induced by the subsequent application of high K^+ (10–50 mM) (Fig. 2). This effect of PMA and congeners (1) is not due to changes of the extent of K^+-induced depolarization; (2) appears rapidly (within 1 min) after the application of PMA; (3) requires concentrations of PMA in the nM range; (4) is accompanied (and most likely caused) by a marked inhibition of the depolarization-induced Ca^{2+} influx, with no change of the Ca^{2+} efflux; and (5) decreases markedly in cells that were long-term treated with protein kinase C activators, a procedure that is known to result in the down-regulation of the enzyme (DiVirgilio et al. 1986 and unpublished results). Taken together, these results strongly suggest that the Ca^{2+} VOCs (of the L type) of PC12 cells operate under the inhibitory control of protein kinase C. This conclusion is further strengthened by results on chick dorsal root ganglion neurons, published by Rane and Dunlap (1986) at the same time as ours in PC12 cells. These authors used patch clamp to demonstrate that a Ca^{2+} current is rapidly inhibited by protein kinase C activators. We conclude therefore that the metabolic control of Ca^{2+} VOCs includes not only cAMP-dependent, but also protein kinase C-dependent phosphorylations. It is interesting to note that in PC12 cells cAMP has no role in Ca^{2+} VOC control (Vicentini et al. 1986). It appears therefore that in different cell types Ca^{2+} VOCs may be regulated by either one of the two phosphorylation systems, although at the present time we cannot exclude that in other cell types, not yet investigated, both systems could cooperate in VOC regulation.

4 Inhibition of N-Tape Ca^{2+} VOCs by Activation of a Muscarinic M_1 Receptor in Rat Sympathetic Neurons

Recently, experiments were carried out on sympathetic neurons prepared from the superior cervical ganglion of newborn rats (Wanke et al. 1987). The cells were loaded with fura-2 and analyzed by a microscopic technique that allows signals to be recorded from perikaria of single neurons. Similar to PC12 cells discussed above, the $[Ca^{2+}]_i$ transients induced in these neurons by the application of high K^+ is composed of an initial spike followed by a plateau, and both these phases are sustained by the activation of Ca^{2+} VOCs. Parallel patch clamp studies revealed however that the nature of the channels involved is not identical in the two types of cells. Indeed, in the sympathetic neurons only about 50% of the Ca^{2+} current measured immediately after depolarization was due to long lasting, L-type channels, while the rest was contributed by rapidly (within 0.1 s) inactivating Ca^{2+} VOCs of the N type. No trace of the small T channels described by Nowicky et al. (1985) was found in these neurons.

Treatment of sympathetic neurons with low (μM) concentrations of acetylcholine failed to show an appreciable effect on the resting $[Ca^{2+}]_i$, but caused the initial spike induced by high K^+ to be markedly inhibited. These results correlate nicely with parallel electrophysiological data, that demonstrated the disappearance of the N channel current after acetylcholine. The inhibitory effect of the neurotransmitter was rapidly eliminated by low concentrations of antimuscarinic drugs, in particular of pirenzepine (concentrations in the 0.01 μM range). This latter result demonstrates that the effect of acetylcholine is mediated by a muscarinic receptor of the M_1 type.

5 Conclusions

The three examples that we have reported demonstrate that intracellularly trappable fluorescent $[Ca^{2+}]_i$ indicators can be conveniently used to investigate the functioning of Ca^{2+} VOCs and their regulation. Compared to the classical techniques, these indicators offer advantages, but also show distinct limitations. For example, time resolution, sensitivity and accuracy of the results obtained with quin-2 and fura-2 are in our experience much better than those obtainable by ^{45}Ca flux experiments. On the other hand, the data with fluorescent indicators concern average $[Ca^{2+}]_i$, and can be interpreted in terms of fluxes only when complemented by additional information. With respect to electrophysiology, in particular to the new technique of patch clamping, the use of fluorescent indicators should be considered as an extremely useful and powerful complement. Clearly, the time resolution of electrophysiology is unrivaled, as it is its potentiality to identify, analyze, and characterize individual currents. However, the electrophysiological analysis (1) does not go beyond the study of membrane currents and potential; (2) is hard to carry out for prolonged periods of time; and (3) makes use of preparations that can be far from physiological (for example, internally dyalized cells and excised membrane patches). Parallel studies, carried out by electrophysiology in these preparations, as well as by fluorescent indicators in intact cells, should therefore not only extend the investigation from currents and voltage to $[Ca^{2+}]_i$, but also strengthen considerably the reliability and biological importance of the results. For this kind of experimental approach the recent development of the fura-2 technique on single cells attached to solid substrates (Grynkiewicz et al. 1985; Meldolesi et al. 1987) is of great value. The possibility of carrying out concomitant measurements by electrophysiology and quantitative microspectrofluorimetry in the same cell represents at the moment a challenging opportunity for the future.

References

DiVirgilio F, Pozzan T, Wollheim CB, Vicentini LM, Meldolesi J (1986) Tumor promoter phorbol miristate acetate inhibits influx through voltage-gated Ca^{++} channels in two secretory cell lines, PC12 and RINm5F. J Biol Chem 261:32–35

DiVirgilio F, Milani D, Leon A, Meldolesi J, Pozzan T (1987) Voltage-dependent activation and inactivation of Ca^{++} channels in PC12 cells: correlation with neurotransmitter release. J Biol Chem 262:9189–9195

Grynkiewicz G, Poenie M, Tsien RY (1985) A new generation of fluorescent Ca^{++} indicators with greatly improved fluorescence properties. J Biol Chem 260:3440–3450

Hagiwara S, Byerly L (1981) Calcium chennel. Annu Rev Neurosci 4:69–125

Kogsamut S, Miller RJ (1986) Nerve growth factor modulates the drug sensitivity of neurotransmitter release from PC12 cells. Proc Natl Acad Sci USA 83:2243–2248

Meldolesi J, Huttner WB, Tsien RY, Pozzan T (1984) Free cytoplasmic Ca^{++} and neurotransmitter release. Studies on PC12 cells and synaptosomes exposed to latrotoxin. Proc Natl Acad Sci USA 81:620–624

Meldolesi J, Malgaroli A, Wollheim CB, Pozzan T (1987) Ca^{++} transients and secretion. Studies with quin2 and other Ca^{++} indicators. In: Poisner A, Trifaro JM (eds) In vitro methods for studying secretion. Elsevier, Amsterdam, pp 283–308

Nowicky MC, Fox PA, Tsien RW (1985) Three types of neuronal calcium channel with different calcium agonist sensitivity. Nature (Lond) 316:440–443

Pozzan T, Gatti G, Dozio N, Vicentini LM, Meldolesi J (1984) Ca^{++}-dependent and –independent release of neurotransmitters from PC12 cells. A role for protein kinase C activation. J Cell Biol 99:623–638

Rabe CS, McGee R (1983) Regulation of depolarization-dependent release of neurotransmitters by adenosine: cyclic AMP-dependent enhanced release from PC12 cells. J Neurochem 41: 1623–1634

Rane SG, Dunlap K (1986) Kinase C activator 1,2 oleoyl acetylglycerol attenuates voltage-dependent calcium current in sensory neurons. Proc Natl Acad Sci USA 83:184–189

Reuter H (1984) Calcium channels modulation by neurotransmitters, enzymes and drugs. Nature (Lond) 301:569–574

Tsien RY, Pozzan T, Rink TJ (1982) Calcium homeostasis in intact lymphocytes: cytoplasmic free calcium monitored by a new, intracellularly trapped fluorescent indicator. J Cell Biol 94: 325–334

Vicentini LM, Ambrosini A, DiVirgilio F, Meldolesi J, Pozzan T (1986) Activation of muscarinic receptors in PC12 cells. Correlation between cytosolic Ca^{++} rise and phosphoinositide hydrolysis. Biochem J 234:555–562

Wanke E, Ferroni A, Malgaroli A, Ambrosini A, Pozzan T, Meldolesi J (1987) A novel type of inhibition of voltage-gated Ca^{++} channels via muscarinic receptors in mammalian sympathetic neurons. Proc Natl Acad Sci USA 84:4313–4317

Calcium Entry Blockers

M. WIBO and T. GODFRAIND[1]

1 Introduction

The concept of calcium antagonism was developed in simultaneous and independent studies on cardiac tissues by Fleckenstein and coworkers (Fleckenstein et al. 1967), and on smooth muscle tissues by Godfraind and coworkers (Godfraind and Polster 1968). Undoubtedly, this concept has opened a new field in cardiovascular therapy, and calcium entry blockers are increasingly used to treat angina pectoris and hypertension, for instance. Concomitantly, these agents have contributed to our understanding of the mechanisms whereby electrical or chemical messages influence cellular activity.

A rise in the cytosolic Ca^{2+} concentration is quite often an early and essential step in the cell response to external stimulation. The activator Ca^{2+} may be released from intracellular stores, and/or flow into the cytoplasm through specific channels in the cell membrane. As illustrated in Fig. 1, two modes of calcium channel activation have been postulated in smooth muscle tissues (Bolton 1979): potential-operated channels (POCs), and receptor-operated channels (ROCs). Potential-operated channels open in response to action potentials or to sustained depolarization. As some agonists may promote calcium entry without concomitant depolarization, it has been proposed that they could act through their receptors to induce the opening of calcium channels. Up to now, however, the evidence in favor of the hypothesis that receptor-operated and potential-operated channels are distinct entities is not compelling.

The passage of Ca^{2+} ions through calcium channels is inhibited by some polyvalent cations, such as Co^{2+} or Mn^{2+}, that compete with Ca^{2+} for coordination site(s) in the channels, but are unable to pass through them. We shall not consider further these inorganic calcium antagonists. Calcium entry blockers may be defined as drugs inhibiting calcium entry through plasmalemmal calcium channels (Godfraind et al. 1986). Their mechanism of action differs from that of inorganic antagonists. Other categories of calcium antagonists have been distinguished, which do not act by blocking calcium channels in the plasma membrane. For instance, some drugs may interfere with other Ca^{2+} transport systems in the plasma membrane or in the endoplasmic (sarcoplasmic) reticulum.

1 Laboratoire de Pharmacodynamie Générale et de Pharmacologie, Université Catholique de Louvain (UCL 73.50), B–1200 Bruxelles, Belgium

Ch. Gerday, R. Gilles, L. Bolis (Eds.)
Calcium and Calcium Binding Proteins

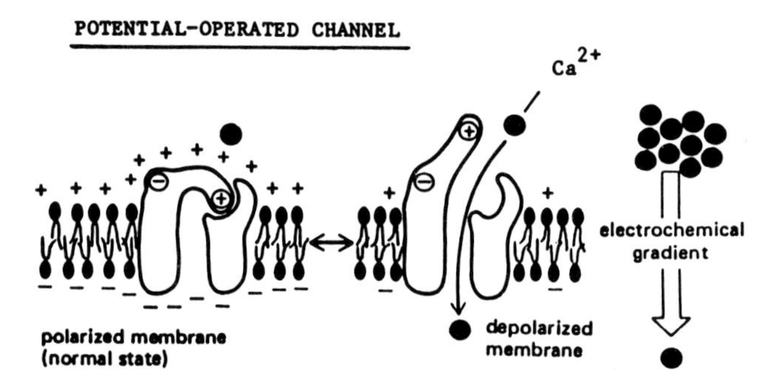

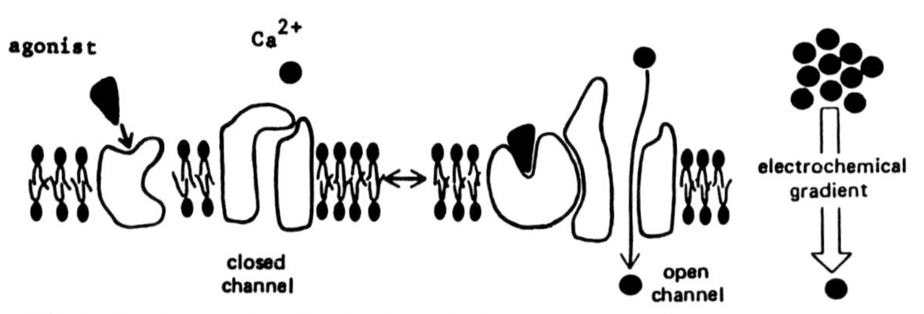

Fig. 1. The two modes of activation of calcium channels. See text for explanations

2 Classification of Calcium Entry Blockers

As shown in Table 1, these drugs may be divided into two groups, selective and nonselective calcium entry blockers. Selective calcium entry blockers act primarily on calcium channels opened by depolarization or other stimuli, whereas nonselective calcium entry blockers act at similar concentrations on calcium and sodium channels, or have a primary site of action that is not the calcium channels.

Two subgroups can be distinguished among selective calcium entry blockers. Subgroup I A includes three chemical classes with verapamil, nifedipine and diltiazem as prototypes. These drugs have been shown to block the slow calcium inward current in voltage-clamped myocardial preparations, without concomitantly affecting the fast sodium channel (Fleckenstein 1983). Cinnarizine and flunarizine (subgroup I B) could not be satisfactorily characterized in electrophysiological experiments on myocardial tissue. However, these drugs behave as selective calcium entry blockers in arterial preparations (Godfraind and Dieu 1981; Godfraind et al. 1982). The subsequent discussion will focus on the selective calcium entry blockers.

Some dihydropyridine derivatives (Bay K 8644, CGP 28392) have been shown to activate, rather than inhibit, calcium channels (Schramm et al. 1983). The discovery of these "calcium agonists" has raised much interest in experimental pharmacology, but up to now these compounds have not found therapeutic applications.

Table 1. Classification of calcium entry blockers

Group I: selective calcium entry blockers

Subgroup I A: selective inhibitors of the slow calcium inward current in myocardium (voltage-clamp)
Phenylalkylamines: verapamil, gallopamil (D 600), desmethoxyverapamil (D 888), falipamil (AQ-A-39)
Dihydropyridines: nifedipine, nicardipine, niludipine, nimodipine, nisoldipine, nitrendipine, ryosidine, dazodipine (PY 108-068), PN 200-110
Benzothiazepines: diltiazem
Subgroup I B: agents with no perceived actions on the calcium inward current in myocardium
Diphenylpiperazines: cinnarizine, flunarizine

Group II: nonselective calcium entry blockers

Subgroup II A: agents acting at similar concentrations on calcium channels and fast sodium channels
Bencyclane, bepridil, caroverine, etafenone, fendiline, lidoflazine, perhexiline, prenylamine, proadifen (SKF 525A), terodiline, tiapamil
Subgroup II B: agents interacting with calcium channels, while having another primary site of action
Local anesthetics, phenytoin, phenoxybenzamine, phenothiazines, pimozide, propranolol, diazepam, loperamide, barbiturates, reserpine, . . .

3 Pharmacological Actions of Calcium Entry Blockers in Cardiovascular Tissues

3.1 Heart

In atrial and ventricular muscles and Purkinje fibers, calcium entry blockers produce excitation-contraction uncoupling, i.e. they markedly reduce the contractile activity while inducing only limited changes in action potentials. The upstroke velocity and height of overshoot, which depend mainly on the activity of sodium channels, remain unaffected, but the plateau phase is abbreviated, due to inhibition of calcium channels, as revealed by voltage-clamp experiments (Kohlardt et al. 1972; Kohlardt and Fleckenstein 1977). Blockade of Ca^{2+} entry prevents the subsequent Ca^{2+} release from the sarcoplasmic reticulum and abolishes contraction. The effects of the calcium entry blockers can be antagonized by increasing the extracellular Ca^{2+} concentration (Fleckenstein 1983).

The order of potency for inhibition of Ca^{2+} current and contractility is nifedipine > verapamil > diltiazem. The inhibitory potency of verapamil and related drugs, and of diltiazem, increases markedly with the rate of stimulation (Bayer et al. 1982). This "use-dependence" has not been observed with nifedipine. At higher concentrations, verapamil and related drugs interact with sodium currents (Bayer et al. 1975) and potassium currents (Kass and Tsien 1975). Nifedipine and other dihydropyridines appear to be more selective inhibitors of calcium channels (Bayer et al. 1982).

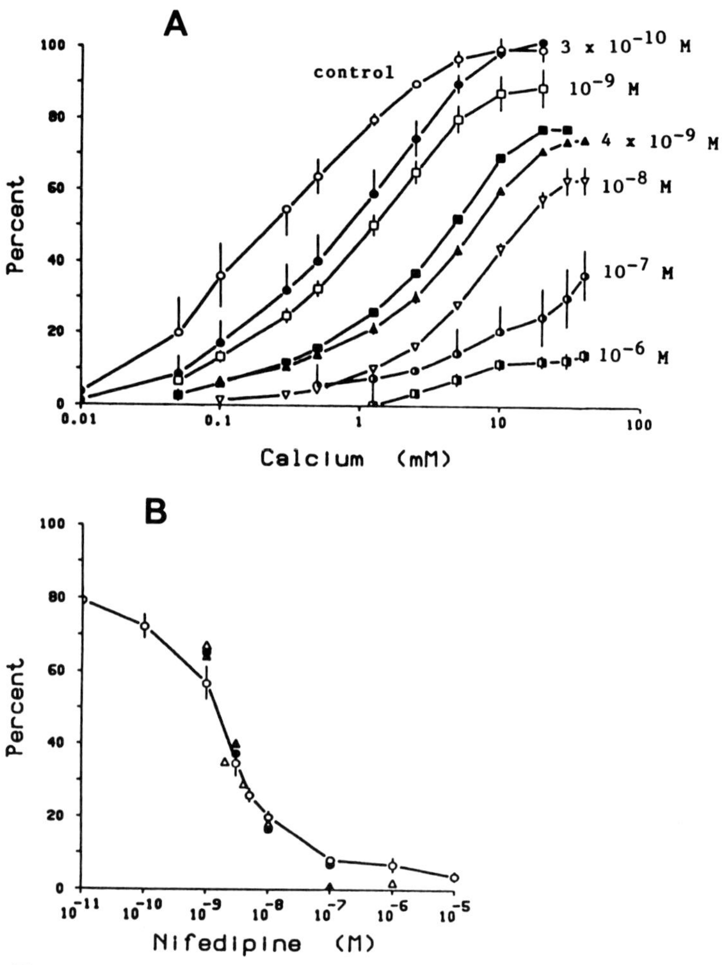

Fig. 2A,B. Effects of nifedipine in K^+-depolarized arteries. **A** Rat aortic rings preincubated in the absence or presence of nifedipine at the indicated concentrations were depolarized in a Ca^{2+}-free solution containing 100 mM KCl and further incubated with increasing Ca^{2+} concentrations. Responses are expressed in percentage of maximal contraction evoked in the absence of nifedipine. For further details, see Godfraind (1983). **B** Nifedipine concentration-effect curves in K^+-depolarized rat aorta (*open symbols*) and mesenteric artery (*closed symbols*). Contractions (*circles*) and ^{45}Ca influxes (*triangles*) are expressed in percentage of control values. For further details, see Godfraind (1983)

Sinoatrial and atrioventricular nodes are characterized by "slow" action potentials for which the inward charge carrier is largely Ca^{2+}. Thus, calcium entry blockers have prominent electrophysiologic effects in these tissues. In vivo, atrioventricular conduction is slowed down by verapamil (Mangiardi et al. 1978) and diltiazem.

3.2 Vascular Smooth Muscle

Figure 2A shows the effect of nifedipine on the contraction induced by calcium in depolarized rat aorta (Godfraind 1983). Similar results have been obtained with other calcium entry blockers, e.g. cinnarizine (Godfraind and Kaba 1969). Aortic rings were first depolarized by immersing them in a Ca^{2+}-free solution containing 100 mM KCl. Addition of Ca^{2+} to the depolarizing solution evoked an increase in tension which was concentration-dependent. In the presence of nifedipine the concentration-effect curve was displaced to the right, in a manner which is reminiscent of the action of competitive antagonists. Higher concentrations, however, also reduced the maximum contraction attainable.

That nifedipine actually inhibits the entry of calcium in depolarized smooth muscle cells is shown in Fig. 2B. When the concentration of nifedipine is varied, there is a close correlation between inhibition of contraction and of ^{45}Ca entry. Moreover, in chemically skinned preparations, in which the permeability barrier of the plasma membrane is suppressed, nifedipine does not inhibit the contraction induced by increasing Ca^{2+} concentrations (Kreye et al. 1983).

Let us now consider the effects of calcium entry blockers on responses to neurotransmitters, in particular norepinephrine. As illustrated in Fig. 3A, nifedipine inhibits the contraction produced by norepinephrine in a non-competitive manner. Micromolar concentrations of nifedipine, which completely blocked contraction evoked by depolarization, reduced by only about 50% the maximal contraction evoked by the agonist in rat aorta (Godfraind 1983). It has been shown that the nifedipine-sensitive part of the contraction is attributable to an influx of Ca^{2+} and that the potency of nifedipine as inhibitor of this influx is about tenfold less than its potency as inhibitor of the Ca^{2+} entry induced by KCl-depolarization.

As shown in Fig. 3B, high concentrations of other calcium entry blockers such as flunarizine or D-600 also leave about 50% of the contraction unaffected (Godfraind and Dieu 1981; Godfraind and Miller 1983). A contraction of a similar magnitude is also evoked by norepinephrine in aortic preparations maintained for 10 min in a Ca^{2+}-free medium (while this treatment completely abolishes the mechanical response to K^+-depolarization). Thus, we may conclude that calcium entry blockers block only that part of the contraction that depends on an influx of Ca^{2+} through receptor-operated channels. The other part, that persists after removal of extracellular Ca^{2+}, can be attributed to a mobilization of Ca^{2+} from internal stores, probably in the endoplasmic reticulum. It is likely that alpha-adrenoceptor activation induces an increased hydrolysis of phosphatidylinositol bisphosphate in the plasma membrane, and that inositol 1,4,5-trisphosphate serves as a messenger from the plasma membrane to the endoplasmic reticulum and triggers the release of Ca^{2+} from this organelle (Somlyo et al. 1985).

The effect of calcium entry blockers on agonist-evoked contractions may be substantially different in other arteries. For example, in rat mesenteric artery, less than 25% of the contraction evoked by norepinephrine is resistant to nifedipine (Godfraind 1983). In rabbit aorta, calcium entry blockers have little effect on norepinephrine-induced contractions (Van Breemen et al. 1980), probably because these contractions are less dependent on extracellular Ca^{2+} in this tissue than they are in rat aorta. It has

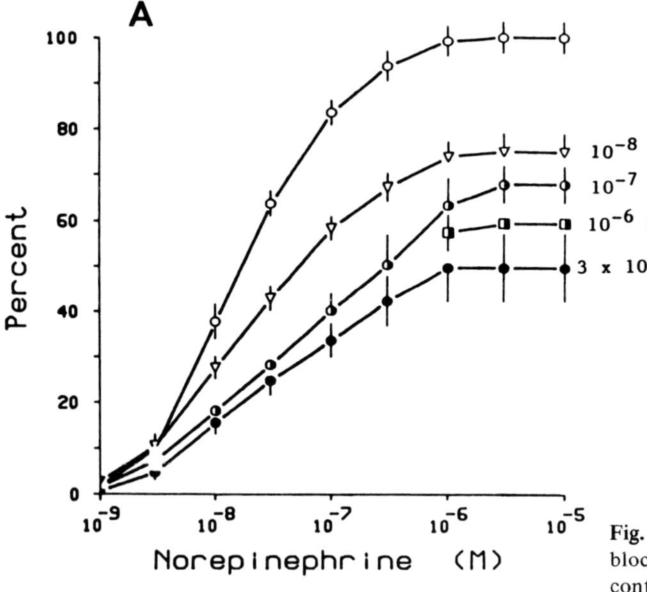

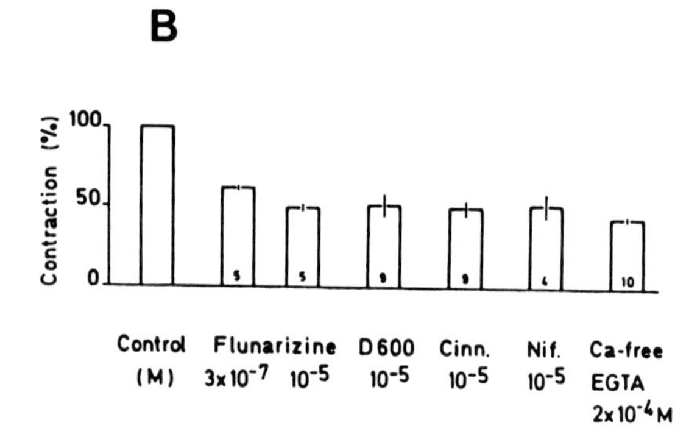

Fig. 3A,B. Effects of calcium entry blockers on norepinephrine-induced contractions in rat aorta. **A** Effect of nifedipine at the indicated concentrations on norepinephrine concentration-effect curves. Contractions are expressed in percentage of the maximal response of control rings to norepinephrine. For further details, see Godfraind (1983). **B** Effect of calcium entry blockers and of Ca^{2+}-free medium on contractions evoked by 10 μM norepinephrine in rat aorta. *numbers in columns* number of determinations; *Cinn.* Cinnarizine; *Nif.* nifedipine. For further details, see Godfraind and Miller (1983)

been reported that, in rabbit small mesenteric vessels, diltiazem is more potent as an inhibitor of agonist- than of K^+-induced contractions and ^{45}Ca influx (Cauvin and Van Breemen 1985).

4 Mode of Action of Calcium Entry Blockers as Revealed by Patch-Clamp Studies

Thanks to the development of patch-clamp analysis, ionic currents flowing through single calcium channels have been studied in several types of excitable cells. Hess et al. (1985) and Kokubun and Reuter (1984) have shown that the inhibitory action of dihydropyridines cannot be explained on the basis of a simple plugging of the channels. In the absence of drug, three modes of calcium channel gating behavior have been recognized, expressed as current records with brief openings (mode 1), with no opening (mode 0), and with long-lasting openings and brief closing periods (mode 2, which appears only rarely). Dihydropyridine inhibitors depress calcium currents by favoring mode 0, while calcium channel "agonists" (Bay K 8644, CGP 28392) enhance calcium currents by favoring mode 2. This modulation does not appear to be related to cyclic AMP-dependent phosphorylation.

In addition, these studies have revealed the existence of distinct subtypes of potential-operated channels in various cell types (Nilius et al. 1985; Nowycky et al. 1985). In cardiac cells, two types of unitary conductances have been recognized. The L-type (long-lasting) channel is characterized by currents that inactivate slowly. The T-type (transient) channel opens at much more negative membrane potentials than the L-type and produces inward currents that inactivate more rapidly. Only the L-type channel is influenced by dihydropyridines.

5 Specific Binding Sites for Calcium Entry Blockers

Over the last few years, specific binding sites for three chemical classes of calcium entry blockers have been characterized in membranes isolated from various tissues (Triggle and Janis 1984; Godfraind et al. 1986). Dihydropyridine binding sites have been more thoroughly studied and they show very similar properties in all excitable tissues, with the exception of skeletal muscle. Most studies have concluded to the existence of only one class of pharmacologically relevant binding sites of high affinity (K_d of [^{3}H]-nitrendipine in the 0.1–1 nM range), localized in the plasma membrane (Godfraind and Wibo 1985). These sites are different from, but allosterically linked to, those that bind verapamil and diltiazem (see Fig. 4). Recently, Galizzi et al. (1986) have obtained evidence suggesting that the binding sites for dihydropyridines, diltiazem and bepridil are all associated with the same polypeptide chain.

The pharmacological relevance of dihydropyridine binding sites in intestinal smooth muscle has been established by Bolger et al. (1983). Using a series of nifedipine analogs, these authors found a good correlation between binding and inhibition of mechanical response to K^+-depolarization. Using a similar approach, we have compared in rat aorta the potencies of a series of dihydropyridines as inhibitors of [^{3}H]-(+)PN 200-110 binding in isolated membranes (Fig. 5) with their potencies as inhibitors of the K^+-evoked contraction. There is obviously a very good 1 to 1 correlation between the two sets of data (Fig. 6). In particular, one observes the same stereoselectivity in binding and contractility studies.

Although dihydropyridine binding sites are very similar in cardiac and smooth muscle membranes, much higher concentrations of dihydropyridines are required to inhibit cardiac contractility than to inhibit smooth muscle contraction. There is a difference of 2 to 3 orders of magnitude between potencies in pharmacological and binding experiments in the case of myocardial tissue. Sanguinetti and Kass (1984) and Bean (1984) have proposed a plausible interpretation for these observations. These authors studied calcium currents in cardiac cells by the whole-cell variant of the patch-clamp technique. When cells were held at a potential of −80 mV, calcium

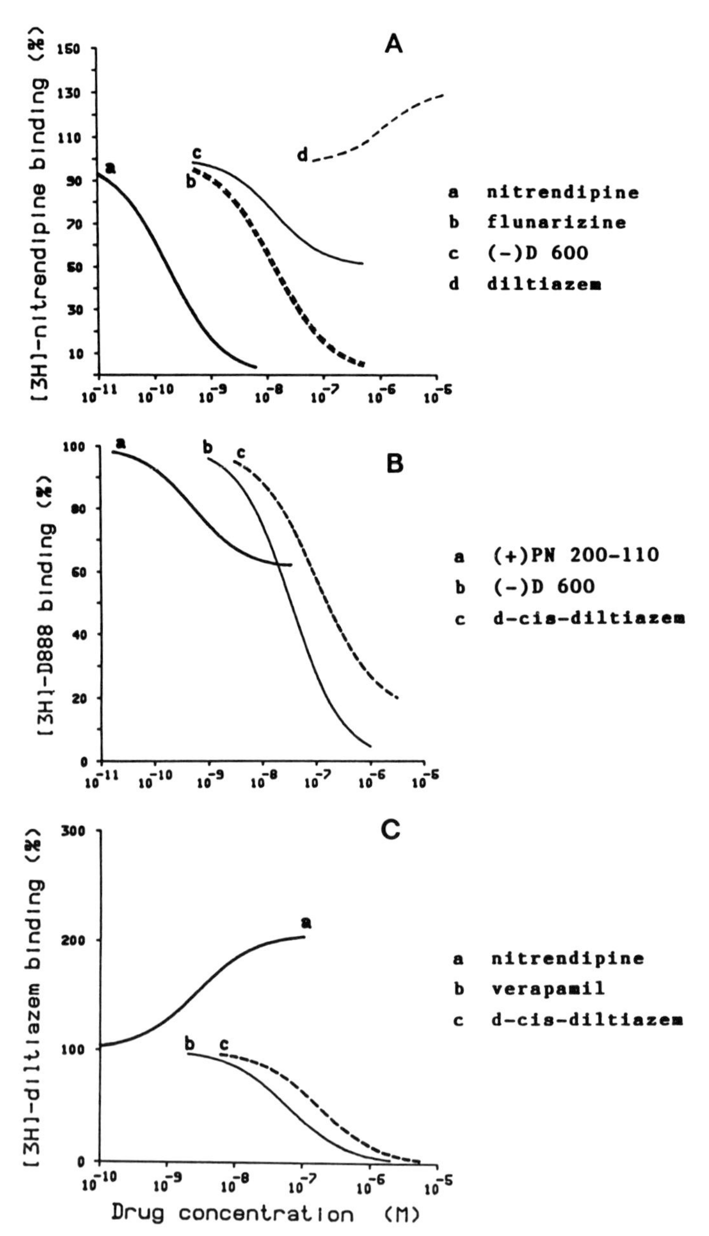

Fig. 4A–C. Allosteric interactions of the various chemical classes of calcium entry blockers at their respective binding sites. **A** [^{3}H]-Nitrendipine binding inhibition curves with membranes from guinea-pig ileal smooth muscle. Curves were drawn from data reported by Bolger et al. (1983). While curves obtained with nitrendipine and flunarizine were compatible with competitive interactions, (−)D 600 was unable to displace more than 50% of bound [^{3}H]-nitrendipine, and diltiazem increased [^{3}H]-nitrendipine binding, indicating allosteric interactions with the dihydropyridine binding site. **B** [^{3}H]-D 888 binding inhibition curves with membranes from guinea-pig hippocampus. Curves were drawn from data reported by Ferry et al. (1984). (+)PN 200-110 (a dihydropyridine) and d-cis-diltiazem (the pharmacologically active isomer) were unable to completely displace bound [^{3}H]-D 888 (a verapamil analog). **C** [^{3}H]-Diltiazem binding inhibition curves with membranes from rat cerebral cortex. Curves were drawn from data reported by Schoemaker and Langer (1985). Nitrendipine stimulated [^{3}H]-diltiazem binding, whereas curves obtained with verapamil and diltiazem were compatible with competitive interactions

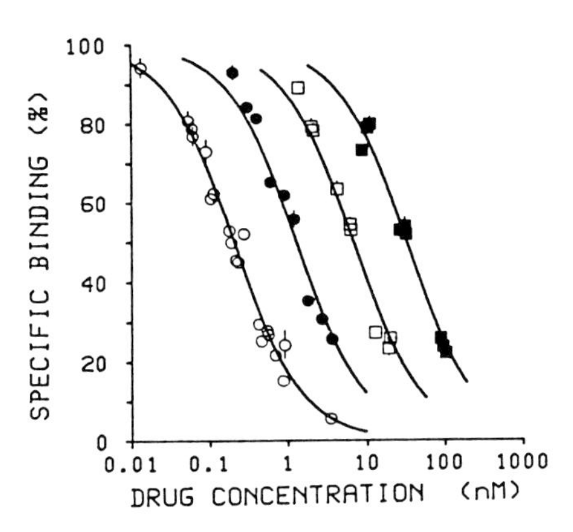

Fig. 5. [^{3}H]-(+)PN 200-110 binding inhibition curves with membranes from rat aorta (unpublished results). Membranes were incubated at 37°C for 60 min with 0.11 nM [^{3}H]-(+)PN 200-110 and various concentrations of unlabeled drugs [○ (+)PN 200-110; ● (−)nimodipine; □ (+)-nimodipine; ■ (−)PN 200-110]. Curves indicate competitive interactions with a single class of binding site

currents elicited by depolarizing pulses were half-maximally inhibited by 1 μM nitrendipine. In contrast, nitrendipine was much more potent when cells were held at a potential of −10 mV, and a concentration of 1 nM then blocked the current almost completely (Bean 1984). This voltage-dependence has been explained on the basis of a modulated receptor model, in which the calcium channel can be either in a resting state (R) (channel closed but available to be opened by depolarization), or in an inactivated state (I) (channel closed and not available to be opened). The distribution of channels between states R and I depends on the holding potential. At −80 mV all channels are in state R and the low potency of nitrendipine as inhibitor of the calcium current would be due to a low affinity for the R state. At −10 mV, about 70% of the channels are in state I, and the high potency of nitrendipine would correspond to a high affinity for the I state. This high-affinity state of the channel in depolarized cells would be consistent with the existence of high-affinity binding sites in isolated membranes, which are obviously nonpolarized. Recently, some binding

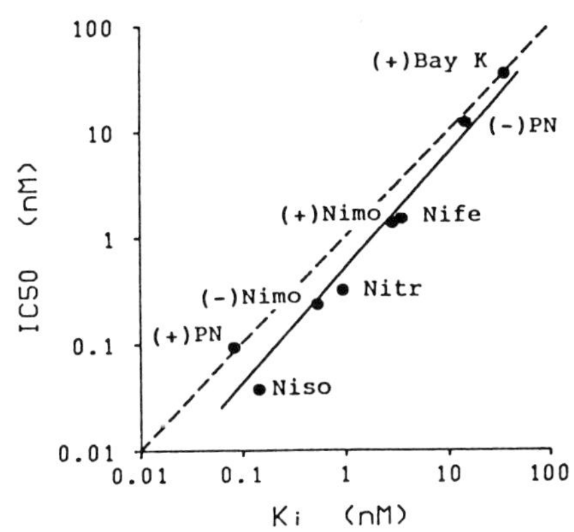

Fig. 6. Correlation between inhibition of K^+-evoked contraction and inhibition of [^{3}H]-(+)PN 200-110 binding in rat aorta. IC_{50} values (50% inhibitory concentrations) for inhibition of contraction were determined from concentration-effect curves (see Fig. 2B) (data from Godfraind 1983; Godfraind et al. 1985; and unpublished results). K_i values (unpublished results) were obtained from binding competition experiments (illustrated in Fig. 5), as described by Godfraind and Wibo (1985). *PN* PN 200-110; *Niso* nisoldipine, *Nimo* nimodipine; *Nitr* nitrendipine; *Nife* nifedipine; *Bay K* Bay K 8644

studies on intact cells have shown that K^+-depolarization may indeed increase the affinity of dihydropyridine binding (Kokubun et al. 1987).

6 Tissue Selectivity of Calcium Entry Blockers

The tissular selectivity of calcium entry blockers is of major importance for their therapeutic indications. Dihydropyridines are especially potent as inhibitors of smooth muscle contraction and they are mainly used for their vasorelaxant properties. Several factors may contribute to tissue selectivity (Godfraind et al. 1986).

a) Calcium entry blockers can only be effective inhibitors if activation of the tissue is dependent to a significant extent on entry of extracellular Ca^{2+}. In some cell types (platelets, mast cells, skeletal muscle, exocrine pancreas), the dependence on extracellular Ca^{2+} appears doubtful, and depressing effects of calcium entry blockers have generally been observed only at high concentrations at which nonspecific effects are likely. In other tissues, the dependence on external Ca^{2+} may vary according to the stimulus, as illustrated above for vascular smooth muscle.

b) At least two modes of calcium channel activation (POCs and ROCs) and several subtypes of POCs have been demonstrated, and they appear to be differently sensitive to calcium entry blockers. For instance, in chick dorsal root ganglion cells, three types of Ca^{2+} conductances have been detected by patch-clamp techniques, only one of them being depressed by dihydropyridine blockers (Nowycky et al. 1985). Thus, one can imagine that at least some neuronal activities that depend on Ca^{2+} entry will not be affected by these drugs.

c) As for potential-operated channels that are sensitive to calcium entry blockers, their modulation by these drugs may be profoundly influenced by the local tissue conditions and by the features of the stimulus. As mentioned previously, the frequency of stimulation markedly affects the potency of some calcium entry blockers (use-dependence). The voltage-dependence of dihydropyridine binding offers a simple explanation of why nanomolar concentrations of these drugs do not affect cardiac contractility under usual conditions. Indeed, although cardiac calcium channels become inactivated during the action potential, the time spent in the inactivated state is less than 1 s and does not allow dihydropyridines to bind (Bean 1984). In contrast, low concentrations of dihydropyridines have pronounced effects on arterial smooth muscle cells, and this may be related to a lower resting membrane potential and/or to longer depolarizing responses to physiological (or pathological) stimuli (Klöckner and Isenberg 1986).

7 Conclusion

Calcium entry blockers have been invaluable tools in studies devoted to the physiology and pharmacology of excitable tissues and they will undoubtedly be essential for a deeper understanding of the functioning of calcium channels at the molecular level.

Concomitantly, they have opened the way towards purification and chemical characterization of calcium channels. Their clinical applications are growing steadily, and it may be expected that the development of novel, more selective derivatives will still improve their therapeutic effectiveness.

References

Bayer R, Kalusche D, Kaufmann R, Mannhold R (1975) Inotropic and electrophysiological actions of verapamil and D 600 in mammalian myocardium. III. Effects of the optical isomers on transmembrane action potentials. Naunyn-Schmiedeberg's Arch Pharmakol 290:81–97

Bayer R, Kaufmann R, Mannhold R, Rodenkirchen R (1982) The action of specific Ca antagonists on cardiac electrical activity. Prog Pharmacol 5:83–85

Bean BP (1984) Nitrendipine block of cardiac calcium channels: High-affinity binding to the inactivated state. Proc Natl Acad Sci USA 81:6388–6392

Bolger GT, Gengo P, Klockowski R, Luchowski E, Siegel H, Janis RA, Triggle AM, Triggle DJ (1983) Characterization of binding of the Ca^{2+} channel antagonist, [^{3}H]nitrendipine, to guinea-pig ileal smooth muscle. J Pharmacol Exp Ther 225:291–309

Bolton TB (1979) Mechanisms of action of transmitters and other substances on smooth muscle. Physiol Rev 59:606-718

Cauvin C, Van Breemen C (1985) Different Ca^{2+} channels along the arterial tree. J Cardiovasc Pharmacol 7:S4–S10

Ferry DR, Goll A, Gadow D, Glossmann H (1984) (–)-^{3}H-desmethoxyverapamil labelling of putative calcium channels in brain: autoradiographic distribution and allosteric coupling to 1,4-dihydropyridine and diltiazem binding sites. Naunyn-Schmiedeberg's Arch Pharmakol 327: 183–187

Fleckenstein A (1983) Calcium antagonism in heart and smooth muscle. Experimental facts and therapeutic prospects. Wiley & Sons, New York

Fleckenstein A, Kammermeier H, Döring HJ, Freund HJ (1967) Zum Wirkungsmechanismus neuartiger Koronardilatatoren mit gleichzeitig Sauerstoff-einsparenden Myokard-Effekten, Prenylamin und Iproveratril. Z Kreislaufforsch 56:839–853

Galizzi J-P, Borsotto M, Barhanin J, Fosset M, Lazdunski M (1986) Characterization and photoaffinity labeling of receptor sites for the Ca^{2+} channel inhibitors *d-cis*-diltiazem, (±)-bepridil, desmethoxyverapamil, and (+)-PN 200-110 in skeletal muscle transverse tubule membranes. J Biol Chem 261:1393–1397

Godfraind T (1983) Actions of nifedipine on calcium fluxes and contraction in isolated rat arteries. J Pharmacol Exp Ther 224:443–450

Godfraind T (1986) Calcium entry blockade and excitation contraction coupling in the cardiovascular system (with an attempt of pharmacological classification). Acta Pharmacol Toxicol 58:(Suppl II) 5–30

Godfraind T, Dieu D (1981) The inhibition by flunarizine of the norepinephrine-evoked contraction and calcium influx in rat aorta and mesenteric arteries. J Pharmacol Exp Ther 217:510–515

Godfraind T, Kaba A (1969) Blockade or reversal of the contraction induced by calcium and adrenaline in depolarized arterial smooth muscle. Br J Pharmacol 36:549–560

Godfraind T, Miller RC (1983) Specificity of action of Ca^{++} entry blockers. A comparison of their actions in rat arteries and in human coronary arteries. Circ Res 52:(Suppl I) 81–91

Godfraind T, Polster P (1968) Etude comparative de médicaments inhibant la réponse contractile de vaisseaux isolés d'origine humaine ou animale. Thérapie 23:1209–1220

Godfraind T, Wibo M (1985) Subcellular localization of [^{3}H]-nitrendipine binding sites in guinea-pig ileal smooth muscle. Br J Pharmacol 85:335–340

Godfraind T, Miller R, Socrates Lima J (1982) Selective α_1- and α_2-adrenoceptor agonist-induced contractions and ^{45}Ca fluxes in the rat isolated aorta. Br J Phamacol 77:597–604

Godfraind T, Wibo M, Egleme C, Wauquaire J (1985) The interaction of nimodipine with calcium channels in rat isolated aorta and in human neuroblastoma cells. In: Betz E, Deck K, Hoffmeister F (eds) Nimodipine. Pharmacological and clinical properties. Schattauer, Stuttgart New York, pp 217–228

Godfraind T, Miller R, Wibo M (1986) Calcium antagonism and calcium entry blockade. Pharmacol Rev 38:321–416

Hess P, Lansman JB, Tsien RW (1985) Different modes of Ca channel gating behaviour favoured by dihydropyridine Ca agonists and antagonists. Nature (Lond) 311:538–544

Kass RS, Tsien RW (1975) Multiple effects of calcium antagonists on plateau currents in cardiac Purkinje fibers. J Gen Physiol 66:169–192

Klöckner U, Isenberg G (1986) Tiapamil reduces the calcium inward current of isolated smooth muscle cells. Dependence on holding potential and pulse frequency. Eur J Pharmacol 127:165–171

Kohlardt M, Fleckenstein A (1977) Inhibition of the slow inward current by nifedipine in mammalian ventricular myocardium. Naunyn-Schmiedeberg's Arch Pharmakol 298:267–272

Kohlardt M, Bauer B, Krause H, Fleckenstein A (1972) Differentiation of the transmembrane Na and Ca channel in mammalian cardiac fibres by the use of specific inhibitors. Pflügers Arch 335: 309–322

Kokubun S, Reuter H (1984) Dihydropyridine derivatives prolong the open state of Ca channels in cultured cardiac cells. Proc Natl Acad Sci USA 81:4824–4827

Kokubun S, Prod'hom B, Becker C, Porzig H, Reuter H (1987) Studies on Ca channels in intact cardiac cells: Voltage-dependent effects and cooperative interactions of dihydropyridine enantiomers. Mol Pharmacol 30:571–584

Kreye VAW, Ruegg JC, Hofmann F (1983) Effect of calcium-antagonist and calmodulin-antagonist drugs on calmodulin-dependent contractions of chemically skinned vascular smooth muscle from rabbit renal arteries. Naunyn-Schmiedeberg's Arch Pharmakol 323:85–89

Mangiardi LM, Hariman RJ, McAllister RG, Bhargava V, Surawicz B, Shabetai R (1978) Electrophysiologic and hemodynamic effects of verapamil. Correlations with plasma drug concentrations. Circulation 57:366–372

Nilius B, Hess P, Lansman JB, Tsien RW (1985) A novel type of cardiac calcium channel in ventricular cells. Nature (Lond) 316:443–446

Nowycky MC, Fox AP, Tsien RW 1985) Thre types of neuronal calcium channel with different calcium agonist sensitivity. Nature (Lond) 316:440–443

Sanguinetti MC, Kass RS (1984) Voltage-dependent block of calcium channel current in the calf cardiac Purkinje fiber by dihydropyridine calcium channel antagonists. Circ Res 55:336–348

Schoemaker H, Langer SZ (1985) [^{3}H]-Diltiazem binding to calcium channel antagonists recognition sites in rat cerebral cortex. Eur J Pharmacol 111:273–277

Schramm M, Thomas G, Towart R, Franckowiak G (1983) Novel dihydropyridines with positive inotropic action through activation of Ca^{2+} channels. Nature (Lond) 303:535–537

Somlyo AV, Bond M, Somlyo AP, Scarpa A (1985) Inositol trisphosphate-induced calcium release and contraction in vascular smooth muscle. Proc Natl Acad Sci USA 82:5231–5235

Triggle DJ, Janis RA (1984) Calcium channel antagonists: new perspectives from the radioligand binding assay. In: Back N, Spector S (eds) Modern methods in pharmacology. Liss, New York, pp 1–28

Van Breemen C, Aaronson P, Loutzenhiser R, Meisheri KD (1980) Ca^{2+} movements in smooth muscle. Chest 78:(Suppl) 157–165

Subject Index